WITHDRAWN
UTSA LIBRARIES

Optical Communications

Optical Communications

ROBERT M. GAGLIARDI
University of Southern California

and

SHERMAN KARP
Naval Electronics Laboratory Center

A Wiley-Interscience Publication

JOHN WILEY & SONS

New York · London · Sydney · Toronto

Library of Congress Cataloging in Publication Data:

Gagliardi, Robert M   1934 –
　Optical communications.

　Includes bibliographical references and index.
　1.　Optical communications.　I.　Karp, Sherman,
joint author.　II.　Title.

TK5103.59.G33　　　　621.38'0414　　　　75–26509
ISBN 0–471–28915–9

Printed in the United States of America

10 9 8 7 6 5 4 3 2 1

*To Our Parents, Wives,
and Children*

Preface

This book is devoted to an analytical study of optical communications. Its primary purpose is to compliment earlier books dealing with optical technology and device description by emphasizing the theoretical aspects of system design. Mathematical models of optical components are developed and used to determine design equations, limitations to system performance, and technological trade-offs involved in an optical communication system. The material is presented in a textbook format, with homework problems included to aid chapter discussions. We have used these problems to fill in mathematical operations, carry out specific examples and applications, and investigate interesting digressions. We think that the book is suitable for classroom instruction, for self learning, or as a reference book for system analysis and design.

The bulk of the text is oriented toward theory and systems, with little attention devoted to hardware design. Other books in this area have concentrated heavily on this device aspect. To a system engineer, however, it often matters very little what the actual technology entails, so long as he has component models that are accurate and can identify satisfactorily with key parameters. We therefore do not consider devices per se but concentrate instead on the fundamental and invariant features of optical systems, especially those that are clearly distinct from the other regions of the electromagnetic spectrum. Our objective is to attempt to unify the large body of work that has appeared in the literature during the past decade, which we feel forms the basis of the area of optical communications.

The material is presented primarily in the language of the communication engineer. We avoid extensive dealings with the fields of optics, quantum mechanics, and electrophysics, examining these areas only sufficiently to derive the necessary models and examine specific examples. In Chapter 1 we have made a special effort to show how the diverse areas of Fourier optics, electromagnetic theory, noise theory, and quantum electronics interrelate in the overall system model. In addition to Fourier analysis, a reader should have a working knowledge of probability theory and field expansions. As an aid, we have included in appendices a short review of the important aspects of these latter areas. However, we feel that a thorough understanding is only an advantage and not a necessity. A user of the book may work around the statistical manipulations and concentrate only on their implications, which we have tried to emphasize throughout. For these reasons the book is aimed at several types of readers—the student in electrical engineering or electrophysics, the communication engineer who may wish to become familiar with optical technology, and the optical physicist who is familiar with the field of optics but perhaps has not considered the theoretical implications of system interfacing for transmitting information over an optical link. We consider the book to be appropriate for a one or two semester course in optical communications at a senior or graduate level. The first seven chapters would make a suitable one semester course. The material in the remaining chapters, being more specialized in its application, would be better suited as a follow on, or special topic, type of course.

The organization of the material developed from several courses on optical communications in which we were involved. In essence, the content of the text can be divided into four parts. The first part—Chapters 1 to 4—concentrates on the development of the system model of the optical transmitter, channel, and receiver. The second part consists of Chapters 5 and 6, both devoted to analog communications. Chapters 7 and 8 consider digital systems as a third part, and the fourth part is formed by the remaining material—Chapters 9 through 11, which investigate the advanced topics of estimation, synchronization, and pointing. The optical communication system differs from the conventional microwave system in two primary areas—the spatial effect of the optical beam and the operation of the quantum detector. For this reason there is emphasis on model development pertaining to these aspects of the total system.

Chapter 1 discusses the overall optical system from source to receiver, setting the framework for the ensuing analysis. An instructor of a course must use care to properly match this material to the background of the student. Chapter 2 explores the optical field and optical detection, enumerating the importance of the photoelectron count process. The main focus of this chapter is on an understanding of the statistics of the count process for a detector operating under the influence of an impinging electromagnetic field. The physical model of the detection operation leads to the development of a count process governed by conditional poisson statistics. In Chapter 3 the latter process is rigorously examined, exhibiting its relation to the time and spatial properties of the received field. In Chapter 4 the time process of the detector output is intrinsically tied to the count process and the received field, establishing a statistical model of the conversion from optical field to detector output during optical reception. This detected time function now forms the basis for the analysis of the processing in the remaining portion of the receiver, the design of which is strictly in the hands of the communication engineer and therefore becomes the objective of the remaining parts of the book.

Chapters 5 and 6 consider the problem of data transmission in terms of output signal to noise ratios. This topic naturally divides into two chapters— incoherent, or direct detection, systems and coherent, or heterodyned, systems. In the next two chapters we focus on digital communications, essentially from the vantage point of detection theory. Digital processing, signal design, and resulting system performance is presented. Chapter 7 considers binary operation, while Chapter 8 extends to block encoded systems. The primary objective of the latter chapter is to demonstrate some of the advantages of using somewhat complicated digital encoding in optical digital systems. Chapter 9 is devoted to the use of parameter estimation in optical system design. Particular emphasis is placed on its application to acquisition and synchronization subsystems. Chapter 10 uses these results to examine the time synchronization problem in detail. In Chapter 11 the problems of pointing, spatially acquiring, and spatially tracking a narrow optical beam is investigated. Short appendices are included as a review of probability theory and random fields, field expansions, for tabulation of properties of the Laguerre polynomial, and for a discussion of random optical channels.

Although the material presented was selected to be as basic as possible it is difficult for authors not to succumb to a personal bias. For example, we have chosen not to include a wealth of material on quantum detection theory and optical channel measurements and modeling. This was not due to a lack of regard for this material but rather reflects our own perspective on these topics. We have tried to present this material to people who, like ourselves, are conversant with communication and noise theories from the continuous or classical point of view. We felt that to digress and present this from a quantum viewpoint would require background that most communication readers may not yet possess. We have also not included a great deal of material that has developed due to experimental measurements on the optical channel, which have nonetheless produced widely accepted results. Here again our personal bias in trying to present the fundamental, invariant material is reflected.

This book of course could not have been assembled without the technical contributions of our many colleagues and fellow workers in the field. Over the years we have benefited from technical discussions and informal conversations with many friends in the academic, research, and applied areas of optical communications, for which we are gratefully indebted. We especially want to acknowledge Professors R. Kennedy, C. Helstrom, E. O'Neill, J. Shapiro, J. Clark, W. Pratt, A. Sawchuck, D. Synder, R. Harger, along with Drs. E. Hoversten, S. Personick, N. Mohanty, M. Fitzmaurice, D. Premo, H. Plotkin, W. Bridges, D. Fried, J. Liu, H. Heggestad, P. Livingston, D. Arnush, R. Lutomirski, J. Randall, F. Goodwin, G. Mooradian, R. Anderson, and Lts. R. Driscoll and R. Giannaris, USN. We also want to personally acknowledge Miss Corinne Leslie for patiently typing our manuscript and all its revisions.

ROBERT M. GAGLIARDI
SHERMAN KARP

Los Angeles, California
San Diego, California
August 1975

Contents

Conversion Formulas

Physical Constants

Speed of light, $c = 2.998 \times 10^8$ m/sec

Electron charge, e $= 1.601 \times 10^{-19}$ C

Planck's constant, $h = 6.624 \times 10^{-34}$ W-sec/Hz $= (-335.4$ dBW/Hz$^2)$

Boltzman's constant, $\kappa = 1.379 \times 10^{-23}$ W/°K-Hz

$$\frac{h}{\kappa} = 4.82 \times 10^{-11} \text{ °K/Hz}$$

Conversion Factors

1 micron $= 10^{-6}$ meters $= 10^{-4}$ cm

1 Å $= 10^{-4}$ microns $= 10^{-10}$ meters

1 arc sec $= 2.78 \times 10^{-4}$ degrees $= 4.89 \times 10^{-6}$ radians

Frequency in Hz $= 3 \times 10^{14}$/wavelength in microns

Bandwidth in Hz at center wavelength $\lambda = (c/\lambda^2)$ [bandwidth in wavelength]

Optical Frequencies and Wavelengths

Violet $\approx 7 \times 10^{14}$ Hz	0.38–0.48 microns
Blue $\approx 6 \times 10^{14}$ Hz	0.48–0.52 microns
Green $\approx 5.6 \times 10^{14}$ Hz	0.52–0.56 microns
Yellow $\approx 5.1 \times 10^{14}$ Hz	0.56–0.62 microns
Orange $\approx 4.8 \times 10^{14}$ Hz	0.62–0.64 microns
Red $\approx 4.4 \times 10^{14}$ Hz	0.64–0.72 microns
Infrared $\approx 3 \times 10^{14}$ Hz	0.7–100 microns

1 The Optical Communication System

The objective of any communication system is the transfer of information from one point to another. This information transfer most often is accomplished by superimposing (modulating) the information onto an electromagnetic wave (carrier). The modulated carrier is then transmitted (propagated) to the destination, where the electromagnetic wave is received and the information recovered (demodulated). Such systems are often designated by the location of the carrier frequency in the electromagnetic spectrum (Figure 1.1). In radio systems the electromagnetic carrier wave is selected with a frequency from the radio frequency (RF) portion of the spectrum. Microwave or millimeter systems have carrier frequencies from those portions of the spectrum. In an optical communication system, the carrier is selected from the optical region, which includes the infrared, visible, and ultraviolet frequencies.

The principal advantage in communicating with optical frequencies is the potential increase in information and power that can be transmitted. In any communication system the amount of information transmitted is directly related to the bandwidth (frequency extent) of the modulated carrier, which is generally limited to a fixed portion of the carrier frequency itself. Thus, increasing the carrier frequency theoretically increases the available transmission bandwidth, and therefore the information capacity of the overall system. This means frequencies in the optical range will have a usable bandwidth about 10^5 times that of a carrier in the RF range. This available improvement is extremely inviting to a communication engineer vitally concerned with transmitting large amounts of information. In addition, the ability to concentrate available transmitter power within the transmitted electromagnetic wave also increases with carrier frequency. Thus, using higher carrier frequencies increases the capability of the system to achieve higher power densities, which generally leads to improved performance. For both of these reasons optical communications has emerged as a field of special technological interest.

2

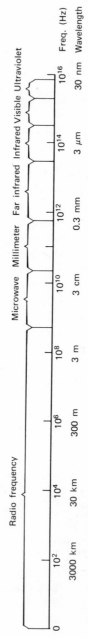

Figure 1.1. The electromagnetic spectrum.

Unfortunately, communicating with optical carrier frequencies has several major difficulties. Since optical frequencies are accompanied by extremely small wavelengths,† optical component design requires essentially its own technology, completely different from design techniques associated with RF, microwave, and millimeter devices. The development of satisfactory optical components has been a primary impairment to optical communication in the past. A significant advance was made by the advent of the laser, a relatively high-powered optical carrier source available in both the infrared and visible frequency range. Further progress was made by development of wide band optical modulators and efficient detectors. Some excellent discussions of optical components—their description, development, and characteristics—are included in recent texts by Pratt [1], Ross [2], and Yariv [3]. Many other engineering discussions of optical components are also available in the literature (see Reference 4 for an extensive bibliography).

Another serious drawback to optical communications is the effect of the propagation path on the optical carrier wave. This is because optical wavelengths are commensurate with molecule and particle sizes, and propagation effects are generated that are uncommon to radio and microwave frequencies. Furthermore, these effects tend to be stochastic and time varying in nature, which hinders accurate propagation modeling. A vast amount of experimental data has been collected to aid in understanding this optical propagation phenomenon and, although certain models have been established, continued exploration is required for refinement and further justification.

The development of optical components and the derivation of propagation models, however, are only part of the overall system design problem. A communication engineer must also be concerned with choice of components, the selection of system operations, and finally the interfacing (interconnecting) of these operations in the best possible manner. These interfacing decisions require reasonably accurate mathematical models, which indicate component performance, anomalies, and degradations, knowledge of which can be used to advantage in system design. It is this aspect of optical communications at which this book is aimed. Our objective is to understand system capability and to formulate system design procedures and performance characteristics for the implementation of an overall optical communication system.

1.1 THE SYSTEM MODEL

The block diagram of a typical optical communication system is shown in Figure 1.2. The diagram is composed of standard communication blocks, endemic to any communication system. A source producing some type of

---

† Wavelength equals the speed of light times the reciprocal of the frequency in hertz.

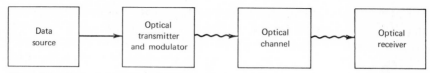

Figure 1.2. Optical communication system block diagram.

information (waveforms in time, digital symbols, etc.) is to be transmitted to some remote destination. This source has its output modulated onto an optical carrier (a carrier frequency in the optical portion of the electromagnetic spectrum). This carrier is then transmitted as an optical light field, or beam, through the optical channel (free space, turbulent atmosphere, fiberoptic waveguide, etc.). At the receiver the field is optically collected and processed (photodetected), generally in the presence of all interference and inherent background radiation (undesired light fields or other electromagnetic radiation). Of course, except for the fact that the transmission is accomplished at the optical range of carrier frequencies, the operations just mentioned describe any communication system using modulated carriers. Nevertheless, the optical system employs devices somewhat uncommon to the standard components of the RF system. These devices have significant differences in their operation and associated characteristics, often requiring variations in design procedures.

The modulation of the source information onto the optical carrier can be in the form of frequency modulation (FM), phase modulation (PM), or possibly amplitude modulation (AM), each of which can be theoretically implemented at any carrier frequency in the electromagnetic range. In addition, however, several other less conventional modulation schemes are also often utilized with optical sources. These include intensity modulation (IM), in which information is used to modulate the intensity (to be defined subsequently) of the optical carrier, and polarization modulation (PLM), in which spatial characteristics of the optical field are modulated.

The optical receiver in Figure 1.2 collects the incident optical field and processes it to recover the transmitted information. A typical optical receiver can be represented by the three basic blocks shown in Figure 1.3,

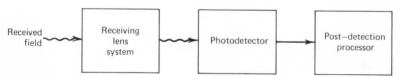

Figure 1.3. The optical receiver.

consisting of an optical receiving lens system (referred to as the receiver front end), an optical photodetector, and a postdetection processor. The lens system filters and focuses the received field onto the photodetector, where the optical signal is converted to an electronic signal. The processor performs the necessary amplification and filtering operations to recover the desired information from the detector output.

Optical receivers can be divided into two basic types—power detecting receivers and heterodyning receivers. Power detecting receivers (often called direct detection, or noncoherent, receivers) have the front end system shown in Figure 1.4a. The lens system and photodetector operate to detect the instantaneous power in the collected field as it arrives at the receiver. Such receivers represent the simplest type for implementation, and can be used whenever the transmitted information occurs in the power variation

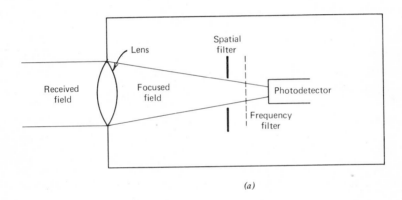

(a)

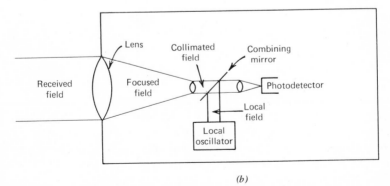

(b)

Figure 1.4. (a) Direct detection receiver. (b) Heterodyne detection receiver.

of the received field. Heterodyning receivers have the front end system shown in Figure 1.4b. A locally generated optical field is electromagnetically mixed through a front end mirror, and the combined wave is photodetected. Such receivers are used whenever information is amplitude modulated, frequency modulated, or phase modulated on to the optical carrier. Heterodyning receivers are more difficult to implement and require close tolerances on the spatial coherence of the two optical fields being mixed. For this reason, heterodyned receivers are often called (spatially) coherent receivers. For either type of receiver, the front end lens system has the role of focusing the received or mixed field onto the photodetector surface. This focusing allows the photodetector area to be much smaller than that of the receiving lens.

Since the receiver front end is essentially an optical receiving antenna, it can be described most simply by its effective receiver area and field of view. The receiver area is the collecting area presented to the impinging field, and, ideally, corresponds to its physical area. Often, however, receiver losses are accounted for by defining an effective receiver area which is less than its actual size. The field of view of a receiver antenna defines the various directions of arrival of electromagnetic waves that the photodetector will observe. In later work these parameters are made more concrete by formal definitions.

The receiver front end, in addition to focusing the optical field onto the photodetector, also provides some degree of filtering, as shown in Figure 1.4. These filters are employed prior to photodetection to reduce the amount of undesired background radiation. Optical filters may operate on the spatial properties of the focused fields (polarization filters, field stops, etc.) or may filter in the frequency domain; that is, pass certain bands of frequencies and reject others. The latter filters determine the bandwidth of the resulting optical field subsequently photodetected. Typical frequency filters are on the order of hundreds of gigahertz.

Photodetectors convert the focused optical field into an electrical signal for processing. Although there are several types of detectors available all behave according to quantum mechanical principles, utilizing photosensitive materials to produce current or voltage responses to changes in impinging optical field power. The basic model defining this interaction for all photodetectors is well accepted, although detectors may differ in their output response characteristics. This basic model, which is examined in detail in Section 1.8, is extremely important to the communication engineer, since it generates the inherent statistics that must be utilized in design of the post-detection processing. The most common type of photodetectors are the photodiodes, photomultipliers, and photoconductors.

The detection of optical fields is hampered by the various noise sources present throughout the receiver. The most predominant in long-distance communication is the interference radiation that is collected at the receiver lens along with the desired optical field. Although this radiation may be reduced by proper spatial filtering, it still represents the most significant interference in the detection operation. The background effect, can be essentially eliminated over short distances where direct-coupled fiberoptic waveguides can be used for the transmission path. A second noise source is the circuit and electronic noise generated in the processing operations. Such noise is accurately modeled as additive white Gaussian thermal noise, whose voltage spectral level is directly related to the receiver temperature. The thermal noise, is the primary noise source in any RF or microwave communication receiver. Lastly, the photodetector itself, not being a purely ideal device, produces internal interference during the photodetection operation. This generally takes the form of a detector "dark current"—an output response being produced even when no impinging field is applied to the input. Each of these noise sources must be properly accounted for in any receiver analysis. We point out that other types of effective "noise" often appear in optical analysis (quantum noise, fluctuation noise, etc.). As we shall see, these are derived from various causes of error in the detection of desired fields, but in reality are not physical noise sources.

The remainder of this chapter addresses the modeling and analysis necessary to understand optical communications. By and large all communication components have counterparts at optical frequencies. But, as previously stated, sometimes these counterparts take a different form from that which experience dictates. Again, this is primarily because of the large difference in wavelength. Certain components, such as optical modulators, while performing the same mathematical functions, require novel means of implementation when compared to past RF experience.

1.2 OPTICAL TRANSMITTERS

The optical transmitter in Figure 1.2 can be essentially subdivided into a cascade connection of an optical source, a modulator, and an optical antenna, as shown in Figure 1.5. Optical sources (lasers, photodiodes, light bulbs) produce optical electromagnetic fields, while having provisions for modulating data waveforms onto its output. Depending on the source, the modulation may be impressed during the generation of light, or after the light has been generated.

The optical antenna is a lens system that focuses the modulated source output into an optical beam (electromagnetic field) for transmission. The antenna may be physically separated from the source, or it may be incorporated within it. Any type of electromagnetic transmitting antenna is described

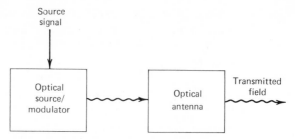

Figure 1.5. The optical modulator/transmitter.

basically by its radiation pattern, the latter describing the power density (power per unit solid angle) transmitted in each direction. These patterns can be summarized by their *gain* and *beam angle*. The gain of an antenna is related to the maximum power density of the radiation pattern. Formally, the antenna gain G_a is defined as†

$$G_a = \frac{\text{maximum power density of the radiation pattern}}{\text{power density due to an isotropic antenna}} \quad (1.2.1)$$

An isotropic antenna is one that transmits electromagnetic energy uniformly in all directions. If a total antenna power of P W is available for transmission, an isotropic antenna will transmit a power density of $P/4\pi$ W/unit solid angle. This becomes the denominator of (1.2.1). Thus, G_a is the maximum power density of the radiated field normalized by the factor $P/4\pi$. Alternatively, G_a can be interpreted as the factor by which $P/4\pi$ should be multiplied to give the maximum power density within the beam.

The antenna beam angle Ω_a is defined as the solid angle in steradians, measured from the antenna, into which the maximum power density must be concentrated in order to have the same total power. Thus,

$$G_a\left(\frac{P}{4\pi}\right)\Omega_a = P \quad (1.2.2)$$

or

$$\Omega_a = \frac{4\pi}{G_a} \qquad \text{sr} \quad (1.2.3)$$

The beam angle therefore indicates the solid angle into which most of the transmitted power appears to flow, and is inversely related to the antenna gain. Often this solid beam angle is described by its planar angle beamwidth— the projection of the solid angle onto a plane through its center, as shown in Figure 1.6. For a circular lens antenna of diameter d, transmitting an

† Antenna gains are often stated in terms of decibels (dB). G_a in decibels equals $10 \log G_a$.

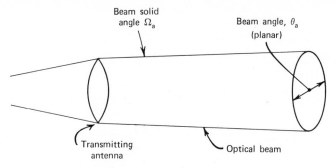

Figure 1.6. Transmitting antenna system.

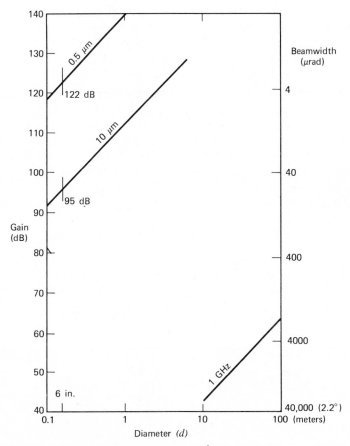

Figure 1.7. Antenna gain and beam width versus diameter.

electromagnetic wave of wavelength λ, the antenna planar beamwidth θ_a and gain G_a are given approximately by

$$\theta_a \cong \frac{\lambda}{d} \tag{1.2.4}$$

$$G_a = \left(\frac{4d}{\lambda}\right)^2 \tag{1.2.5}$$

Figure 1.7 shows a plot of these parameters as a function of diameter for several optical frequencies. The values of these parameters clearly show the power advantage of an optical system. For example, a 6 in. lens antenna at an optical frequency of 6×10^{14} Hz has an effective gain of approximately 122 dB, a sizable improvement over an RF antenna. (A 210 ft antenna generates only about 60 dB gain at RF.) On the other hand, the corresponding beamwidth is on the order of 3×10^{-6} rad—about 0.6 arcsecond. Thus, large gain is achieved in an extremely narrow beamwidth and it is evident that extreme pointing problems (aiming the transmitter at the receiver) are generated. This latter problem, which immediately becomes an integral part of the overall system design, is considered in Chapter 11.

1.3 THE TRANSMITTED OPTICAL FIELD

Since the transmitted optical field is an electromagnetic wave it is described at any spatial point by solutions to Maxwell's equations. Let ξ represent a point of a selected coordinate system in which the field source is located at the origin, as shown in Figure 1.8. At any time t, the electrical field is described at ξ by

$$\text{electric field} = \text{Real}\{f(t, \xi)\} \qquad \text{V/length} \tag{1.3.1}$$

where Real{ } means "real part of" and $f(t, \xi)$ is referred to as the complex field. At each such point ξ, the field has an *intensity* given by

$$I(t, \xi) = \frac{1}{2}\left(\frac{1}{Z_w}\right)|f(t, \xi)|^2 \qquad \text{W/area} \tag{1.3.2}$$

where Z_w is the wave impedance (i.e., the impedance of the medium; in free space, $Z_w \approx 377$ ohms). The *instantaneous power* of the field over an arbitrary area $\mathscr{A}$ at time t is given by the surface integral of the intensity over the area,

$$P_{\mathscr{A}}(t) = \int_{\mathscr{A}} I(t, \xi)(\mathbf{i} \cdot d\mathbf{a}) \tag{1.3.3}$$

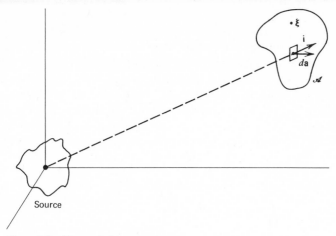

Figure 1.8. Transmission geometry.

where the integration is over all ξ in $\mathscr{A}$, and the dot product indicates the directional cosine between the unit vector **i** in the direction of power flow† and $d\mathbf{a}$, the normal vector to the surface area at ξ (Figure 1.8). Equations 1.3.1 to 1.3.3 are key equations relating optical field, intensity, and power, and are important to subsequent communication analysis. This is because optical detectors inherently respond to the intensity of the impinging fields.

If the source producing the field in Figure 1.8 is a point source (which is the usual model for an electromagnetic transmitter) operating in a free space medium, the field propagates as a plane wave in directions specified by the antenna gain pattern. At a point ξ, the complex field of a plane wave can be written specifically as

$$f(t, \xi) = \mathscr{E}(t, \xi) \exp\left[j\left(\omega t - \frac{2\pi}{\lambda} |\xi| \right) \right] \tag{1.3.4}$$

where $|\xi|$ is the distance to the point ξ, $\mathscr{E}(t, \xi)$ is the instantaneous field amplitude vector at ξ, and ω, λ are the wave radian frequency and wavelength, respectively. The amplitude vector $\mathscr{E}$ describes the field polarization in a plane (x, y) normal to the direction of propagation. This can be expanded as

$$\mathscr{E}(t, \xi) = \mathscr{E}_x(t, \xi)\mathbf{1}_x + \mathscr{E}_y(t, \xi)\mathbf{1}_y \tag{1.3.5}$$

† The vector $I(t, \xi)\mathbf{i}$ through ξ is called the Poynting vector. Equation 1.3.3 is the integration of the Poynting vector over the surface area.

where $\mathscr{E}_x(t, \xi)$ and $\mathscr{E}_y(t, \xi)$ are the complex polarization components, and 1_x, 1_y are unit coordinate vectors in the (x, y) plane. The polarization components determine the polarization state of the plane wave. If $|\mathscr{E}_x(t, \xi)| = |\mathscr{E}_y(t, \xi)|$ and are 90° out of phase, the field is circularly polarized. If both are in phase or one is zero, the field is linearly polarized. In polarization modulation systems, information is transmitted by varying the polarization vector of the plane wave. Unless otherwise stated, we assume linearly polarized fields throughout subsequent discussions. For a point ξ located within the transmitting antenna beamwidth, the linearly polarized field can be simplified to

$$f(t, \xi) = a(t, \xi) \exp[j\omega t] \tag{1.3.6}$$

where

$$a(t, \xi) = \frac{C}{|\xi|} \mathscr{E}(t) \exp[-j2\pi|\xi|/\lambda] \tag{1.3.7}$$

and

$$C = \left[\frac{G_a}{4\pi}\right]^{1/2} \tag{1.3.8}$$

Here $\mathscr{E}(t)$ is the transmitted field amplitude variation of the point source and G_a is the antenna gain. (If ξ is not located within the antenna beamwidth, G_a must be replaced by the actual antenna gain determined from the radiation pattern.) The exponential term in (1.3.7) accounts for the time delay in propagating the distance $|\xi|$. The complex field in (1.3.6) has the intensity given by (1.3.2),

$$I(t, \xi) = \frac{C^2}{2Z_w|\xi|^2} |\mathscr{E}(t)|^2 \tag{1.3.9}$$

It is convenient to define

$$P_s(t) \triangleq \frac{|\mathscr{E}(t)|^2}{2Z_w} \tag{1.3.10}$$

as the transmitted source power function, and rewrite (1.3.9) as simply

$$I(t, \xi) = \frac{G_a}{4\pi|\xi|^2} P_s(t) \tag{1.3.11}$$

Thus, for a plane wave, the instantaneous field intensity at a point ξ depends directly on the transmitted field power and antenna gain, and varies indirectly with the square of the distance to ξ. The function $a(t, \xi)$ in (1.3.6) is called the *complex envelope* of the plane wave field at ξ. If $a(t, \xi)$ does not

Table 1.1   Modulated Fields

Modulation type	$\mathscr{E}(t)$
Amplitude modulation (AM)	$1 + d(t)$
Phase modulation (PM)	$\exp[j\,d(t)]$
Frequency modulation (FM)	$\exp[j\int d(t)\,dt]$
Intensity modulation	$[1 + d(t)]^{1/2}$

$d(t) = $ real modulating waveform
$\mathscr{E}(t) = $ complex field envelope

depend on t, the field is said to be a *monochromatic field*, and the electric field in (1.3.6) corresponds to a pure sine wave in time at any ξ. If $a(t, \xi)$ has time variation with a narrow frequency band relative to the carrier frequency ω, the field is said to be *quasi monochromatic*. Any time variation in the complex envelope $a(t, \xi)$ must be due to the amplitude function $\mathscr{E}(t)$ in (1.3.7). Thus, any type of modulation imposed on the optical carrier at the transmitter must be exhibited in the complex amplitude $\mathscr{E}(t)$. In Table 1.1 we summarize the various mathematical forms for $\mathscr{E}(t)$, dependent on the type of carrier modulation used.

We are often interested in determining the plane wave field power over an area $\mathscr{A}$, due to a remote point source. This can be obtained directly from (1.3.3). Consider the area $\mathscr{A}$ in Figure 1.9 containing points $\mathbf{r}$ of a planar coordinate system having its origin at the center of $\mathscr{A}$. Let a point source a distance L away transmit a plane wave in the direction of a ray vector $\mathbf{z}$ through the origin of $\mathscr{A}$, as shown. If we assume $L \gg |\mathbf{r}|$ (the source is far

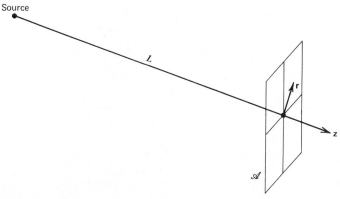

Figure 1.9.   Rotated transmission geometry.

away), then all points in $\mathscr{A}$ are essentially a distance L away from the source. Neglecting the propagation delay (i.e., reorienting the time axis), the field at a point $\mathbf{r}$ in $\mathscr{A}$ due to the point source can be written

$$f(t, \mathbf{r}) = a(t) \exp[\, j(\omega t - \mathbf{z} \cdot \mathbf{r})] \tag{1.3.12}$$

where

$$a(t) = \left[\frac{G_a}{4\pi L^2} \right]^{1/2} \mathscr{E}(t)$$

and the vector $\mathbf{z}$ has magnitude $2\pi/\lambda$. The dot product accounts for the phase variation over the surface of $\mathscr{A}$. The power over $\mathscr{A}$ is then given by (1.3.3),

$$
\begin{aligned}
P_{\mathscr{A}}(t) &= \frac{1}{2Z_w} \int_{\mathscr{A}} |a(t)|^2 \left(\frac{\mathbf{z}}{|\mathbf{z}|} \cdot d\mathbf{a} \right) \\
&= \frac{G_a P_s(t)}{4\pi L^2} \int_{\mathscr{A}} \frac{\mathbf{z}}{|\mathbf{z}|} \cdot d\mathbf{a}
\end{aligned} \tag{1.3.13}
$$

If the area $\mathscr{A}$ is normal to the direction of arrival of the plane wave, $\mathbf{z}/|\mathbf{z}| \cdot d\mathbf{a} = d\mathbf{a}$, and

$$P_{\mathscr{A}}(t) = \frac{G_a A}{4\pi L^2} P_s(t) \tag{1.3.14}$$

where A is the integrated area of $\mathscr{A}$. Thus the power collected over the normal area $\mathscr{A}$, due to a plane wave from a remote point source, is directly proportional to the area A. The factor

$$\mathscr{L}_p \triangleq \frac{1}{4\pi L^2} \tag{1.3.15}$$

is called the free space propagation loss in transmitting power over a distance L. (The units of L must be identical to those of A.) Note that this free space loss does not depend on frequency, and therefore is the same for any carrier frequency of the electromagnetic spectrum.†

Let us extend (1.3.12) to the case of several point sources. Suppose each is approximately a distance L away, each transmitting toward $\mathscr{A}$ from a different direction $\mathbf{z}_i$, with the same carrier frequency ω (Figure 1.10a). The field at $\mathbf{r}$ in $\mathscr{A}$ now becomes the combined fields from each point,

$$f(t, \mathbf{r}) = a(t, \mathbf{r})e^{j\omega t} \tag{1.3.16}$$

---

† Only when the receiving area A is replaced by an equivalent receiving antenna gain will a wavelength parameter enter into (1.3.15). This tends to be misleading, since it implies that propagation losses are different in RF and optical systems.

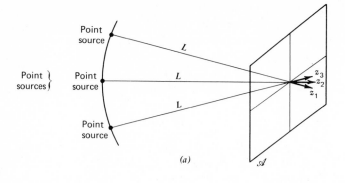

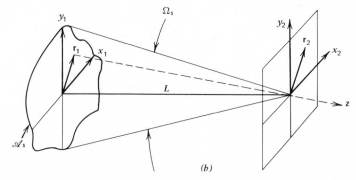

Figure 1.10. (a) Point source geometry. (b) Extended source geometry.

where now

$$a(t, \mathbf{r}) = \sum_i a_i(t) e^{-j(\mathbf{z}_i \cdot \mathbf{r})} \qquad (1.3.17)$$

Here $\{a_i(t)\}$ are the complex field envelope variations, and the sum is over all source points. Thus each point $\mathbf{r}$ receives the superposition of the individual plane waves. Note that each plane wave will in general produce a different phase variable at each point $\mathbf{r}$ of $\mathscr{A}$. We can also consider (1.3.17) as an expansion of the complex envelope $a(t, r)$ into a sum of functions where the functions correspond to arriving plane waves. In later work, we investigate this interpretation further.

Equation 1.3.17 suggests an extension to an integral over an extended source amplitude function. Let $F_s(t, \mathbf{r}_1)$ describe the source field envelope at time t and at point $\mathbf{r}_1$ in the source plane a distance L away from the receiver plane, as shown in Figure 1.10b. The resulting field that will be produced at point $\mathbf{r}_2$ in the receiver plane is obtained by use of Fresnel–Kirchhoff diffraction [5–7]

$$f_r(t, \mathbf{r}_2) = \frac{\exp[(j\pi/\lambda L)|\mathbf{r}_2|^2 + j(2\pi L/\lambda)]}{j\lambda L}$$

$$\cdot \int_{\mathcal{A}_s} F_s(t, \mathbf{r}_1) \exp\left(j\frac{\pi}{\lambda L}|\mathbf{r}_1|^2\right) \exp\left[-j\frac{2\pi}{\lambda L}(x_1 x_2 + y_1 y_2)\right] d\mathbf{r}_1$$

(1.3.18)

where (x, y) are the coordinates of the corresponding points $\mathbf{r}_1$ and $\mathbf{r}_2$, and $\mathcal{A}_s$ is the area of the source. Thus a source amplitude function produces a receiver field at a distance L according to (1.3.18). When $L\lambda \gg |\mathbf{r}_1|^2$, it is convenient to deal with the source angular spectrum rather than its amplitude function. The angular spectrum describes the source amplitude along ray lines $\mathbf{z}$ emitted from the source plane and passing through the center of the receiver, as shown in Figure 1.10b. If we consider a ray $\mathbf{z}$ with arrival angles (θ_x, θ_y) with respect to the normal to the receiver area, we can then define the source angular spectrum $B_s(t, \mathbf{z})$ as the source amplitude in the direction of arrival of $\mathbf{z}$. Thus, $B_s(t, \mathbf{z})$ is obtained by evaluating $F_s(t, \mathbf{r}_1)$ at the point $\mathbf{r}_1 = (L\theta_x, L\theta_y)$. If in addition we let $|\mathbf{z}| = 2\pi/\lambda$, then (1.3.18) can be rewritten as

$$f_r(t, \mathbf{r}_2) = \frac{\exp[(j\pi/\lambda L)|\mathbf{r}_2|^2 + j(2\pi L/\lambda)]}{j(\lambda/L)} \iint_{\Omega_s} B_s(t, \mathbf{z}) e^{-j\mathbf{z} \cdot \mathbf{r}_2} \, d\theta_x \, d\theta_y \quad (1.3.19)$$

where Ω_s is the solid angle subtended by the source area $\mathcal{A}_s$. When compared with (1.3.17), (1.3.19) appears as an integration over a continuum of point sources, with $B_s(t, \mathbf{z})$ describing the source emission from $\mathcal{A}_s$ in the direction of $-\mathbf{z}$.

In later analysis of optical fields, it is often convenient to consider orthogonal expansions of the complex envelope into orthonormal functions. That is, we consider the field over a spatial area $\mathcal{A}$ and a time interval $(0, T)$, and expand the field envelope within these regions as an infinite series:

$$a(t, \mathbf{r}) = \sum_{i=1}^{\infty} a_i \phi_i(t, \mathbf{r}) \qquad 0 \le t \le T, \quad \mathbf{r} \in \mathcal{A} \qquad (1.3.20)$$

where

$$a_i = \int_{\mathcal{A}} \int_0^T a(t, \mathbf{r}) \phi_i^*(t, \mathbf{r}) \, dt \, d\mathbf{r} \qquad (1.3.21)$$

Here $\{a_i\}$ are the complex expansion coefficients, and $\{\phi_i(t, \mathbf{r})\}$ represents a complete set of complex orthonormal basis functions over $\mathcal{A}$ and $(0, T)$.

That is,

$$\int_{\mathscr{A}} \int_T \phi_i(t, \mathbf{r})\phi_j^*(t, \mathbf{r}) \, dt \, d\mathbf{r} = \begin{cases} 1, & i = j \\ 0, & i \neq j \end{cases} \qquad (1.3.22)$$

where the asterisk denotes complex conjugate. The equality in (1.3.20) is in a squared integrable sense, and the convergence of the sum on the right to the envelope function on the left requires only a bounded energy constraint on the radiation field.† The advantage of an expansion of this form is primarily for mathematical convenience, although such expansions often yield physical insight into the related optical processing. It is natural to consider the functions $\{\phi_i(t, \mathbf{r})\}$ as defining the "modes" of the field over the area $\mathscr{A}$ and interval $(0, T)$. The coefficients $\{a_i\}$ then become the modal coefficients, describing the envelope in each mode. Since different orthonormal sets are available, mode descriptions of a given field are not necessarily unique, each corresponding to a different expansion set. This mode expansion procedure is examined in more detail in Chapter 3.

1.4 STOCHASTIC FIELDS

In optical system modeling we are often forced to deal with stochastic, or random, fields. Such fields arise when dealing with transmission effects (turbulence, scattering, fading, etc.) which cause random variations to occur in the transmitted beam. These random variations lead to statistical fluctuations in the received field intensity which can only be analyzed after associating proper statistics with the field itself. A large amount of optical communication research has been devoted to this aspect of field modeling, with a significant portion concerned with reducing large amounts of empirical field data to convenient mathematical forms.

The complex envelope of a stochastic field must be considered random at each point t and $\mathbf{r}$ describing the field over a designated area. As such, random fields are completely described by their probability densities, that is, the probability densities associated with each point, or set of points, of the field. These densities are often difficult to model exactly, and often assumptions must be imposed upon these statistics. For this reason, stochastic field analysis is often confined to second-order statistics associated with the field, in particular, its coherence function. In this regard the *time-space (mutual) coherence function* of a stochastic field $f(t, \mathbf{r})$ at points $(t_1, \mathbf{r}_1)$ and $(t_2, \mathbf{r}_2)$ is formally defined as

$$R_f(t_1, \mathbf{r}_1; t_2, \mathbf{r}_2) \triangleq E[f(t_1, \mathbf{r}_1)f^*(t_2, \mathbf{r}_2)] \qquad (1.4.1)$$

† A brief review of field expansions with orthonormal functions is presented in Appendix B.

where E is the expectation operator over the joint field densities at the points involved.† The mean squared value of the field at $(t, \mathbf{r})$ then follows as $R_f(t, \mathbf{r}; t, \mathbf{r})$, and can be evaluated directly from the coherence function at any t and $\mathbf{r}$.

Stochastic fields are often described by their inherent coherence properties. A stochastic field is said to be *temporally stationary* if the time dependence in its coherence function depends only on the time difference $t_1 - t_2$. The field is *spatially homogeneous* if the spatial dependence in the coherence function depends only on the spatial distance $(\mathbf{r}_1 - \mathbf{r}_2)$. A field is completely *homogeneous* if it is both temporally *stationary* and spatially *homogeneous*. A stochastic field is said to be *coherence-separable* if its coherence function factors as

$$R_f(t_1, \mathbf{r}_1; t_2, \mathbf{r}_2) = R_t(t_1, t_2)R_s(\mathbf{r}_1, \mathbf{r}_2) \qquad (1.4.2)$$

The factor $R_s(\mathbf{r}_1, \mathbf{r}_2)$ is called the *spatial coherence function* while $R_t(t_1, t_2)$ is called the *temporal correlation* of the field. The function $R_s(\mathbf{r}, \mathbf{r})$ is called the field *irradiance function*. Often the spatial coherence function is normalized as

$$\tilde{R}_s(\mathbf{r}_1, \mathbf{r}_2) \triangleq \frac{R_s(\mathbf{r}_1, \mathbf{r}_2)}{[R_s(\mathbf{r}_1, \mathbf{r}_1)R_s(\mathbf{r}_2, \mathbf{r}_2)]^{1/2}} \qquad (1.4.3)$$

The normalized space coherence function is therefore bounded in magnitude by 1. A field that is coherence-separable and homogeneous is said to be *spectrally pure*, and its coherence function can always be written as $R_t(\tau)\tilde{R}_s(\boldsymbol{\rho})$ where $\tau = t_1 - t_2$ and $\boldsymbol{\rho} = \mathbf{r}_1 - \mathbf{r}_2$. Spectrally pure fields therefore have $\tilde{R}_s(0) = 1$ and a mean squared value of $R_t(0)$ at all $\mathbf{r}$. The Fourier transform of $R_t(\tau)$ is called the *intensity spectrum* of the field.

A stochastic field is *space coherent over an area* $\mathscr{A}_0$ if $\tilde{R}_s(\mathbf{r}_1, \mathbf{r}_2) = 1$ for all $\mathbf{r}_1, \mathbf{r}_2$ in $\mathscr{A}_0$. Note that in this case its mutual coherence is given by $R_t(t_1, t_2)$, and therefore only the temporal randomness is exhibited over $\mathscr{A}_0$. In this sense, the field is identical to the plane wave in (1.3.7), except it has its field amplitude $\mathscr{E}(t)$ as a random process in t. Hence, random fields coherent over an area behave as random plane waves over that area. A stochastic field is *space incoherent* if $\tilde{R}_s(\mathbf{r}_1, \mathbf{r}_2) = 0$, $\mathbf{r}_1 \neq \mathbf{r}_2$. Otherwise, it is *partially space coherent*.

Stochastic field envelopes also have infinite series expansions into orthonormal functions as in (1.3.20), except the coefficients in (1.3.21) are now random variables.‡ The convergence of the sum is now in a mean squared sense, and requires bounded average energy in the field over the expansion

---

† The reader may wish to review statistical averaging and random fields in Appendix A before further discussions of the random field case.

‡ See the discussion of random field expansions in Appendix B.

area and time interval. Again, any set of complete orthonormal functions can be used for the expansion. If the coefficients $\{a_i\}$ are uncorrelated variables, the expansion is called a *Karhunen–Loeve* (KL) expansion. A KL expansion will occur if the orthonormal functions $\phi_i(t, \mathbf{r})$ are such that they each satisfy the integral equation

$$\int_{\mathscr{A}} \int_0^T R_f(t_1, \mathbf{r}_1; t_2, \mathbf{r}_2) \, \phi_i(t_2, \mathbf{r}_2) \, dt_2 \, d\mathbf{r}_2 = \gamma_i \phi_i(t_1, \mathbf{r}_1) \qquad (1.4.4)$$

for some set of constants $\{\gamma_i\}$. The members of the set $\{\phi_i(t, \mathbf{r})\}$ satisfying this equation are called *eigenfunctions*, and the $\{\gamma_i\}$ are its *eigenvalues*. It can be shown (Problem 1.6) that the eigenvalues are also the mean squared value of the corresponding random coefficients $\{a_i\}$. Note that (1.4.4), and therefore the resulting eigenfunctions, depend only on the field mutual coherence function. We therefore see that although any orthonormal set can be used for the expansion, the coefficients will be uncorrelated only if a KL expansion is used; that is, if the expansion functions satisfy (1.4.4).

1.5 THE OPTICAL CHANNEL

When designing a communication system the properties of the propagation path from transmitter to receiver must be taken into account. Proper characterization of this path is equivalent to defining the communication channel in Figure 1.2. Electromagnetic propagation can be roughly divided into guided and unguided transmissions. In an unguided channel the source transmits the field freely into a medium with no attempt to control its propagation other than by its antenna gain pattern. The principal example of an unguided channel is the so-called space channel in which the medium involved may be free space, the atmosphere, or the ocean (underwater). In a guided channel, a waveguide is used to confine the wave propagation from transmitter to receiver. The primary example of a guided system in the optical region is the fiberoptic channel. In this section we attempt to summarize some of the extensive work dealing with the modeling of these channels.

1.5.1 The Unguided (Space) Channel

The operating characteristics of the space channel depend primarily on the properties of the medium involved. The simplest type of unguided channel is the free space channel, in which the medium involved between transmitter and receiver is free space. This would characterize propagation paths outside the Earth's atmosphere, or perhaps in a short-range, controlled-laboratory environment. The principal effect is the propagation loss in

(1.3.15). The transmission is distortion free, and the object of system design is merely to overcome the loss factor by sufficient size transmitters and receivers.

When the propagating medium is not free space, additional effects must be included in the channel model. This is because when propagating through a nonfree space environment, an electromagnetic wave undergoes effects tending to alter the structure of the wave. These alterations depend on the frequency of the wave, and are due primarily to interactions with inhomogeneities and foreign particles comprising the medium. One can expect these effects to predominate whenever the particle size is comparable to or greater than the wavelength of the field. For this reason atmospheric distortion may be quite severe in the optical range, where wavelengths are commensurate with matter as small as molecules. Thus the nonfree space channel presents a major communication hurdle that must be of concern to system designers. Particles in the transmission path primarily cause field absorption and scattering, although the degree of each will depend on the type of channel (underwater, clear air, turbulent atmosphere, etc.). If the absorption is severe, little communication can be accomplished, even with sophisticated techniques. On the other hand, if the field is merely distorted, with little additional power loss, application of communication theoretic techniques can sometimes be applied to recover a sufficient portion of the field modulation so as to maintain a satisfactory communication link.

Absorption and scattering in a space channel manifest themselves in the electromagnetic field through both amplitude and spatial effects. Amplitude effects are exhibited in the time variation of the field, and primarily involve power loss, power fluctuations, and frequency filtering. Spatial effects appear as variations in the beam direction, or as distortion effects across the beam front. In particular, they exhibit effects over points of a receiving surface, and limit the extent to which a transmitted beam can be collected coherently. Since the particle structure of the channel tends to be random in nature (because of temperature gradients in the atmosphere, currents in the ocean, etc.) amplitude and spatial fluctuations tend to be stochastic in nature and can only be described statistically, sometimes only by their coherence properties. An analysis of the behavior of field coherence functions in a random optical channel is presented in Section D.1 of Appendix D.

Amplitude Effects. The propagation of an electromagnetic wave is always accompanied by a reduction in field strength. A portion of this reduction is due to the propagation loss predicted by (1.3.15). Additional loss is caused by the absorption and scattering phenomena. Since these anomalies are critically related to wavelength, their effects are negligible for most of the RF range, but extremely important in the optical range. Careful selection of optical frequencies is therefore necessary to avoid obvious space inhomogeneities (hydrogen, oxygen, etc.) while sustaining a transmission channel. Power losses due to absorption and scattering effects are accounted for at a receiver by multiplying the free space received power by the transmission loss factor (transmissivity)

$$\mathcal{L}_t = e^{-\alpha_t L} \tag{1.5.1}$$

where L is the propagation distance and α_t is a per unit length loss coefficient. This latter term is often written as the sum of separate absorption and scattering coefficients:

$$\alpha_t = \alpha_a + \alpha_{sc} \tag{1.5.2}$$

Absorption coefficients α_a and scattering coefficients α_{sc} are readily available as a function of wavelength and altitude, essentially derived from years of empirical studies [8]. Scattering coefficients α_{sc} are generally separated into two types of scattering. *Rayleigh scattering* is essentially diffuse scattering and tends to be low level in effect, characterizing a quiet, clear air atmospheric channel. *Mie scattering* is that caused by larger particles, such as with fog, smoke, water droplets, and so on, and is more severe. Additional power loss factors are often included to account for amplitude variations due to polarization fluctuations (variation of the polarization vector during transmission).

While (1.5.1) represents an average type of loss factor which can be simply and conveniently applied in power calculations, the instantaneous effect of particle scattering is to produce a multipath phenomenon that appears as a random fading of the received wave power. This fading can be accounted for by a multiplication of the expected received field power by a random channel gain factor G_c. Theoretical studies of wave propagation in the atmosphere, using a perturbation approximation to the wave equation solution (called the Rytov approximation [9]), have shown that in the optical frequency range the normalized channel gain appears to obey lognormal statistics. That is, the probability density of the normalized gain G_c is given by

$$p_{G_c}(G) = \frac{1}{\sqrt{2\pi}\,\sigma_1 G} \exp\left[\frac{\ln G + \sigma_1^2}{2\sigma_1^2}\right] \tag{1.5.3}$$

Note that this density depends only on the one parameter σ_1^2, which is the variance of the log of G_c (called the log-variance of the channel). One can easily show that the lognormal channel gain G_c has a mean and variance given by

$$\text{mean } G_c = e^{-\sigma_1^2/2} \tag{1.5.4a}$$

$$\text{variance } G_c = 1 - e^{-\sigma_1^2} \tag{1.5.4b}$$

The mean gain produces the average attenuation, as in (1.5.1). The variance of the gain can be interpreted as a variation from the expected gain, producing a fluctuation in the received power level. These fluctuations are referred to as *scintillation*, and the variance in (1.5.4b) is often called the scintillation noise power. There have been extensive studies, both theoretical and empirical, to derive equations relating the log variance σ_1^2 to the characteristics of the atmosphere, but such results are hindered by its dependence on a large number of different parameters—wavelength, path length, size of scatters, altitude, and so on [10]. In addition, some discrepancies exist concerning the actual range of validity of the lognormal model and the saturation behavior of the log variance [11].

In addition to scintillation and attenuation, time variations in the electromagnetic field are susceptible to nonlinear phase shifting during transmission. This is because variations of the index of refraction with wavelength produce varying propagation times (delay spreading) of the frequencies of the modulated waveform. This spreading places upper limits on the modulation bandwidth that can be propagated. This effect, though predominant in the optical range, has been reported to be negligible for modulation bandwidths less than several hundred gigahertz.

In a strong scattering channel, a second type of delay distortion occurs when the scattering causes multipath effects. The various multipath signals arrive with different delays which, if sufficient, can cause destructive interference at the receiver. When the maximum delay difference among the paths (referred to as the *delay dispersion* of the channel) is a sizable portion of the period of significant frequencies, the combined multipath signal will no longer resemble the transmitted signal. Hence, delay dispersion limits the coherence bandwidth [12] of the channel. In general, if t_d is the expected channel time dispersion, the channel coherence bandwidth is limited to

$$B_c \approx \frac{1}{2t_d} \text{ Hz} \tag{1.5.5}$$

Typical values of t_d for a clear air turbulent channel are reported to be fractions of a nanosecond, leading to a frequency limitation of hundreds of gigahertz. However, in a strong scattering channel (fog, water, smoke) dispersion times as large as microseconds may occur, which severely limit coherence bandwidths.

Spatial Effects. Particulate scattering in the optical range also produces spatial effects on the transmitted beam.† These spatial effects are generally divided into *weak scattering* effects, producing beam movement while essentially retaining the structure of the beam front, and *strong scattering*, in which the beam is spatially distorted. In weak scattering, the beam is basically refocused by the medium, which tends to produce (a) beam wander

† Analytical models for the optical scattering channel are reviewed in Section D.2 of Appendix D.

and dancing (the direction of propagation of the beam moves about), (b) wave tilt (the orientation of the plane wavefront changes), and (c) beam spreading (the beamwidth is enlarged, effectively diluting its power and causing a smaller portion to be received). Such effects generally occur with quiet channels when the beamwidth is much narrower than the cross section of the inhomogeneities. For example, the effects are predominant during clear air transmission in an uplink from the Earth, where the atmospheric scatterers are relatively close to the transmitter and the optical beam has not had sufficient distance to spread.

Strong scattering occurs when the medium particles are particularly dense (fog, smoke, rain, underwater, etc.) and the beam diameter is much larger than the cross section of the impurities. Each particle now acts as a separate and independent scatterer for different points of the beam front, producing multiple scattering and refocusing of different parts of the field. The overall result is to cause a breakup of the wavefront of the beam during transmission. Since the location, movement, and denseness of the scatterers tend to be random in nature, the distorted beam becomes stochastic in form, characterized by a randomly varying amplitude and phase across the beam. From a communication point of view, there are several ways to interpret this transmission distortion. Consider a point source transmitting a monochromatic plane wave which, in free space, would arrive normal to the receiver plane with envelope a_0 over the receiver area $\mathscr{A}_r$. Each point of the plane wave over the receiver would be in phase, varying identically (sinusoidally) in time. Nonfree space transmission with strong scattering produces instead the field envelope $a_0 | f(\mathbf{r}) | \exp[j\phi(\mathbf{r})]$ where $[| f(\mathbf{r}) |, \phi(\mathbf{r})]$ are the random amplitude and phase factors at $\mathbf{r}$ on the receiver. The random amplitude and phase differences between pairs of points at the receiver can be associated with random amplitude plane waves arriving at random off-axis angles, as in (1.3.17). Thus beam breakup in space channels can be envisioned as the transformation of a normal plane wave into a collection of random plane waves, each arriving at random off-axis directions. We can associate these plane waves with an effective transmission from surface points of a large source, subtending a solid angle which includes all these directions. In effect, the point source has been transformed into an extended source by the channel. In certain cases communication improvement can be achieved by designing the receiver to observe more of this effective source area. This strategy is in fact made more rigorous in Chapter 3.

A measure of the effect of field distortion can be obtained by examining the field spatial coherence function over the receiver area $\mathscr{A}_r$. [$\tilde{R}_s(\mathbf{r}_1, \mathbf{r}_2)$ in (1.4.3)]. If the field were a pure plane wave, its spatial variation, and therefore its coherence function, would be constant over the entire receiver area $\mathscr{A}_r$. A disturbed beam would instead have a coherence function $\tilde{R}_s(\mathbf{r}_1, \mathbf{r}_2)$. If the disturbed field is spatially coherent over a region r_0 [i.e., if $\tilde{R}_s(\mathbf{r}_1, \mathbf{r}_2) \approx 1$, $|\mathbf{r}_1|$, $|\mathbf{r}_2| \in r_0$, and $\tilde{R}_s(\mathbf{r}_1, \mathbf{r}_2) \approx 0$ elsewhere], then r_0 can be interpreted as the coherence radius of the received field. In this case, the channel has reduced the effective coherence area of the receiver from A_r to πr_0^2. We shall find that in a direct detection receiver, the correlation area of the received field amplitude is of paramount importance, whereas in heterodyning receivers, the correlation area of the received field phase is important.† There has been considerable effort, both theoretically and experimentally, devoted to determining values of r_0 in terms of the atmospheric properties of the optical space channel [10, 13, 14. See also Section D.2 of Appendix D]. By introducing the Fourier transform of the spatial coherence as an effective source area, reduction of correlation radius r_0 can in fact be rigorously related to an equivalent source extension, similar to the interpretation stated earlier.

1.5.2 The Guided (Fiberoptic) Channel

In a guided channel the optical beam is confined to a closed path, or "pipe," from transmitter to receiver. Such optical channels have application whenever hard wire lines can be used, such as Earth-based systems, or short-range internal intercommunications. Besides having the inherent bandwidth of the optical beam, the guided channel has the advantage of (1) being completely shielded from background and electromagnetic interference, (2) immunity from turbulence, (3) complete control of the propagating beam by the periodic insertion of refocusing and amplification devices, and (4) interfacing neatly with the rapidly developing field of integrated optics.

Although optical waveguides can be constructed similar to microwave guides (hollow, metallic pipes), the modern techniques use encased glass fibers for lower loss. The waves propagate in a cylindrical glass core, which is surrounded by a dielectric, called a cladding (Figure 1.11a). The core is made to have a slightly higher refractive index than the cladding. Optical beams are transmitted into the core by inserting them at an angle θ (Figure 1.11b). The beam will propagate if the angle θ satisfies

$$\theta \leq \cos^{-1}\left(\frac{n_2}{n_1}\right) \tag{1.5.6}$$

where n_1 and n_2 are the refractive indices of the core and cladding, respectively. Beam directions having steeper angles are absorbed in the cladding and do not propagate. If $n_1 \approx n_2$, then θ will be small, and only small grazing angles are allowed to propagate. These smaller angles have less boundary

---

† It is common practice in field theory to deal with the wave structure function, rather than its coherence function. [See Appendix A, (A.2.28).] The structure function leads directly to the amplitude and phase coherence functions.

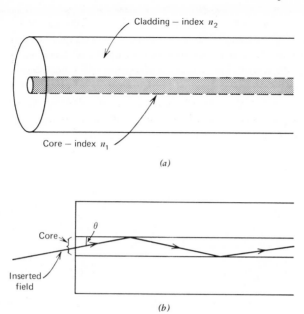

Figure 1.11. (a) Fiberoptic waveguide. (b) Field propagation path.

absorption, and therefore less loss during propagation. The only field attenuation results from absorption by the impurities in the core. In early design 100 dB/km was typical of fiberoptic loss. However, the use of purer glass has reduced this to approximately 10 dB/km, and improved methods of glass formation can reduce this further to about 2 dB/km at selected frequencies. The inability to purify the glass completely places the ultimate limit on achievable loss factors to this order of magnitude.

Mathematically, the electromagnetic field within a waveguide can be described by the same field equations as for free space, except boundary conditions must be satisfied at the guide walls. Thus, guided fields can be represented by orthonormal expansions, just as in (1.3.20). Again the individual terms of the expansions define the modes of the propagating field, as in the space channel. However, in free space the modes tend to be plane waves that separate spatially, whereas in guides the modes essentially superimpose as waves with orthogonal field vectors. (These field vectors correspond to the various types of transverse electric fields similar to those found in microwaveguides.) These additional terms are generated by nonperfect boundary conditions which create cross coupling. Physically this corresponds to imperfections in the core walls, especially at bends and interconnections. Even though a single field mode is inserted into the guide, additional modes are therefore almost unavoidable. These additional modes will each have slightly different grazing angles, and will propagate if they satisfy (1.5.6). We therefore see that the cladding is an important part of the fiberoptic guide. In addition to supporting and shielding the glass core, it allows for careful control of the dielectric index n_2, By proper construction, this will control the grazing angle θ which, in addition to reducing the loss, can limit the number of modes (number of terms in the field expansion) that will propagate. The approximate number of modes D_s of wavelength λ that propagate in a core of diameter d is given by [15]

$$D_s \cong \left(\frac{2\pi d}{\lambda}\right)^2 (n_1{}^2 - n_2{}^2) \qquad (1.5.7)$$

As n_2 is made closer to n_1, the number of modes (propagating field patterns) is reduced. Only a single mode will propagate if the core is narrow enough and if $n_2 \approx n_1$, that is, if

$$d \approx \frac{\lambda}{2\pi(n_1{}^2 - n_2{}^2)^{1/2}} \qquad (1.5.8)$$

Typically, claddings can be constructed with indices within several percent of n_1, and (1.5.8) therefore requires a core diameter on the order of a few microns (Problem 1.13). This obviously presents a nontrivial fabrication problem in fiberoptic construction.

From a communication point of view, having many modes propagate (referred to as an overmoded guide) is generally disadvantageous. This is due to the delay dispersion among the modes and the associated power division. Since each mode propagates at a slightly different angle, they propagate at different velocities, and are therefore delayed with respect to each other. (This is somewhat similar to the multipath effect in the space channel). If the delay dispersion is sufficient, the fields interfere. For a power detection receiver at the end of the guide, a dispersion of t_d sec limits the system coherence bandwidth according to (1.5.5). In a typical guide, delay dispersion may be as large as 50 nsec/km, and a 1 km guide would be limited to a frequency of about 10 MHz (Problem 1.14). Improved methods of fiber construction can often be used to reduce the dispersion. For example, by using a tapered index along the fiber radius (called an index profile) the dispersion can possibly be reduced to fractions of a nanosecond per kilometer, which increases the upper frequency to more than 500 MHz for a 1 km guide. In a heterodyning receiver, only the desired field pattern is used for detection and the overmode condition means a power loss to the unusable modes. Single mode operation avoids these dispersive effects and power losses. Except for the attenuation factor, the only distortion effect in a single mode guide is the nonlinearity of the phase function of the guide. In glass fibers this latter effect is generally not significant until we approach bandwidths of about 50 GHz.

1.6 THE DETECTED OPTICAL FIELD

The objective of the receiving optical antenna in Figure 1.3 is to collect the received field for the photodetecting surface. The receiver therefore responds to the instantaneous field over the receiver area. A detailed optical antenna system is shown in Figure 1.12. The receiving lens is located in the aperture plane, having a receiver area $\mathscr{A}_r$ and focal length f_c. The lens focuses the received field onto the focal plane located a distance f_c behind the lens. The focused field appears as a diffraction pattern in the focal plane. The photosensitive optical detector, with area $\mathscr{A}_d$, is located in the focal plane, and responds to the portion of the diffracted field on its surface.

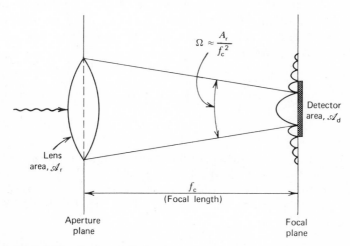

Figure 1.12.   Receiver optical system.

We often must determine the amount of transmitted field power collected over the receiving area $\mathscr{A}_r$. Let the source be a point source transmitting a normal plane wave at a distance L from the receiver, with a transmitter gain G_a in the receiver direction. For a free space channel, the instantaneous power collected over $\mathscr{A}_r$ is given directly by (1.3.14):

$$P_r(t) = G_a \mathscr{L}_p A_r P(t) = \frac{G_a P(t)}{4\pi L^2} A_r \qquad (1.6.1)$$

For the nonfree space channel, absorption and scatter effects $\mathscr{L}_t$ must be included. The received power must then be modified to

$$P_r(t) = (G_a \mathscr{L}_p \mathscr{L}_t A_r)P(t) \tag{1.6.2}$$

where $\mathscr{L}_t$ accounts for these channel effects. In dealing strictly with average power, $\mathscr{L}_t$ is generally taken as an equivalent power loss factor as in (1.5.1). In dealing with instantaneous power, on a slowly varying channel, $\mathscr{L}_t$ is treated as a random gain associated with the channel, and appears as a multiplicative random variable. As such, it introduces an average channel gain, similar to (1.5.4), with an additive power variance (scintillation noise). In later work we shall find it necessary to examine system performance from both of these points of view.

Often we must accurately describe the field wave itself at the detector, rather than simply its power value at the receiver lens. This requires us to account for the conversion of the received field in the aperture plane to the diffracted field in the focal plane. A well-designed receiver lens allows for Fraunhofer diffraction in the focal plane. Thus, if $f_r(t, \mathbf{r})$ is the received field over the aperture lens, and if $f_d(t, \mathbf{r})$ is the diffracted field in the focal plane, then the two are related by [5]

$$f_d(t, u, v) = \frac{\exp[j(\pi/\lambda f_c)(u^2 + v^2)]}{j\lambda f_c} \int_{\mathscr{A}_r} f_r(t, x, y) \exp\left[-j\frac{2\pi}{\lambda f_c}(xu + yv)\right] dx\, dy$$

$$\tag{1.6.3}$$

where $\mathbf{r} = (x, y)$ are the field coordinates in the aperture plane and (u, v) are the field coordinates in the focal plane, as shown in Figure 1.13. Equation 1.6.3 describes the manner in which the received and focal plane fields are related. Note that $f_d(t, \mathbf{r})$ is also related to the two-dimensional Fourier

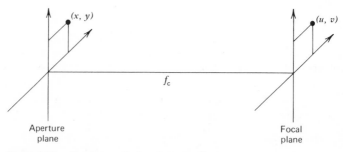

Figure 1.13. Receiver optical system coordinates.

transform of $f_r(t, \mathbf{r})$. That is, if we denote

$$F_r(t, \omega_1, \omega_2) = \int_{\mathscr{A}_r} f_r(t, x, y) \exp[-j(x\omega_1 + y\omega_2)] \, dx \, dy \qquad (1.6.4)$$

as the two-dimensional transform, then

$$f_d(t, u, v) = \frac{[j(\pi/\lambda f_c)(u^2 + v^2)]}{j\lambda f_c} F_r\left(t, \frac{2\pi u}{\lambda f_c}, \frac{2\pi v}{\lambda f_c}\right) \qquad (1.6.5)$$

Thus, diffraction patterns in optical receivers can be generated by simply resorting to transform theory. In communication analyses, this is an extremely useful result, since it means much of our receiver analysis reduces to straightforward linear system theory.

Consider a normal monochromatic plane wave over the receiver area $\mathscr{A}_r$. From our discussion in Section 1.3 the received field is then

$$
\begin{aligned}
f_r(t, x, y) &= a_0 e^{j\omega t} && x, y \in \mathscr{A}_r \\
&= 0 && \text{elsewhere}
\end{aligned}
\qquad (1.6.6)
$$

The resulting diffraction pattern in the focal plane is then obtained directly from (1.6.3). Its magnitude becomes

$$|f_d(t, u, v)| = \frac{a_0}{\lambda f_c} \left| \int_{\mathscr{A}_r} \exp\left[-j \frac{2\pi}{\lambda f_c} (xu + yv)\right] dx \, dy \right| \qquad (1.6.7)$$

If the aperture lens area is assumed rectangular with dimensions (d, b), the limits of integration in (1.6.7) become $|x| \le d/2$, $|y| \le b/2$, and the result integrates to

$$|f_d(t, u, v)| = a_0 \left(\frac{bd}{\lambda f_c}\right) \left| \frac{\sin(\pi \, du/\lambda f_c)}{(\pi \, du/\lambda f_c)} \cdot \frac{\sin(\pi \, bv/\lambda f_c)}{(\pi \, bv/\lambda f_c)} \right| \qquad (1.6.8)$$

The result for the u coordinate is sketched in Figure 1.14a. A similar plot exists along the v coordinate. If a circular lens of diameter d is used, the transform in (1.6.7) can be evaluated by converting to polar coordinates. This yields

$$
\begin{aligned}
|f_d(t, u, v)| &= \left(\frac{a_0}{\lambda f_c}\right) 2\pi \left| \int_0^d r J_0\left(\frac{\pi r \rho}{\lambda f_c}\right) dr \right| \\
&= a_0 \left(\frac{\pi d^2}{4\lambda f_c}\right) \left| \frac{2J_1(\pi \, d\rho/\lambda f_c)}{(\pi \, d\rho/\lambda f_c)} \right|
\end{aligned}
\qquad (1.6.9)
$$

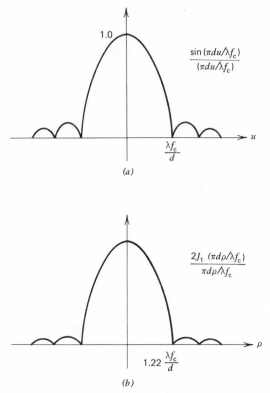

Figure 1.14. (a) Diffraction pattern for rectangular aperture. (b) Diffraction pattern for circular aperture.

where $\rho \triangleq (u^2 + v^2)^{1/2}$ and $J_0(x)$ and $J_1(x)$ are Bessel functions. This diffraction pattern is also sketched in Figure 1.14b, as a function of the circular radial distance ρ, and is the familiar "Airy disc" in optical diffraction theory. Note that in both cases, (1.6.8) and (1.6.9), the diffracted pattern occupies an area of approximately $(\lambda f_c)^2/A_r$ (i.e., the area encompassed by the largest hump) in the focal plane. Since this area is quite small at optical wavelengths, the detector area is generally many times larger than plane wave diffraction patterns.

If the plane wave arrives off the normal, as in Figure 1.15a, the received field over the receiver lens is now described by

$$f_r(t, x, y) = (a_0 e^{j\omega t}) e^{-j\mathbf{z} \cdot \mathbf{r}}$$

$$= (a_0 e^{j\omega t}) \exp\left[-j \frac{2\pi}{\lambda} (x \sin \theta_x + y \sin \theta_y) \right] \qquad (1.6.10)$$

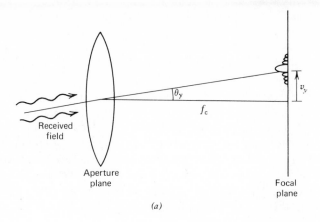

(a)

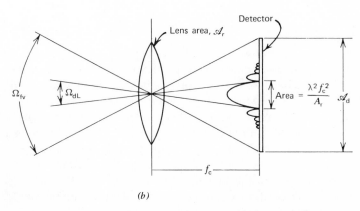

(b)

Figure 1.15. (a) Off-axis reception. (b) Relationship between Ω_{fv} and Ω_{dL}.

where θ_x, θ_y are the arrival angles from the normal. From transform theory, the resulting diffraction pattern magnitude is now

$$|f_d(t, u, v)| = \left(\frac{a_0}{\lambda f_c}\right)\left|F_r\left[t, \frac{2\pi}{\lambda f_c}(u - u_x), \frac{2\pi}{\lambda f_c}(v - v_y)\right]\right| \qquad (1.6.11)$$

where $u_x = f_c \sin \theta_x$ and $v_y = f_c \sin \theta_y$. Thus, off-angle incident plane waves generate position-shifted diffraction patterns in the focal plane. If the arrival angle is too large, the pattern is shifted off the detector surface area, and the field is not detected. This allows for a formal definition of *receiver field of view* as the solid angle, looking out from the receiver, within which all arriving plane waves must occur in order to project their diffraction pattern onto the detector. By standard geometric analysis in Figure 1.15b we see that

$$\Omega_{\mathrm{fv}} \approx \frac{A_{\mathrm{d}}}{f_{\mathrm{c}}^{2}} \tag{1.6.12}$$

where f_{c} is the focal length, A_{d} the detector area, and the approximation assumes that small angles are involved. In addition, we note that all arriving plane waves that superimpose their diffraction patterns are in essence indistinguishable in terms of direction of arrival. Thus the absolute minimal field of view that we can distinguish is that set of arrival angles whose patterns superimpose. This minimal field of view is limited only by the receiver optics (diffraction pattern) and is called *the diffraction limited field of view* of the receiver. Denoting this as Ω_{dL} we see from Figure 1.15b that

$$\Omega_{\mathrm{dL}} \cong \frac{(\lambda f_{\mathrm{c}})^{2}/A_{\mathrm{r}}}{f_{\mathrm{c}}^{2}} = \frac{\lambda^{2}}{A_{\mathrm{r}}} \tag{1.6.13}$$

Thus, the diffraction limited field of view depends on the receiver (aperture lens) area. In practice, of course, detectors cannot easily be constructed whose area yields a receiver field of view equal to the diffraction limited field of view. That is, $\Omega_{\mathrm{fv}} \gg \Omega_{\mathrm{dL}}$, and the receiver field of view is generally many times its diffraction limited angle. However, (1.6.13) does point out the inherent spatial resolution of an optical system (since λ is on the order of microns) as opposed to an RF antenna having centimeter wavelengths.

When the received field is not a plane wave, then we must resort to two-dimensional transform theory to determine the focal plane field. Consider an extended source at a distance L with an arbitrary amplitude spectrum $F_{\mathrm{s}}(t, \mathbf{r})$ producing at the receiver aperture plane the field $f_{\mathrm{s}}(t, x, y)$, according to (1.3.18). The lens focuses the portion of this on its surface onto the focal plane. If we define the lens pupil function

$$
\begin{aligned}
G(x, y) &= 1, && (x, y) \in \mathscr{A}_{\mathrm{r}} \\
&= 0, && \text{elsewhere}
\end{aligned}
\tag{1.6.14}
$$

then the field at the receiver lens due to the source is

$$f_{\mathrm{r}}(t, x, y) = G(x, y) f_{\mathrm{s}}(t, x, y) \tag{1.6.15}$$

The focal plane field is then obtained by substituting (1.6.15) into (1.6.3). However, it also can be written using the complex convolution of the transforms of each term in (1.6.15). The inverse Fourier transform of $f_{\mathrm{s}}(t, x, y)$ is related to the source spectrum $F_{\mathrm{s}}(t, x, y)$ (Problem 1.16). If we let $g(\mu, v)$ be the Fourier transform of the pupil function $G(x, y)$, then the focal plane field magnitude becomes

$$|f_d(t, -u, -v)| = \frac{L}{f_c}\left|\int_{\mathscr{A}_s} F_s(t, L\lambda x, L\lambda y)g\left(\frac{u}{\lambda f_c} - x, \frac{v}{\lambda f_c} - y\right)dx\,dy\right| \quad (1.6.16)$$

Thus the optical receiver reproduces the source amplitude function in the focal plane through its convolution with the function g.[†] Since $g(u, v)$ is the transform of $G(x, y)$ in (1.6.14), it is related to the diffraction pattern produced by a normal plane wave over the receiver lens, and therefore is similar to (1.6.8) or (1.6.9). Another interpretation can be made by considering the case when $F_s(t, x, y) = \delta(x)\,\delta(y)$, corresponding to a monochromatic point source at $x = 0$, $y = 0$. Equation 1.6.16 then yields $|f_d| = g(u/\lambda f_c, v/\lambda f_c) \cdot (1/\lambda^2 L f_c)$. This can be interpreted as the focal plane field resulting from a point source at distance L centered in front of the lens. The function $g(u, v)$ therefore indicates the manner in which the receiver optics "spreads" the point in the focal plane, and is often called the *point spread* function of the receiver. Hence, extended sources are focused as convolutions of source amplitude and point spread functions. Similar interpretations can also be associated with the reception of stochastic fields in random media (see Appendix D).

An important point in our later discussions involves the application of Parceval's theorem[‡] to the integrated receiver and detector fields. This will yield the equality

$$\int_{\mathscr{A}_d} |f_d(t, \mathbf{r})|^2\,d\mathbf{r} = \left(\frac{1}{2\pi}\right)^2 \int\!\!\!\int_{-\infty}^{\infty} |F_r(t, \omega_1, \omega_2)|^2\,d\omega_1\,d\omega_2$$

$$= \int_{\mathscr{A}_r} |f_r(t, \mathbf{r})|^2\,d\mathbf{r} \quad (1.6.17)$$

Since the integrands are the field intensities [we have canceled the $2Z_w$ factor in (1.3.2) which is common to both sides] we see that the integrals are proportional to the power collected over the receiver and detector areas. This means the power collected over the detector area due to the focal plane field is equal to the power collected over the receiver surface due to the received field. Hence, power is preserved in focusing the received field onto the detector (provided the detector is large enough to encompass the entire focused field). This means detector power levels can be computed directly at the receiver lens without the necessity of computing the actual diffracted field.

† The negative signs mean the source is inverted and reversed.
‡ Parceval's theorem states that if $f \leftrightarrow F$ and $g \leftrightarrow G$ are Fourier transform pairs, then $\int f \cdot g^* = \int F \cdot G^*/2\pi$. When $f = g$ this becomes $\int |f|^2 = (1/2\pi)\int |F|^2$.

If the received field has an orthonormal expansion as in (1.3.20),

$$f_r(t, \mathbf{r}) = \sum_i f_i \phi_i(t, \mathbf{r}); \qquad 0 \le t \le T, \qquad \mathbf{r} \in \mathscr{A}_r \qquad (1.6.18)$$

then the detector field has the associated expansion

$$f_d(t, \mathbf{r}) = \sum_i f_i \Phi_i(t, \mathbf{r}); \qquad 0 \le t \le T, \qquad \mathbf{r} \in \mathscr{A}_d \qquad (1.6.19)$$

where

$$\Phi_i(t, u, v) \equiv \frac{\exp[j(\pi/\lambda f_c)(u^2 + v^2)]}{j\lambda f_c} \int_{\mathscr{A}_r} \phi_i(t, x, y) \exp\left[-j \frac{2\pi}{\lambda f_c} (xu + yv) \right] dx \, dy$$

$$(1.6.20)$$

The detector functions $\{\Phi_i\}$ are the diffracted versions of the receiver functions $\{\phi_i\}$, and therefore are related to its Fourier transform. By use of Parceval's theorem, we can easily show that these detector functions are themselves orthonormal (Problem 1.4). Hence, (1.6.19) itself represents an orthonormal expansion of the detected field over $\mathscr{A}_d$ having the same coefficients as the receiver field expansion. This means the detector field expansion can be derived from the received field expansion by using the same coefficients and modifying the orthonormal functions according to (1.6.20). Note that since the orthonormal functions are different, the associated field intensity will be distributed differently at the detector from that at the receiver lens, even though its power (integral) has been preserved. The principal point here is that the optical field can be expanded either at the receiver area or at the detector area with a straightforward conversion between the two. The former is obviously more convenient when discussing the received field, whereas the latter is better suited for analysis at the detector surface. If the fields are random, the fact that the coefficients are preserved means that a Karhunen–Loeve expansion at the receiver is transformed directly into a KL expansion at the detector with the same eigenvalues (mean squared coordinate values; Problem 1.5).

In this section we have presented several basic equations from Fourier optics that will be of use in our subsequent receiver analysis. We choose not to digress further in this area, since the material presented is sufficient for a basic understanding of optical communication systems. The reader interested in pursuing further the areas of diffraction theory, optics, and Fourier transform theory, is referred to any of the recent texts in this area [5–7].

1.7 BACKGROUND RADIATION

In addition to the desired signal power, a receiving system viewing an atmospheric background also collects undesirable background radiation falling within the spatial and frequency ranges of the detector. This collected background radiation is processed along with the desired signal background, and presents a basic degradation to the overall system performance. Of particular importance is the actual amount of background radiation power that is collected. The determination of the power, however, requires an accurate model for the source of this radiation. The accepted model is to consider the background to be generated from uniformly radiating sources. These sources divide into two basic types—(1) the diffuse sky background, assumed to occupy the whole hemisphere, and therefore is always present in any antenna field of view, and (2) discrete, or point, sources, such as stars, planets, sun, and the like, that are more localized but more intense, and may or may not be in the antenna field of view. In this section we review the analysis of background noise radiators.

Uniform background radiators are most often described by their *spectral radiance function*, $N(f)$, defined as the power radiated at frequency f per cycle bandwidth into a unit solid angle per unit of source area. If the receiving antenna occupies an area A_r at distance L from the source, it represents a solid angle, measured from the source, of approximately A_r/L^2 sr. If the radiation source has area A_s, then the total power collected depends on the portion of source area lying within the receiver field of view. Thus the background power collected at the receiver in a bandwidth of B Hz around the frequency f is

$$P_b = N(f)BA_s\left(\frac{A_r}{L^2}\right)\left(\frac{\Omega_{fv}}{\Omega_s}\right) \qquad \text{if} \quad \Omega_{fv} \in \Omega_s$$

$$= N(f)BA_s\left(\frac{A_r}{L^2}\right) \qquad \text{if} \quad \Omega_s \in \Omega_{fv} \qquad (1.7.1)$$

where Ω_{fv} is the receiver field of view and Ω_s is the solid angle subtended by the source when viewed from the receiver. Since $\Omega_s = A_s/L^2$, we can rewrite (1.7.1) as

$$P_b = N(f)BA_r\Omega_{fv} \qquad \text{if} \quad \Omega_{fv} \in \Omega_s \qquad (1.7.2a)$$

$$= N(f)BA_r\Omega_s \qquad \text{if} \quad \Omega_s \in \Omega_{fv} \qquad (1.7.2b)$$

Thus, if the background source is extended (as with a sky background), the background power is given by (1.7.2a) and depends only on the receiver area and field of view. If we assume $\Omega_{fv} = \Omega_{dL}$ in (1.6.13), then (1.7.2a) becomes

$$P_b = P_{b0} \triangleq \lambda^2 N(f)B; \qquad \Omega_{fv} = \Omega_{dL} \qquad (1.7.3)$$

where λ is the wavelength at frequency f. Since the bandwidth B is generally much smaller than frequency f, we often denote

$$N_{0b} \triangleq \lambda^2 N(f) \qquad (1.7.4)$$

as the effective one-sided background spectral level at f. The diffraction limited power is then simply

$$P_{b0} = N_{0b} B \qquad (1.7.5)$$

over a bandwidth B.

When the field of view is not diffraction limited as is usually the case, we must resort to (1.7.2a) to determine noise power. However, it is convenient to write this result in terms of the diffraction limited result in (1.7.5). Multiplying and dividing in (1.7.2a) by Ω_{dL} allows us to write

$$P_b = P_{b0}\left(\frac{\Omega_{fv}}{\Omega_{dL}}\right) = N_{0b} B\left(\frac{\Omega_{fv}}{\Omega_{dL}}\right) \qquad (1.7.6)$$

The ratio Ω_{fv}/Ω_{dL} appears as a multiplying factor by which the diffraction limited power is multiplied to get total power. The diffraction limited power P_{b0} is often called the power per spatial "mode," and the ratio Ω_{fv}/Ω_{dL} is taken as the "number of modes" of the optical field. (This notion is made more meaningful in Chapter 3.) We should point out that in certain situations the desired source itself may take on the appearance of an extended background noise source. This would occur, for example, if a point source were behind an extended scattering medium (cloud) producing diffuse light throughout. A receiver in front of the cloud would observe optical transmissions over the entire surface of the cloud, the latter appearing as an extended source to the receiver. If the source radiance of the cloud, $N(f)$, is properly described, (1.7.2a) can be used to determine the amount of source power collected at the receiver. In this case, however, the collected power is desired signal power rather than interference. It would now be important to collect as much source radiance as possible, which often means selecting the receiver field of view to match the subtended solid angle of the extended source.

If the source is localized (e.g., point sources as with stars, planets, etc.), then (1.7.2b) must be used, requiring knowledge of the source solid angle. For this reason when dealing with point sources, we often define instead a *spectral irradiance function* of the source, $H(f) = \Omega_s N(f)$, having units of power per hertz per unit receiver area. The latter function can be easily measured without explicit knowledge of Ω_s. Data on irradiance functions, $H(f)$, are readily available for the most common point sources in the hemisphere [1, Chap. 6]. Thus, their power contribution for a particular bandwidth B and receiving area A_r can be easily determined as $H(f)BA_r$, without specific knowledge of their spatial solid angle, provided they fall within the receiver field of view.

The most important background interference is that due to the sky, and (1.7.2a) or (1.7.6) must be used to determine its effect. The typical model is to assume the sky appears as an ideal *blackbody* radiator. For an ideal blackbody source at temperature $\mathcal{T}$ degrees Kelvin, the linearly polarized radiance function is known to be [1, Chap. 6; 16]

$$N(f) = \frac{hf^3}{c^2} \left[\frac{1}{\exp[hf/\kappa\mathcal{T}] - 1} \right] \qquad (1.7.7)$$

where h is Planck's constant, κ is Boltzmann's constant, and c is the speed of light. The resulting effective spectral level in (1.7.4) for a blackbody radiator is then

$$N_{0b} = \lambda^2 N(f) = \frac{hf}{e^{hf/\kappa\mathcal{T}} - 1} \qquad (1.7.8)$$

and the power collected in a diffraction limited receiver would be

$$P_{b0} = N_{0b}B = \frac{hfB}{e^{hf/\kappa\mathcal{T}} - 1} \qquad (1.7.9)$$

The function N_{0b} is plotted in Figure 1.16 as a function of f. Note that for $hf/\kappa\mathcal{T}^\circ \ll 1$, we have

$$N_{0b} \approx \frac{hf}{(1 + hf/\kappa\mathcal{T}) - 1} = \kappa\mathcal{T} \qquad (1.7.10)$$

and $P_{b0} = N_{0b}B = \kappa\mathcal{T}B$. In this case the background noise appears to be caused by a flat spectrum of level $\kappa\mathcal{T}$ W/Hz. This is the usual model for background radiation noise in RF communication systems. At extremely high frequencies, the blackbody spectral function falls to zero. The break point in this spectral function occurs at about $(4 \times 10^{10})\mathcal{T}$, which is

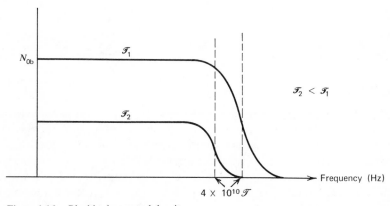

Figure 1.16. Blackbody spectral density.

generally well below optical frequencies. Thus in the optical range $N_{0b} \ll 1$, for a typical sky temperature, and the diffraction limited optical system is susceptible to much less sky background radiation per mode than a comparable RF system.

1.8 PHOTODETECTION

Photodetection of the light field represents the key operation in the receiver, converting the collected field to a current or voltage waveform for subsequent post detecting processing. For optimal design of this latter processor it is important that the system designer be cognizant of the characteristics of the photodetecting element. This becomes particularly significant when the actual statistics of the detector waveform are necessary for optimal design procedures. As we shall see in Chapters 7 to 9, this occurs in detection and estimation systems where performance depends explicitly on the mathematical model of the photodetector.

An optical photodetector has the basic structure shown in Figure 1.17. A photosensitive material responds to incident light over its surface by releasing free electrons from its inner surface. These electrons are susceptible to an electric force field that pulls the released electrons to a collecting anode.

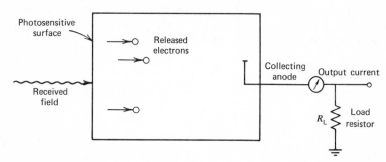

Figure 1.17. Photodetector model.

If *photomultiplication* is desired, a series of dynode surfaces are used, each one responding to incident electrons by releasing more electrons to the next dynode. As each dynode releases more electrons than it collects, a form of electron multiplication occurs, the overall effect to produce an inherent gain in the detector operation. The flow of electrons from photosensitive surface to final anode is manifested at the output terminals as an output current. (Recall current is the movement of electron charge.) Thus each electron movement is exhibited as a current contribution to the detector output, and the cumulation of such effects represents the observed output current. This current is converted to a voltage by passing through a detector load resistor. Since the conversion of optical field to electron flow and the reproduction of electrons by anode impingement, are both probabilistic in nature, the photodetector output always evolves as a random process in time. The overall effect is to induce an inherent randomness to the photo-detection operation when responding to any optical field, whether stochastic or not. This detector randomness must be properly accounted for in system models.

The photodetected output current process is described mathematically by the superposition of the individual current effects of each released electron. As a single electron moves from the photosensitive material to the final collecting anode, it produces a current response function $h(t)$. (This is the time response observed in an ideal ammeter connected to the output, when an electron is released at $t = 0$.) Physically, each electron moves for a short time period, ideally coming to rest at the anode. Hence, its current response $h(t)$ will also be of a finite time duration, existing only while the electron is in motion. The actual shape of $h(t)$ depends on the electron velocity during transition. Specifically, $h(t)$ is the rate of movement of charge. Thus an electron moving at constant velocity to the anode produces the response function† in Figure 1.18a, whereas a decelerating electron is described by Figure 1.18b. In all cases, however, the area under the response function is a fixed constant, since the integral of $h(t)$ is the change in electric charge during electron motion. Thus,

$$\int_0^\infty h(t)\, dt = \text{charge of a single electron}$$

$$\triangleq e = 1.6 \times 10^{-19} \text{ coulombs} \tag{1.8.1}$$

In a photomultiplier, the release of an electron produces the movement of many other electrons to the collecting anode. If the photomultiplier is ideal, exactly G_m electrons are collected for each photoemissive electron released, and (1.8.1) becomes

$$G_m \int_0^\infty h(t)\, dt = G_m e \tag{1.8.2}$$

Thus the effective charge is increased by the factor G_m, the latter called the *gain* of the photomultiplier. In nonideal photomultiplication, the number of

---

† Current in a electronic network is typically defined in the opposite direction of electron flow. We avoid this complication by treating $h(t)$ as a non-negative function, and considering the electron to have a positive charge.

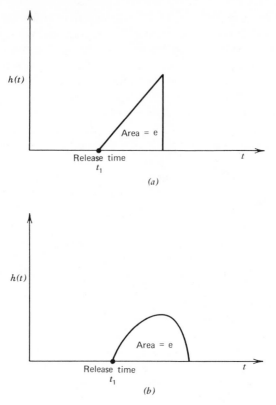

$h(t)$

Area = e

Release time
t_1

(a)

$h(t)$

Area = e

Release time
t_1

(b)

Figure 1.18. (a) Constant velocity electron effect. (b) Decelerating electron effect.

collected electrons, G_m, for each released electron, is itself random, and (1.8.2) must be analyzed with G_m as a random variable.

An electron released at a later time, say t_m, produces the response function $h(t - t_m)$. (That is, the same response function shifted to the point $t = t_m$.) If the optical field is impressed at time $t = 0$, the total cumulative response at time t is the combined response of all electrons released during the interval $(0, t)$. This produces the output current function

$$x(t) = \sum_{m=1}^{k(0,\,t)} h(t - t_m) \qquad (1.8.3)$$

where t_m is the time of release of the jth electron and $k(0, t)$ is the number of electrons released during the time interval $(0, t)$. Note that only electrons released prior to time t contribute to the output current at time t. The number $k(0, t)$ is often called the "count" of the electrons, and $k(0, t)$, considered as a function of t, is often called the *counting process* of the photodetection operation. Since photodetection is statistical in nature, the location times $\{t_m\}$ and the electron count k(0, t) are random variables when used to model the optical detector output. Thus, (1.8.3) represents a sum of a random number of randomly located response functions $h(t)$. Such processes are called *shot noise processes*, and have been used to model more general types of burstlike noise phenomena, such as shot noise in vacuum tubes (from which its name was derived). The functions $h(t)$ are called the *component functions* of the shot noise, and in photodetection are constrained by the electron transit behavior, as stated before.

Since the $\{t_m\}$ are random variables, and the count k(0, t) is a random count process in time, the current $x(t)$ in (1.8.3) is itself generated as a somewhat complicated random process. Clearly, any analysis to determine its inherent statistics (e.g., probability densities, averages, moments) requires specification of the statistics of the location times and counting process. This is our first objective in this book. Note that the output $x(t)$ does not explicitly contain the received field in any obvious way—yet we expect the photodetector to exhibit the properties of the received field. (Indeed, the optical system is based on the ability to faithfully recover the modulation imposed on the optical beam.) Thus we would expect the optical field to be embedded within the counting process in some manner.

The relationship of the received electromagnetic field and the number of released electrons is governed by the interaction between the radiation field and electrons of the photosensitive material [17–25]. There are two accepted ways to treat this relationship. In the purely quantum treatment, the field is quantized into photons, the average number of which is related to the field energy. Each field photon gives rise to an electron with probability η (called the quantum efficiency). The electrons released are therefore a statistical measurement of the photon occupancy in the field, and electron counting is often called photon, or photoelectron, counting. The alternate treatment (and the one we use) is the semiclassical approach, which is actually a consequence of the quantum treatment. This model treats the field classically (i.e., as a wave) and prescribes a probabilistic relation to account for its interaction with the atomic structure of the detector surface. Although a complete description of the emission and absorption of light by an atom is well beyond our interest here, an outline of the approach is as follows [17]. The semiclassical procedure begins with a charged particle in an electromagnetic field. It is then assumed that the combined system of atom plus field begins in some initial state, and a set of coupling equations is derived for the state transition probabilities. From these one determines the probability rate of finding the combined system in a particular state. Summing over all final states, and making some simplifying assumptions, one derives Fermi's rule for the probability per second for a state transition over a differential area $\Delta\mathbf{r}$ located at point $\mathbf{r}$ on the detector surface. This probability rate has the form

$$\frac{dP_t}{dt} = \alpha I(t, \mathbf{r})\, \Delta\mathbf{r} \tag{1.8.4}$$

where P_t can be interpreted as the probability of an electron release from $\Delta\mathbf{r}$ at t, α is a proportionality constant, and $I(t, \mathbf{r})$ is again the field intensity at time t and point $\mathbf{r}$ on the detector surface. The primary consequence of Fermi's rule is that it implies that in a short time interval Δt, the probability of ejecting an electron from an atom at the elemental surface area $\Delta\mathbf{r}$ is proportional to the incident field intensity over $\Delta\mathbf{r}$ and Δt. That is,

$$\begin{bmatrix} \text{probability that an} \\ \text{electron is released} \\ \text{from an area } \Delta\mathbf{r} \\ \text{during the time } \Delta t \end{bmatrix} \cong \alpha I(t, \mathbf{r})\, \Delta t\, \Delta\mathbf{r} \tag{1.8.5}$$

for sufficiently small $\Delta\mathbf{r}$ and Δt. In addition, (1.8.5) implies that the probability of more than one electron being released must go to zero as $(\Delta\mathbf{r}\, \Delta t)^2$, which means

$$\begin{bmatrix} \text{probability of no} \\ \text{electron being released} \\ \text{from } \Delta\mathbf{r} \text{ during } \Delta t \end{bmatrix} \cong 1 - \alpha I(t, \mathbf{r})\, \Delta t\, \Delta\mathbf{r} \tag{1.8.6}$$

as $\Delta\mathbf{r}\, \Delta t \to 0$. Note that (1.8.5) states that the release of an electron from any elemental area at point $\mathbf{r}$ at any time t depends only on the field intensity at that time and point. This implies that the release of electrons from disjoint differential areas on the surface, and from disjoint intervals in time, can be treated as independent events when given a particular intensity function. This assumption, along with (1.8.5) and (1.8.6), describes the mathematical model of the photodetecting surface, and is of primary importance in the subsequent derivation of the electron count model.

Our discussion has been limited to a single photodetector in the focal plane, which produces an output shot noise process in response to the total diffracted field at the receiver. The natural extension from a single photodetector is to a collection of smaller photodetectors placed side by side in the focal plane, as shown in Figure 1.19, with each detector producing its own output. Such a combination of detectors is called a photodetector *array*. Detector arrays are characterized by the parallel set of response functions

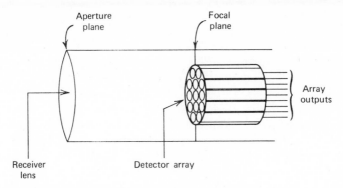

Figure 1.19. Receiver with focal plane detector array.

from each of its elements. Since off-axis arriving plane waves concentrate their diffraction patterns at different points in the focal plane, each detector of the array detects a different angle of the receiver field of view. Thus arrays effectively spatially "sample" the field of view simultaneously, producing separate response processes from each sample in parallel. Detector arrays, therefore, give the optical receiver a form of diversity reception. As the array size is increased the degree of diversity is increased at the expense of added parallel processing. The availability of a high degree of parallel diversity processing in a single receiver makes the optical detector array a theoretically powerful device in optical communications. At various points in the text we attempt to point out this capability.

In summary, this first chapter is used to present an overview of a typical optical communication system, from information source to final destination. Our prime objective was to review the overall system structure, and indicate the physical and mathematical interconnection of the various components. Many of the important results stated are rederived more rigorously in later studies, where the impact on design can be emphasized. A secondary objective was to focus attention on the photodetection operation, which is examined in detail in the next three chapters. This detector signal becomes the demodulated optical signal used for receiver processing, and reflects the properties of the modulation, the channel, and the detection operation. Therefore, the design of the transmission system and receiver processor becomes intimately related to the properties of the detector output. When the latter has been satisfactorily modeled, one can begin considering overall system format and processor design from the point of view of signal-to-noise ratios (Chapters 5 and 6), detection theory (Chapters 7 and 8), or estimation theory (Chapters 9–11), depending on the mission of the communication link.

REFERENCES

[1] Pratt, W., *Laser Communication Systems*, John Wiley and Sons, New York, 1969.

[2] Ross, M., *Laser Receivers*, John Wiley and Sons, New York, 1966.

[3] Yariv, A., *Quantum Electronics*, 2nd ed., John Wiley and Sons, New York, 1975.

[4] Special Issue on Optical Communication, *Proc. IEEE*, **58**, No. 10 (October 1970).

[5] Goodman, J., *Introduction to Fourier Optics*. McGraw-Hill Book Co., New York, 1968.

[6] Papoulis, A., *Systems and Transforms with Applications in Optics*, McGraw-Hill Book Co., New York, 1968.

[7] Born, M. and Wolf, E., *Principles of Optics*, 4th ed., Pergamon Press, London, 1970.

[8] Jamieson, J., et al., *Infrared Physics and Engineering*, McGraw-Hill Book Co., New York, 1963.

[9] Chernov, L. A., *Wave Propagation in a Random Medium*, Dover Publications, New York, 1967.

[10] Brookner, E., "Atmospheric Propagation and Communication Channel Models for Laser Wavelengths," *IEEE Trans. Commun. Tech.*, **COM-18**, 396 (August 1970).

[11] ———, "Atmospheric Propagation for Laser Wavelengths—An Update," *IEEE Trans. Communications*, **COM-22**, 265 (February 1974).

[12] Kennedy, R. S., *Fading Dispersive Communication Channels*. John Wiley and Sons, New York, 1969, Chap. 3.

[13] Fried, D. L., "Atmospheric Modulation Noise in an Optical Heterodyne Receiver," *IEEE J. Quantum Electron.*, **AE-3** (6), 213 (June 1967).

[14] Fried, D. L., "Statistics of a Geometric Representation of Wavefront Distortion," *J. Opt. Soc. Am.*, **55**, No. 11, 1427 (November 1965).

[15] Gloge, D., "Optical Waveguide Transmission," Special Issue on Optical Communications, [4] above, p. 1513.

[16] Kapeika, N. and Bordogna, J., "Background Noise in Optical Systems," Special Issue on Optical Communications, [4] above, p. 1571.

[17] Karp, S., O'Neill, E., and Gagliardi, R., "Communication Theory for the Free Space Optical Channel," Special Issue on Optical Communications, [4] above, p. 1611.

[18] O'Neill, E., *Introduction to Statistical Optics*, Addison-Wessley, Reading, Massachusetts, 1963.

[19] Glauber, R. J., "The Quantum Theory of Optical Coherence," *Phys. Rev.*, **130**, No. 6, 2529–2539 (June 1963).

[20] Glauber, R. J., "Coherent and Incoherent States of the Radiation Field," *Phys. Rev.*, **131**, No. 6 (September 1963).

[21] Glauber, R. J., "Optical Coherence and Photon Statistics," *Quantum Optics and Electronics* (C. de Witt, et al., eds.), Gordon & Breach, New York, 1965, pp. 65–185.

[22] Mandel, L. and Wolf, E., "Coherence Properties of Optical Fields," *Rev. Mod. Phys.*, **37**, No. 2, 231–287 (April 1965).

[23] Klauder, J. R. and Sudarshan, E. C. G., *Fundamentals of Quantum Optics*, W. A. Benjamin, New York, 1968.

[24] Troup, G. J., *Optical Coherence Theory*, Methuen, London, 1967.

[25] Louisell, W. H., *Radiation and Noise in Quantum Electronics*, McGraw-Hill Book Co., New York, 1964.

PROBLEMS

1. Communication engineers deal with bandwidth in terms of frequency. Physicists usually deal with bandwidth in terms of wavelengths.
(a) Determine the conversion between these two bandwidths. (That is, find the frequency difference corresponding to a given wavelength difference.)
(b) Use this result to sketch a plot of frequency bandwidth versus optical (center) frequency for a 1 Å wavelength filter.

2. A stationary satellite, located 22,000 miles above the Earth, is to transmit a green light beam to cover a 1 mile radial distance on the Earth. What is the approximate transmitting lens size needed on the satellite and the associated antenna gain?

3. An Earth–Moon link is to communicate at 10.6 microns. The Earth beam is to have a solid angle beamwidth just encompassing the Moon. The Moon beam must illuminate the United States (3000 miles diameter on Earth). The same beams are used for transmitting and receiving, and the total gain must be at least 150 dB. Design a transmitting and receiving antenna pair that satisfies these specifications.

4. Show that if the eigenfunctions are orthonormal in the aperture plane, then their Fraunhofer diffraction patterns are orthonormal in the focal plane.

5. Show that the eigenvalue associated with a particular eigenfunction in the Karhunen–Loeve expansion in the aperture plan is identical to that of the transformed eigenfunction when expanded in the focal plane.

6. Show that the eigenvalues are the mean squared values of the Fourier coefficients.

7. Show that if the covariance function of a noise field is delta distributed [i.e., $R_b(t_1, \mathbf{r}_1; t_2, \mathbf{r}_2) = N_0\, \delta(t_1 - t_2)\delta(\mathbf{r}_1 - \mathbf{r}_2)$] its Karhunen–Loeve expansion is satisfied by any orthogonal set of functions.

8. A channel has a nonlinear phase function $\phi(\omega)$. The bandwidth of the channel is often defined as the maximum frequency that can be sent such that the delay variation $d^2\phi/d\omega^2$ is 30% of the wavelength. What will be the channel bandwidth when the phase function is given by

$$\phi(\omega) = -\tan^{-1}\left(\frac{\omega}{Q}\right)$$

where Q is a parameter of the channel

9. Determine the effective loss in power in decibels in a beam that is transmitted as a 5 μrad beam but is spread to 20 μrad by the atmosphere. Assume a point receiver (beam power density constant over the receiver area).

10. The size of scattering inhomogeneities in the atmosphere is often modeled as

$$l = (10^{-9} h)^{1/3}, \qquad 10 \le h \le 1000 \text{ meters}$$
$$= 1 \text{ cm}, \qquad\qquad 1000M \le h \le 10 \text{ miles}$$
$$= 0, \qquad\qquad\quad \text{elsewhere}$$

where h is the height above Earth.
(a) If an optical beamwidth of 10^{-5} rad is transmitted from Earth, determine the height at which the beam area is approximately equal to the size of the inhomogeneities.
(b) Determine the ratio of inhomogeneity size to beam area as a function of h for an optical beam with a 10^{-5} rad beamwidth transmitted to the Earth from an orbiting satellite 400 miles above the Earth.
(c) From (a) and (b) explain what the prime beam distortion effects will be in an Earth-satellite up and down link.

11. Beam scattering is often described by a "billiard ball" model. The beam, considered as a ray, undergoes a series of collisions with amplitude being multiplied by the gain constant $\exp(-a_i)$ at the ith collision, where a_i is a random variable. Show by the central limit theorem (Equation A.2.17) that if the number of collisions is large, the resulting beam amplitude will have a log-normal probability density.

12. (a) Derive the log-normal moments in (1.5.4) from the density in (1.5.3).
(b) Show that if G is a lognormal gain, then

$$\text{Prob}(G < G_0) = \int_0^{\ln G/\sigma_1} \frac{e^{-u^2/2}}{\sqrt{2\pi}} \, du$$

13. A fiberoptic guide uses a core of diameter 1 mm and index 1.5, and a cladding of index 1.485. For a beam wavelength of 0.6 micron, find
(a) the maximum grazing angle for propagation,
(b) number of propagating modes,
(c) how small the core diameter should be for a single propagating mode,
(d) repeat (a), (b), and (c) if the cladding is removed (index of air is 1).

14. Time dispersion in a fiberoptic guide is obtained by determining the delay difference between a propagating mode at grazing angle $\theta = 0$ and maximum grazing angle.
(a) Determine an expression for delay dispersion for a guide of length L, wavelength λ, and core and cladding indices n_1 and n_2.
(b) Use the parameters in Problem 1.13 to find the dispersion and resulting coherence bandwidth in (1.5.5).

15. A fiberoptic guide uses an index that is tapered from the guide center, according to the equation

$$n(r) = n_0\left[1 - \Delta\left(\frac{r}{d/2}\right)^2\right]$$

where r is the radial distance, d is the diameter, and Δ is the rate of taper. For such a taper the maximum grazing angle θ is given by

$$\cos\theta = \left(\frac{1 - 3\Delta}{1 - \Delta}\right)^{1/2}$$

Derive the fiber delay dispersion as in Problem 1.14.

16. Show that in (1.3.18) $f_r(t, \mathbf{r}_2)$ is proportional to the two-dimensional Fourier transform of the modified source amplitude function

$$F_s(t, \mathbf{r}_1)\exp[j(\pi/\lambda L)|\mathbf{r}_1|^2]$$

at point $\mathbf{r}_1 = (x, y)$ in the source plane.

17. Show that a set of plane waves arriving at different angles is essentially orthogonal over an infinite (large) square detector area.

18. A laser emits 5 mW at 5×10^{14} Hz with a diffraction limited beam angle of 1 mrad, and a 1 GHz bandwidth. What would be the temperature of a blackbody yielding this same intensity? What if the laser bandwidth can be reduced to 1 MHz

19. Consider an isotropic radiator in Figure 1.20 which is square in shape with dimension d. Consider the coordinate system shown to be centered on

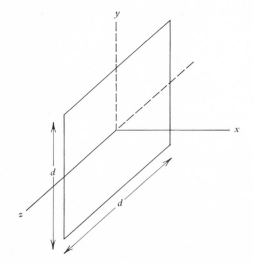

Figure 1.20. Isotropic radiating surface.

the square. Consider a receiver with a field that is conical in shape with half-angle α and located on the x axis at point x_0 ($x_0 \gg d$). If the spectral radiance of the radiator is N_0, how much power is received at the receiver as a function of α?

20. Several models for optical detector current response functions $h(t)$ have been suggested. In the following give a physical interpretation and a possible application for the stated forms of $h(t)$:

(a) $h(t) = \delta(t)$ (unit impulse),

(b) $h(t) =$ Fourier transform of the detector frequency response,

(c) $h(t) = a_i g(t)$ ($\{a_i\}$ a set of independent, identically distributed, random variables),

(d) $h(t) = g(t/\sigma)$ (σ a random variable).

2 Counting Statistics

The overall model of an optical communication system is discussed in Chapter 1. It is pointed out that the receiver and its associated signal processing are directly related to the ability to detect, or respond to, the received optical radiation. Photodetecting surfaces used for this purpose produce randomly emitted electrons in response to the received radiation. Any communication processing following photodetection must be inherently linked to this detector response. In this and the next two chapters we investigate the statistical relationships between the received intensity and the resulting electron flow at the photodetector output. Our ultimate objective is to determine statistical properties of the detector output that will have a direct bearing on the study of optical communication system design and its theoretical performance.

2.1 DERIVATION OF COUNT STATISTICS

We start with a derivation of the actual statistics (probability densities) concerning the number of electrons that flow during a particular time interval when optically detecting a given radiation field with a photoresponsive surface. The number is called the electron *count*, and its associated statistics are referred to as *counting statistics*. It is shown in the first chapter that the fundamental property of photodetection, from the point of view of counting statistics, is that the probability of releasing an electron from an area of the detector surface can be related to the squared envelope of the classical electromagnetic field over that area. We labeled this squared envelope as the intensity of the received field. This relation between the electron probabilities and received field is the basis of our investigation of count statistics.

We first determine the probability density of the number of electrons produced during a specific time interval, say $(t, t + T)$ from the photodetecting surface area $\mathscr{A}$. We denote this random counting variable as k. To facilitate this derivation we find it convenient to make use of the concept of a time-space domain. This domain contains vectors whose components correspond to time and spatial coordinates associated with these regions. We denote these vectors as $\mathbf{v} = (t, \mathbf{r})$, where t is the scalar time component and $\mathbf{r}$ represents the two-dimensional spatial coordinates of the detector surface. Unless otherwise stated, the origin of the coordinate axis for $\mathbf{r}$ is taken to be the detector center. We define the volume V in this domain to be composed of all vectors $\mathbf{v} = (t', \mathbf{r})$ such that $t \leq t' \leq t + T$ and $\mathbf{r} \in \mathscr{A}$. This volume is shown in Figure 2.1a. This notation allows us to denote the normal electromagnetic field intensity at point $\mathbf{r}$ on the detector surface at time t' by $I(t', \mathbf{r}) \equiv I(\mathbf{v})$. The volume V is therefore the set of all points in the time-space domain over which we observe the radiation field with a given detector area in a given time interval.

Now consider the partition of the volume V into disjoint subvolumes, or cells, $\Delta \mathbf{v} = \Delta \mathbf{r} \, \Delta t$, as shown in Figure 2.1b. We assume $\Delta \mathbf{r}$ and Δt are smaller than the spatial and temporal variations in $I(t, \mathbf{r})$ so that within each

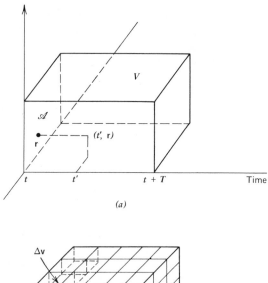

(a)

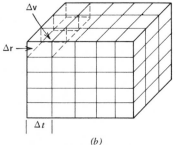

(b)

Figure 2.1. (a) Observation volume. (b) Observation cells.

Δv, $I(v)$ is approximately constant. (This is certainly possible with continuous fields.) Let ΔV be the volume of the cells Δv, and let q be the total number of cells in V after partitioning. The ensemble of q disjoint cells Δv can now be ordered to form the sequence $\{\Delta v_1, \Delta v_2, \ldots, \Delta v_q\}$, where each Δv_i is centered at some point v_i in V. Each Δv_i can be interpreted as an observation cell corresponding to an elemental surface area and elemental time interval over which we observe the radiation field. In this notation, the probability model of electron emissions in Section 1.8 becomes, for $\Delta V \to 0$,

$$\begin{bmatrix} \text{probability of an electron} \\ \text{emitted from } \Delta v_i \text{ at point } v_i \end{bmatrix} = \alpha I(v_i)\, \Delta V, \tag{2.1.1a}$$

$$\begin{bmatrix} \text{probability of no} \\ \text{electrons emitted} \end{bmatrix} = 1 - \alpha I(v_i)\, \Delta V, \tag{2.1.1b}$$

where α is again the proportionality constant. We now consider the probability of the detector releasing k total electrons from the total surface area $\mathscr{A}$ during the time interval $(t, t + T)$. This is equivalent to the compound probability that k electrons will be emitted from the totality of all cells $\{\Delta v_i\}$ spanning V. For small volumes ΔV, this can be written as

$$\begin{bmatrix} \text{probability of} \\ k \text{ emitted} \\ \text{electrons from } V \end{bmatrix} = \frac{1}{k!} \sum_{\substack{\text{all} \\ \text{orderings}}} \begin{bmatrix} \text{probability of} \\ \text{one electron} \\ \text{from } k \text{ different} \\ \text{ordered cells} \end{bmatrix} \begin{bmatrix} \text{probability of} \\ \text{no electrons in} \\ \text{the } q - k \text{ remaining} \\ \text{ordered cells} \end{bmatrix}$$

$$= \frac{\alpha^k}{k!} \sum_{\substack{\text{all} \\ \text{orderings}}} I(v_{i_1}) \cdots I(v_{i_k})(\Delta V)^k \prod_{j=k+1}^{q} [1 - \alpha I(v_{i_j})\, \Delta V]$$

$$\tag{2.1.2}$$

where $(i_1, i_2, \ldots, i_q)$ is a particular index ordering of the integers 1 to q. The summation considers all possible orderings, and therefore all possible arrangements of k and $(q - k)$ index groupings. The division by $k!$ is necessary since different arrangements of the same k cells need only be considered once. Note that (2.1.2) has used (2.1.1) and the assumption of independent electron emissions from disjoint cells. However, as long as q is finite, (2.1.2) must be considered only an approximation.

Now consider the limit of (2.1.2) as $\Delta V \to 0$ and $q \to \infty$, so that the approximation approaches a true equality. Since the limits of sums and products are equal to the sum and products of the limits, we can investigate the limit of the individual terms in (2.1.2) and then recombine. We first show that the product term has the same limit for all orderings. This can be seen by considering the limit of the logarithm of the product. For finite q this log is

$$\log \prod_{j=k+1}^{q} [1 - \alpha I(\mathbf{v}_{i_j}) \Delta V] = \sum_{j=k+1}^{q} \log[1 - \alpha I(\mathbf{v}_{i_j}) \Delta V] \qquad (2.1.3)$$

Adding and subtracting the k terms not included in the summation allows us to write the right side of (2.1.3) always as

$$\sum_{j=1}^{q} \log[1 - \alpha I(\mathbf{v}_j) \Delta V] - \sum_{j=1}^{k} \log[1 - \alpha I(\mathbf{v}_{i_j}) \Delta V] \qquad (2.1.4)$$

Now, in the limit as $\Delta V \to 0$, we have $q \to \infty$, and

$$\lim_{\Delta V \to 0} \log[1 - \alpha I(\mathbf{v}_j) \Delta V] \to -\alpha I(\mathbf{v}_j) \Delta V \qquad (2.1.5)$$

The first summation in (2.1.4) therefore has the limit

$$\lim_{\substack{\Delta V \to 0 \\ q \to \infty}} \sum_{j=1}^{q} \log[1 - \alpha I(\mathbf{v}_j) \Delta V] \to \lim_{\substack{\Delta V \to 0 \\ q \to \infty}} \left[-\alpha \sum_{j=1}^{q} I(\mathbf{v}_j) \Delta V \right]$$

$$= -\alpha \int_{V} I(\mathbf{v}) \, d\mathbf{v} \qquad (2.1.6)$$

where the integration is over the volume V. However, the second summation in (2.1.4) involves only a finite number of terms and therefore has the limit

$$\lim_{\substack{\Delta V \to 0 \\ q \to \infty}} \sum_{j=1}^{k} -\alpha I(\mathbf{v}_{i_j}) \Delta V = 0 \qquad (2.1.7)$$

for any ordering. Hence, in (2.1.2)

$$\lim_{\Delta V \to 0} \prod_{j=k+1}^{q} [1 - \alpha I(\mathbf{v}_{i_j}) \Delta V] = \exp\left[-\alpha \int_{V} I(\mathbf{v}) \, d\mathbf{v} \right] \qquad (2.1.8)$$

Now consider the summation term in (2.1.2). We want to evaluate the limit

$$\lim_{\Delta V \to 0} \alpha^k \sum_{\substack{\text{all} \\ \text{orderings}}} [I(\mathbf{v}_{i_1}) \cdots I(\mathbf{v}_{i_k})(\Delta V)^k] \qquad (2.1.9)$$

Since each ordering above requires $i_1 \neq i_2 \neq \cdots \neq i_k$, the summation can be equivalently written as

$$\alpha^k \sum_{i_k=1}^{q} \cdots \sum_{i_1=1}^{q} [I(\mathbf{v}_{i_1}) \cdots I(\mathbf{v}_{i_k})] \Delta V^k - \begin{bmatrix} \text{sum over} \\ \text{all orderings} \\ \text{in which} \\ \text{at least} \\ \text{two } i_j \text{ are} \\ \text{equal} \end{bmatrix} \qquad (2.1.10)$$

The second term involves q^{k-1} terms of order $(\Delta V)^k$. Since q behaves as $1/\Delta V$, the limit of the second term will be zero as $\Delta V \to 0$. The first term, on the other hand, has the limit

$$\lim_{\substack{\Delta V \to 0 \\ q \to \infty}} \left(\alpha \sum_{i=1}^{q} I(\mathbf{v}_i) \, \Delta V \right)^k = \left(\alpha \int_V I(\mathbf{v}) \, d\mathbf{v} \right)^k \qquad (2.1.11)$$

Therefore, using (2.1.6), (2.1.7), and (2.1.11) in (2.1.2), we derive the final limiting form of the probability of k emissions over V as†

$$P_k(k) = \frac{(m_V)^k}{k!} \exp[-m_V], \qquad k \geq 0 \qquad (2.1.12)$$

where

$$m_V \triangleq \alpha \int_V I(\mathbf{v}) \, d\mathbf{v} \qquad (2.1.13)$$

Recalling our definition of the volume V, we can also write

$$m_V = \alpha \int_{\mathscr{A}} \int_t^{t+T} I(\rho, \mathbf{r}) \, d\rho \, d\mathbf{r} \qquad (2.1.14)$$

The probability in (2.1.12) is the probability that exactly k electrons will be emitted over the observation volume V, that is, from the spatial area $\mathscr{A}$ during the time interval $(t, t + T)$ defining V. Note that the probability depends on the integral of the field intensity over the volume V. We have explicitly indicated this dependence by subscripting m. Since m_V depends on the location and size of the area $\mathscr{A}$ and the counting interval $(t, t + T)$, the probability is in general nonstationary in time and nonhomogeneous in space. That is, the probabilities may be different over different volumes.

Since k can represent any nonnegative integer, (2.1.12) represents a probability over all nonnegative integers. This probability is called a Poisson probability, and the parameter m_V is called the *level* of the probability. A sketch of this probability over the integers is shown in Figure 2.2 for several values of m_V. For convenience, we subsequently denote the Poisson probability with level m_V in (2.1.12) as $\text{Pos}[k, m_V]$. That is,

$$\text{Pos}(k, m_V) \triangleq \frac{(m_V)^k}{k!} \exp[-m_V] \qquad (2.1.15)$$

The count probability density associated with these probabilities is then a discrete density over all nonnegative integers, with probability weight given

† Subscripts on probabilities and probability densities refer to the random variable involved. Here k refers to the random count variable.

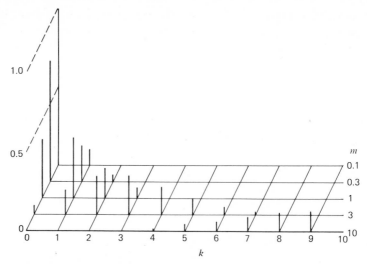

Figure 2.2. Poisson probabilities for various levels, m.

by (2.1.15). Hence, for a specific m_V, the random count variable k has the probability density

$$p_k(x) = \sum_{j=0}^{\infty} \text{Pos}(j, m_V)\, \delta(x - j) \tag{2.1.16}$$

where $\delta(x)$ is the delta function [Appendix A, (A.1.14)]. The probability density in (2.1.16) is called a Poisson density, and random variables having this density are Poisson random variables. Thus the count variable k is a Poisson random variable whose density depends on m_V.

Note that m_V, being a parameter of the count probability (2.1.12), must be dimensionless. Since the integrated field intensity in (2.1.13) has units of energy, α must have reciprocal units of energy. It is therefore convenient to define a normalized field intensity

$$n(t, \mathbf{r}) \triangleq \alpha I(t, \mathbf{r}) \tag{2.1.17}$$

whose integral is dimensionless and yields directly the level m_V. The function $n(t, \mathbf{r})$ is called the *count intensity* (as distinguished from the field intensity) and we write

$$m_V = \int_{\mathscr{A}} \int_t^{t+T} n(\rho, \mathbf{r})\, d\rho\, d\mathbf{r} \tag{2.1.18}$$

Often we further simplify by defining the spatially integrated count intensity as

$$n(t) \triangleq \int_{\mathcal{A}} n(t, \mathbf{r}) \, d\mathbf{r} = \alpha \int_{\mathcal{A}} I(t, \mathbf{r}) \, d\mathbf{r} \qquad (2.1.19)$$

so that m_V has the more compact form

$$m_V = \int_t^{t+T} n(\rho) \, d\rho \qquad (2.1.20)$$

It is convenient in later communication analysis to deal exclusively with the integrated count intensity $n(t)$. We point out that even though m_V is dimensionless, it is nevertheless often referred to as an energy parameter, and sometimes called the *count energy*. For the same reason $n(t)$ is referred to as the *count power*. If the detected field $f(t, \mathbf{r})$ is coherent over $\mathcal{A}$ at any t [i.e., $I(t, \mathbf{r}) = I(t), \mathbf{r} \in \mathcal{A}$], then the spatial effects can be completely suppressed in dealing with any volume V containing $\mathcal{A}$. Equation 2.1.19 becomes

$$n(t) = \alpha \int_{\mathcal{A}} I(t) \, d\mathbf{r} = \alpha A I(t) \qquad (2.1.21)$$

where A is the integrated area of $\mathcal{A}$. Photodetectors operating over coherent areas, and thereby having count intensities given by (2.1.21), are often called *point detectors*. Basically, point detectors collect optical intensity as if they were located at a single spatial point in the coherent region.

2.2 THE POISSON RANDOM VARIABLE

In the preceding section we showed the development of the Poisson counting variable, k. It is clear that further examination of the photodetection model requires properties of this random variable. In this section we derive pertinent statistical characterizations which are useful in subsequent analyses. The following derivations involve straightforward manipulations associated with random variables, and are therefore presented somewhat tersely. (The reader may be interested in first reviewing Appendix A.)

Mean Value. The mean (average, or expected) value of the discrete random variable k is denoted as $E_k(k)$, where $E_k[f(k)]$ is the expectation operator defined by

$$E_k[f(k)] = \sum_{k=-\infty}^{\infty} f(k) P_k(k) \qquad (2.2.1)$$

and $P_k(k)$ is the probability that the count variable k has the value k. For the Poisson count variable,

$$E_k[k] = \sum_{k=0}^{\infty} k P_k(k)$$

$$= \sum_{k=1}^{\infty} k \, \text{Pos}(k, m_V) \qquad (2.2.2)$$

We now make the integer substitution $i = k - 1$ and sum over all i from $(0, \infty)$, resulting in

$$E_k[k] = m_V \sum_{i=0}^{\infty} \frac{(m_V)^i}{i!} \exp[-m_V]$$

$$= m_V \qquad (2.2.3)$$

Thus, the level of the Poisson probability is also the mean value of the electron count over V. This immediately gives new meaning to the spatially integrated count intensity $n(t)$. Since m_V can now be interpreted as the average number of electrons produced in a given volume, the function $n(t)$ in (2.1.19) now plays the role of the average count rate. That is, it can be considered as the average rate at which electrons are being emitted at time t from the surface area $\mathscr{A}$.

As a specific example, consider the case of photodetecting a coherent monochromatic field. Let the constant amplitude field have intensity I and arrive normal to the spatial area $\mathscr{A}$. A detector operating over $\mathscr{A}$ collects the power $P = IA$, where A is the area of $\mathscr{A}$. The count intensity associated with this field is then $n(t) = \alpha I A = \alpha P$ and the level of the Poisson counting probability is αPT. The counts are therefore statistically described by the probabilities

$$P_k(k) = \text{Pos}(k, \alpha PT)$$

$$= \frac{(\alpha PT)^k}{k!} \exp[-\alpha PT] \qquad (2.2.4)$$

Note that for this case the count intensity is a constant, or equivalently, the electrons are being emitted at a uniform rate. We also note that the probability in (2.2.4) does not depend on t, and the statistics are therefore stationary in time. Thus, point detection of a monochromatic optical field is a special case where the count statistics do not change with time, and depend only on the time length of the observation interval and the size of the detector area.

The fact that the average count equals the Poisson level m_V also allows us to evaluate specifically the proportionality constant α. This is accomplished by recalling that the photoelectric effect requires the absorption of an amount of energy hf from the field in order to release an electron, where f is the frequency of the field and h is Planck's constant. This means that the mean number of emitted electrons multiplied by hf must be equal to the average energy absorbed from the field at frequency f. Hence, we can equate

$$\eta \int_{\mathscr{A}} \int_t^{t+T} I(\rho, \mathbf{r}) \, d\rho \, d\mathbf{r} = hf \, E_k[k] \qquad (2.2.5)$$

where η is the ratio of energy absorbed to incident energy. (η is often called the detector quantum efficiency and can also be considered as the probability that an electron will be released given incident energy hf. See Problem 2.1.) Substituting for $E_k(k)$ from (2.2.3) and (2.1.13) yields the formal definition

$$\alpha = \frac{\eta}{hf} \qquad (2.2.6)$$

Note that α above has the proper units (energy)$^{-1}$, and effectively converts field intensity to count intensity.

Mean Square Value. The mean square value of the Poisson variable k is given by $E_k(k^2)$. Thus

$$E_k[k^2] = \sum_{k=0}^{\infty} k^2 \, \text{Pos}(k, m_V)$$

$$= m_V \sum_{k=0}^{\infty} (k + 1) \frac{m_V{}^k}{k!} \exp[-m_V]$$

$$= m_V(m_V + 1) = m_V{}^2 + m_V \qquad (2.2.7)$$

The variance of the count is defined by

$$\text{Var}[k] = E_k[(k - E_k[k])^2] = E_k[k^2] - (E_k(k))^2 \qquad (2.2.8)$$

Use of (2.2.7) and (2.2.3) shows that

$$\text{Var}[k] = m_V \qquad (2.2.9)$$

Thus the level of the Poisson probability also gives the variance of the count. Since the variance essentially describes the mean square variation from the mean value, we see that the Poisson count variable inherently has a variation from the mean as large as the mean itself. Later we relate this variation to an effective noise fluctuation associated with the ability of a photodetector to count electrons accurately. We emphasize again the dependence of the moments above on the volume V, further illustrating the general non-stationarity of the counting statistic.

Characteristic Function. The characteristic function $\psi_k(\omega)$ of a discrete random variable k is the average of the function $\exp(j\omega k)$:

$$\psi_k(\omega) = E_k[e^{j\omega k}]$$

$$= \sum_{k=0}^{\infty} e^{j\omega k} P_k(k) \qquad (2.2.10)$$

For the Poisson random variable this becomes

$$\psi_k(\omega) = \sum_{k=0}^{\infty} e^{j\omega k} \, \text{Pos}(k, m_V)$$

$$= e^{-m_V} \sum_{k=0}^{\infty} \frac{(e^{j\omega} m_V)^k}{k!}$$

$$= \exp[m_V(e^{j\omega} - 1)] \qquad (2.2.11)$$

We recognize the characteristic function in (2.2.10) as the discrete Fourier transform of $P_k(k)$, and as such it can always be inverse transformed back to the probabilities. However, with discrete densities it is not actually necessary to perform the complex integration usually required by the inverse transform. This can be seen if we make the substitution $z = 1 - e^{j\omega}$ in $\psi_k(\omega)$, which is the same as generating the new function

$$g_k(z) = \sum_{k=0}^{\infty} (1 - z)^k P_k(k) \qquad (2.2.12)$$

Notice that by taking q derivatives of $g_k(z)$ and setting z equal to one we obtain

$$\left[\frac{\partial^q}{\partial z^q} g_k(z)\right]\bigg|_{z=1} = \sum_{k=0}^{\infty} P_k(k) \frac{\partial^q}{\partial z^q} [1 - z]^i \bigg|_{z=1}$$

$$= (-1)^q P_k(q) q! \qquad (2.2.13)$$

Hence,

$$P_k(k) = \frac{(\partial^k/\partial z^k)[g_k(z)]|_{z=1}}{(-1)^k k!} \qquad (2.2.14)$$

Thus, a fairly simple inversion exists for computing the discrete probabilities from $g_k(z)$. The latter function is often called the probability transform of $P_k(k)$.

The derivatives of a characteristic function are directly related to the moments of the corresponding random variable through the identity

$$E_k[k^q] = \frac{1}{j^q} \left\{ \frac{d^q}{d\omega^q} [\psi_k(\omega)] \right\}_{\omega=0} \qquad (2.2.15)$$

Therefore all the moments of the Poisson random variable can be obtained by differentiating the characteristic function in (2.2.11). In particular, the first two moments follow from

$$\frac{d\psi_k(\omega)}{d\omega} = [jm_V e^{j\omega}] \exp[m_V(e^{j\omega} - 1)]$$

$$\frac{d^2\psi_k(\omega)}{d\omega^2} = j^2 \exp[m_V(e^{j\omega} - 1)][(m_V e^{j\omega})^2 + m_V e^{j\omega}]$$

(2.2.16)

and from (2.2.15)

$$E_k(k) = m_V$$
$$E_k(k^2) = m_V^2 + m_V$$

(2.2.17)

as derived before. Higher moments can be easily generated in this way.

The characteristic function is also useful in dealing with sums of independent random variables, since the characteristic function of the sum is equal to the product of the individual characteristic functions. That is, if a random variable y is related to a sequence of independent random variables $x_1, x_2, \ldots, x_q$ by

$$y = x_1 + x_2 + \cdots + x_q$$

(2.2.18)

then

$$\psi_y(\omega) = \prod_{i=1}^{q} \psi_{x_i}(\omega)$$

(2.2.19)

As a specific example, consider a binary random variable x_i defined with the following probabilities:

$$\text{Prob}[x_i = 1] = n(\mathbf{v}) \, \Delta V$$
$$\text{Prob}[x_i = 0] = 1 - n(\mathbf{v}) \, \Delta V$$

(2.2.20)

The characteristic function for this variable is then

$$\psi_{x_i}(\omega) = E_{x_i}[e^{j\omega x_i}]$$
$$= 1 + (e^{j\omega} - 1)n(\mathbf{v}) \, \Delta V$$

(2.2.21)

Now extend to the sum of q such independent binary variables, denoted by y in (2.2.18). Note that y corresponds to the number of occurrences of $x_i = 1$ among the q variables. We then have from (2.2.19) and (2.2.21)

$$\psi_y(\omega) = \prod_{i=1}^{q} \{1 + (e^{j\omega} - 1)n(\mathbf{v}) \, \Delta V\}$$

(2.2.22)

If we now interpret q as the number of cells of volume ΔV in a total volume V, then y is precisely the count variable k in (2.1.2). Making these substitutions, taking the log of this function, and passing to the limit yields

$$
\begin{aligned}
\lim_{\Delta V \to 0} \log \psi_y(\omega) &= \lim_{\substack{\Delta V \to 0 \\ q \to \infty}} \log \prod_{i=1}^{q} \{1 + (e^{j\omega} - 1)n(\mathbf{v})\,\Delta V\} \\
&= \lim_{\substack{\Delta V \to 0 \\ q \to \infty}} \sum_{i=1}^{q} \log\{1 + (e^{j\omega} - 1)n(\mathbf{v})\,\Delta V\} \\
&= \left[\int_V n(\mathbf{v})\,d\mathbf{v} \right] (e^{j\omega} - 1)
\end{aligned}
\tag{2.2.23}
$$

Upon exponentiation this becomes

$$
\psi_y(\omega) = \exp[m_V(e^{j\omega} - 1)]
\tag{2.2.24}
$$

which is identical to the characteristic function of a Poisson counting variable with level m_V. This example serves as an alternative, and in fact a simpler, derivation of the Poisson density from the basic probability assumption of (2.1.1).

As a second example, consider (2.2.19) when each x_i is itself a Poisson random variable having level m_i. Then

$$
\begin{aligned}
\psi_y(\omega) &= \prod_{i=1}^{q} \exp[m_i(e^{j\omega} - 1)] \\
&= \exp\left[\left(\sum_{i=1}^{q} m_i \right)(e^{j\omega} - 1) \right]
\end{aligned}
\tag{2.2.25}
$$

which is the characteristic function of a Poisson random variable with level $\sum_{i=1}^{q} m_i$. Hence, the sum of independent Poisson variables is always a Poisson variable whose level is given by the sum of the individual levels.

Second Characteristic Function. It is often convenient in statistical analysis to use the logarithm of the characteristic function, referred to as the second characteristic function. When this is expanded to a Taylor series in the variable $j\omega$, the Taylor coefficients are referred to as semiinvariants. For the Poisson random variable, the second characteristic function becomes

$$
\log \psi_k(\omega) = m_V(e^{j\omega} - 1)
\tag{2.2.26}
$$

Since we can immediately expand

$$
m_V(e^{j\omega} - 1) = \sum_{i=1}^{\infty} \frac{m_V}{i!} (j\omega)^i
\tag{2.2.27}
$$

it is easy to see that all the semiinvariants are equal to the level m_V. The moments of a random variable are related to its semiinvariants $\{\chi_i\}$ by

$$E_k[k] = \chi_1$$
$$E_k[k^2] = \chi_1^2 + \chi_2$$
$$E_k[k^3] = \chi_1^3 + 3\chi_1\chi_2 + \chi_3$$
$$\vdots$$

$$(2.2.28)$$

$$E_k[k^q] = \sum_{i=1}^{q} \frac{1}{i!} \underbrace{\sum_{\substack{i \\ \sum_{j=1} q_j = q}}} \left[\frac{q!}{q_1! q_2! \cdots q_i!} \right] \chi_{q_1} \cdots \chi_{q_i}$$

Substituting the Poisson semiinvariants $\chi_j = m_V$ for all j rederives the moments in (2.2.15). Similarly, the semiinvariants can be related to the moments by the inverse of (2.2.28). This becomes

$$\chi_1 = E_k[k]$$
$$\chi_2 = E_k[k^2] - [E_k[k]]^2 \qquad\qquad (2.2.29)$$
$$\chi_3 = E_k[k^3] - 3E_k[k]E_k[k^2] + 2(E_k[k])^2$$

for the first three semiinvariants. In some of our later work we find situations in which semiinvariants of a random variable can be determined more conveniently than the actual probability density itself, and (2.2.28) is useful for generating moments.

2.3 CONDITIONAL POISSON COUNTING

We have emphasized the fact that the count variable k is Poisson, depending on the mean count over V. In effect, this means the count probability $Pos(k, m_V)$ is in fact only conditionally Poisson, conditioned on a known value for m_V. When attempting to extend count analysis to stochastic fields, we can no longer assume the intensity is deterministic, but must instead treat it as a stochastic process. This means that if $n(t, \mathbf{r})$ is a random intensity, then m_V becomes a random variable (integral of a random field) and the probability of the count variable k requires an additional average of the conditional Poisson count over the probability density on m_V. Formally, if we consider the volume V fixed, then m_V becomes a random variable, having probability density $p_{m_V}(m)$ ($0 \le m < \infty$), and the counting probability of the random variable k is then

$$P_k(k) = \int_0^{\infty} Pos(k, m) p_{m_V}(m)\, dm$$

$$= \int_0^{\infty} p_{m_V}(m) \left[\frac{m^k}{k!} e^{-m} \right] dm \qquad (2.3.1)$$

This counting probability for k is no longer a Poisson probability, and obviously depends on the probability density of m_V induced by the stochastic field. Since the conditional probability in the integrand is Poisson, we call the class of probabilities $P_k(k)$ generated from (2.3.1) *conditional Poisson* (CP) probabilities. In the literature they are also called *doubly stochastic Poisson* probabilities. Thus, the count probability resulting from the photo-detection of random fields belongs to the class of CP probabilities. We call the count k a conditional Poisson (CP) random variable. We emphasize again that the density of m_V in general depends on V, and therefore CP densities are generally functions of V, that is, nonstationary in time and space.

The computation of the CP probability requires performing the integration in (2.3.1). Fortunately, however, many of the statistical properties of the CP random variable can be inferred directly from the preceding results concerning the Poisson random variable. This is due to the fact that the average of any function $f(k)$ of the CP count k is

$$
E_k[f(k)] = \sum_{k=0}^{\infty} f(k) P_k(k)
$$

$$
= \sum_{k=0}^{\infty} f(k) \int_0^{\infty} p_{m_V}(m) \left[\frac{m^k}{k!} e^{-m} \right] dm
$$

$$
= \int_0^{\infty} \left[\sum_{k=0}^{\infty} f(k) \frac{m^k}{k!} e^{-m} \right] p_{m_V}(m) \, dm
$$

$$
= E_{m_V}[E_k[f(k)|m]] \tag{2.3.2}
$$

where we have introduced the conditional average over k as

$$
E_k[f(k)|m] \triangleq \sum_{k=0}^{\infty} f(k) \, \text{Pos}[k, m] \tag{2.3.3a}
$$

and the average over m_V as

$$
E_{m_V}[u(m)] \triangleq \int_{-\infty}^{\infty} u(m) p_{m_V}(m) \, dm \tag{2.3.3b}
$$

We recognize (2.3.2) as the average over m_V of the conditional average of $f(k)$ over a Poisson density with level m_V. This means the moments and characteristic functions of k can be derived from the moments and characteristic functions previously determined for the (conditional) Poisson density. For example,

$$
E_k[k] = E_{m_V}[m]
$$
$$
E_k[k^2] = E_{m_V}[m^2 + m] = E_{m_V}[m^2] + E_{m_V}[m] \tag{2.3.4}
$$
$$
\text{Var}[k] = \text{Var}[m_V] + E_{m_V}[m]
$$

The characteristic function of CP counts has an interesting relation to that of the level m_V, when evaluated by (2.3.2). This becomes

$$
\begin{aligned}
\psi_k(\omega) &= E_{m_V}[E(e^{j\omega k}|m)] \\
&= E_{m_V}[\exp\{m(e^{j\omega} - 1)\}] \\
&= \psi_{m_V}[-j(e^{j\omega} - 1)]
\end{aligned}
\tag{2.3.5}
$$

where $\psi_{m_V}(\omega)$ is the characteristic function of m_V. Hence, $\psi_k(\omega)$ is obtained simply by replacing ω by $-j(1 - e^{j\omega})$ in the characteristic function of m_V. Since $P_k(k)$ can always be obtained by inverse transforming its characteristic function, the CP probability can be alternately written as

$$
\begin{aligned}
P_k(k) &= \frac{1}{2\pi} \int_{-\infty}^{\infty} \psi_k(\omega) e^{-j\omega k} \, d\omega \\
&= \frac{1}{2\pi} \int_{-\infty}^{\infty} \psi_{m_V}[-j(1 - e^{-j\omega})] e^{-j\omega k} \, d\omega
\end{aligned}
\tag{2.3.6}
$$

In terms of the probability transform in (2.2.12), this has the simpler inversion

$$
P_k(k) = \frac{1}{(-1)^k k!} \left\{ \frac{\partial^k}{\partial z^k} \psi_{m_V}[-jz] \right\} \Bigg|_{z=1}
\tag{2.3.7}
$$

In some cases these inversions may be more tractable than the straightforward computation in (2.3.1).

2.4 SHORT-TERM CP COUNTING

When determining the probability densities of the CP counting, using the relation in (2.3.1), we have the initial task of determining the probability density of m_V from the (assumed) known statistics of the intensity. This means, mathematically, we must transform the density of the count intensity process $n(t, \mathbf{r})$ into those of the level process m_V through (2.3.1). In the next chapter we present a general approach to this procedure. However, some degree of simplification, avoiding these general methods, can be used when dealing with "short term" counting intervals and point detectors.

Let us denote our point detector coherence area as A and again let T be the counting interval in time. We define a short-term counting time as a value of T for which the integral of the intensity process $n(t)$ over $(t, t + T)$ is approximately equal to $n(t)T$. That is, we have a short-term counting condition if T is such that

$$
m_V = \int_t^{t+T} n(\rho) \, d\rho \approx n(t)T
\tag{2.4.1}
$$

In the general approach of the next chapter, this approximation and meaning of short term can be somewhat rigorized, so that the validity of the above can be subjectively evaluated. For the present, however, let us proceed under the assumption that (2.4.1) can be treated as an equality. [Certainly, as long as $n(t)$ has continuous sample functions, we know we can always select a T small enough so that any degree of approximation can be reached.] The advantage of the short-term assumption is that the probability density of m_V is now simply related to that of $n(t)$. Since $n(t)$ is proportional to the amplitude squared of the received radiation field, level statistics can be obtained directly from the field statistics at each t. Formally, we have

$$m_V = n(t)T = \alpha A T |a(t)|^2 \tag{2.4.2}$$

By the conversion of probability densities,

$$p_{m_V}(m) = \left(\frac{1}{2\alpha A T} \right) \frac{p_{|a_t|}(a)}{a} \bigg|_{a = (m/\alpha T A)^{1/2}} \tag{2.4.3}$$

where $P_{|a_t|}(a)$ is the probability density of the field magnitude $|a(t)|$ at time t. Thus the level probability density requires only knowledge of the first-order density of the envelope magnitude of the received field. In the following we present examples of this analysis procedure for computing short-term CP densities, using (2.4.3) and (2.3.1). The results of the examples are somewhat important in their own right, and are useful in later discussions of communication systems involving photodetection of stochastic fields.

Example 2.1   Monochromatic Fields

Let the received field be a simple monochromatic field with constant intensity as in Section 2.2. The intensity and level follow from (2.1.19) and (2.1.21) as $n(t) = \alpha I A$ and $m_V = \alpha I A T$. The latter parameters are of course not random. However, we can account for this in our present analysis by using a point mass probability density for m_V

$$p_{m_V}(m) = \delta(m - \alpha I A T) \tag{2.4.4}$$

implying that $m_V = \alpha I A T$ with a probability of one. The corresponding CP count probability in (2.3.1) is then

$$P_k(k) = \int_0^\infty \left[\frac{(m)^k}{k!} e^{-m} \right] \delta(m - \alpha I A T)$$

$$= \frac{(\alpha I A T)^k}{k!} e^{-\alpha I A T} \tag{2.4.5}$$

which is the Poisson count probability with level $\alpha I A T$. Thus we may interpret the Poisson probability with constant level as a special case of the

CP probability, corresponding to a nonrandom, constant intensity, monochromatic field.

Example 2.2   Narrow Band Gaussian Fields

Consider a zero mean, narrow band stationary Gaussian random field whose mean squared intensity over area A is σ^2. It is known that the envelope of this process is Rayleigh distributed in the variable a according to the density

$$p_{|a_t|}(a) = \frac{a}{\sigma^2} \exp\left[\frac{-a^2}{2\sigma^2}\right], \qquad a > 0$$

$$= 0 \qquad \text{elsewhere} \tag{2.4.6}$$

If we now make the substitution in (2.4.3), we can easily see that the probability density $p_{m_V}(m)$ can be written as

$$p_{m_V}(m) = \frac{1}{2\sigma^2 \alpha T A} \exp\left[\frac{-m}{2\sigma^2 \alpha T A}\right], \qquad m \geq 0$$

$$= 0 \qquad \text{elsewhere} \tag{2.4.7}$$

and

$$P_k(k) = \int_0^\infty \frac{m^k}{k!} e^{-m} \left[\frac{\exp[-m/2\sigma^2 \alpha T A]}{2\sigma^2 \alpha T A}\right] dm$$

$$= \frac{1}{2\sigma^2 \alpha T A k!} \int_0^\infty m^k \exp\left[\frac{-m(2\sigma^2 \alpha T A + 1)}{2\sigma^2 \alpha T A}\right] dm$$

$$= \frac{1}{1 + 2\alpha\sigma^2 T A}\left[\frac{2\alpha\sigma^2 T A}{1 + 2\alpha\sigma^2 T A}\right]^k \tag{2.4.8}$$

This count probability is referred to as the Bosé–Einstein probability, and is shown in Figure 2.3. A discrete probability density composed of these probabilities at the nonnegative integers is called a Bosé–Einstein density, and the associated CP count is said to be a Bosé–Einstein random variable. Thus the count associated with the short-term photodetection of narrow band Gaussian noise fields always obeys Bosé–Einstein statistics.

We might digress temporarily to examine this particular probability. Its characteristic function is

$$\psi_k(\omega) = \sum_{k=0}^\infty \left[\frac{1}{1 + 2\alpha\sigma^2 T A}\right]\left[\frac{2\alpha\sigma^2 T A}{1 + 2\alpha\sigma^2 T A}\right]^k e^{j\omega k}$$

$$= \frac{1}{1 + 2\alpha\sigma^2 T A}\left\{\frac{1}{1 - [(2\alpha\sigma^2 T A)e^{j\omega}/(1 + 2\alpha\sigma^2 T A)]}\right\}$$

$$= \frac{1}{1 + 2\alpha\sigma^2 T A(1 - e^{j\omega})} \tag{2.4.9}$$

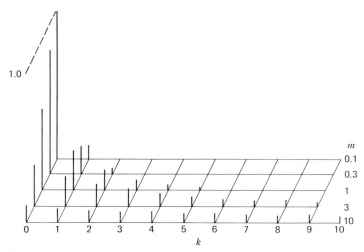

Figure 2.3. Bosé–Einstein probabilities; m = mean count.

Its moments can be calculated directly from (2.4.8) or by use of the characteristic function in (2.4.9). This leads to

$$E_k(k) = 2\alpha\sigma^2 TA$$
$$E_k(k^2) = 2\alpha\sigma^2 TA[1 + 4\alpha\sigma^2 TA] \qquad (2.4.10)$$
$$\text{Var}[k] = 2\alpha\sigma^2 TA[1 + 2\alpha\sigma^2 TA]$$

Notice that the mean of the Bosé–Einstein count is directly related to $\sigma^2 TA$, the amount of energy in the field collected over T sec. The variance of the Bosé–Einstein count appears as the sum of two terms, $2\alpha\sigma^2 TA + (2\alpha\sigma^2 TA)^2$. The first term equals the mean, and therefore would be the variance if the counting were Poisson, as indicated by (2.2.9). The second term is identical to the variance of the intensity probability density in (2.4.7). Thus the Bosé–Einstein count variance appears as the sum of an "input" variance (of the field itself) plus an additive variance due to the counting (detection operation). The addition of these variances occurs in a manner usually associated with the summing of independent random variables. Thus photodetection appears to add an independent "noise" to the inherent "noise" of the received field. This interpretation of the count variance for this example is of course always valid for CP counting, as is evident from (2.3.4). In Chapter 4 we relate this detection variance to an effective additive "quantum" noise.

Example 2.3   Monochromatic Field Plus Narrow Band Gaussian Noise

Let us now extend the results just obtained by considering the field to be the sum of a monochromatic field with constant intensity I, and random phase angle, uniformly distributed $(0, 2\pi)$, plus a zero mean, narrow band Gaussian noise field of variance σ^2. The envelope of this process obeys the Rician density at any t [1]:

$$p_{|a_t|}(a) = \frac{a}{\sigma^2} \exp\left[-\frac{a^2 + I}{2\sigma^2} \right] I_0\left(\frac{a(I)^{1/2}}{\sigma^2} \right), \qquad a \geq 0$$

$$= 0 \qquad \text{elsewhere} \tag{2.4.11}$$

where $I_0(x)$ is the zero-order imaginary Bessel function of argument x. The level density is then

$$p_{m_V}(m) = \frac{1}{2\sigma^2\alpha T A} \exp\left[-\frac{m + \alpha I A T}{2\sigma^2\alpha T A} \right] I_0\left(\frac{(m\alpha I A T)^{1/2}}{\alpha\sigma^2 T A} \right), \qquad m \geq 0$$

$$= 0 \qquad \text{elsewhere} \tag{2.4.12}$$

Expanding $I_0(x)$ as

$$I_0(x) = \sum_{j=0}^{\infty} \frac{(x/2)^{2j}}{(j!)^2} \tag{2.4.13}$$

and substituting into (2.3.1) yields

$$P_k(k) = \left(\frac{1}{2\alpha\sigma^2 T A} \right) \exp\left[-\frac{\alpha I A T}{2\alpha\sigma^2 T A} \right]\left(\frac{1}{k!} \right) \sum_{j=0}^{\infty} \left[\frac{1}{2^{2j}(j!)^2} \right]$$

$$\cdot \int_0^{\infty} m^k \exp\left[-\frac{m(1 + 2\alpha\sigma^2 T A)}{2\alpha\sigma^2 T A} \right]\left[\frac{\alpha m I A T}{(\alpha\sigma^2 T A)^2} \right]^j dm \tag{2.4.14}$$

Performing the integration we obtain

$$P_k(k) = \frac{\exp[-\alpha I A T/2\alpha\sigma^2 T]}{k! 2\alpha\sigma^2 T A}\left[\frac{2\alpha\sigma^2 T A}{1 + 2\alpha\sigma^2 T A} \right]^{k+1} \sum_{j=0}^{\infty} \left[\frac{I/2\sigma^2}{(1 + \alpha\sigma^2 T A)} \right]^j \frac{(k + j)!}{(j!)^2} \tag{2.4.15}$$

This can be written more compactly as

$$P_k(k) = \frac{\exp[-I/2\sigma^2]}{1 + 2\alpha\sigma^2 T A}\left[\frac{2\alpha\sigma^2 T A}{1 + 2\alpha\sigma^2 T A} \right]^k {}_1F_1\left(k + 1, 1, \frac{I/2\sigma^2}{1 + 2\alpha\sigma^2 T A} \right) \tag{2.4.16}$$

where ${}_1F_1(x, y, z)$ is the confluent hypergeometric function of arguments (x, y, z):

$${}_1F_1(x, y, z) = \sum_{j=0}^{\infty} \frac{x_j(z)^j}{y_j j!}$$

$$x_j = x(x + 1) \cdots (x + j - 1)$$

$$y_i = y(y + 1) \cdots (y + j - 1) \tag{2.4.17}$$

It is convenient to use the relationship between the confluent hypergeometric function and the Laguerre polynomial $L_k(x)$†, defined by

$$L_k(x) = \sum_{j=0}^{k} \binom{k}{j} \frac{(-x)^j}{j!} \tag{2.4.18}$$

Using (C.3.1), we rewrite (2.4.16) as

$$P_k(k) = \left[\frac{1}{1 + 2\alpha\sigma^2 TA} \right]\left[\frac{2\alpha\sigma^2 TA}{1 + 2\alpha\sigma^2 TA} \right]^k$$

$$\times \exp\left[-\frac{\alpha IAT}{1 + 2\alpha\sigma^2 TA} \right]L_k\left[\frac{-I/2\sigma^2}{1 + 2\alpha\sigma^2 TA} \right] \tag{2.4.19}$$

This represents the CP counting probability associated with a monochromatic signal field plus a Gaussian field, and is shown in Figure 2.4.

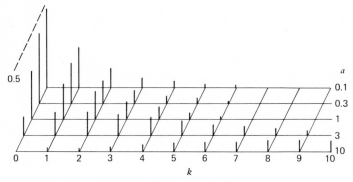

Figure 2.4. Laguerre probabilities, Lag(k, a, 1, 0).

It is referred to as a *Laguerre counting probability*. For convenience in later analysis, we introduce the Laguerre probability function

$$\text{Lag}[k, a, b, c] = \left[\frac{b^k}{(1 + b)^{k+c+1}} \right]\exp\left[\frac{-a}{1 + b} \right]L_k^c\left[\frac{-a}{b(1 + b)} \right] \tag{2.4.20}$$

so that $P_k(k) = \text{Lag}(k, \alpha IAT, 2\alpha\sigma^2 TA, 0)$. We note the following limiting conditions. For the special case of noise alone ($I = 0$)

$$\lim_{I \to 0} P_k(k) = \left[\frac{1}{1 + 2\alpha\sigma^2 TA} \right]\left[\frac{2\alpha\sigma^2 TA}{1 + 2\alpha\sigma^2 TA} \right]^k \tag{2.4.21}$$

† The Laguerre polynomial and its generalizations appear numerous times in subsequent discussions. We have devoted Appendix C to a summary of its definition, properties, and significant identities.

which is the Bosé–Einstein probability calculated in (2.4.8). For the case of a pure monochromatic field only ($\sigma^2 = 0$) we use the fact

$$\lim_{z \to \infty} {}_1F_1(x, y, z) = \frac{(x - 1)!}{(y - 1)!} \exp[z]z^{x-y} \tag{2.4.22}$$

to obtain

$$\lim_{\sigma^2 \to 0} P_k(k) = \frac{(\alpha I A T)^k}{k!} e^{-\alpha I A T} \tag{2.4.23}$$

which is the Poisson density with level $\alpha I A T$. Hence, the density calculated in (2.4.19) is seen to be a generalization of our two previous examples. One can use a Laguerre identity [(C.6.1) in Appendix C] to calculate the characteristic function for the CP probability density in (2.4.19). This becomes

$$\psi_k(\omega) = \left[\frac{1}{1 + 2\alpha\sigma^2 TA(1 - e^{j\omega})} \right] \exp\left[\frac{\alpha I A T(e^{j\omega} - 1)}{1 + 2\alpha\sigma^2 TA(1 - e^{j\omega})} \right] \tag{2.4.24}$$

Under the limiting conditions in (2.4.21) and (2.4.23) previously discussed, this reduces to both (2.2.11) and (2.4.9). It is straightforward to calculate the mean and variance of this density (Problem 2.3). They are, respectively,

$$E_k(k) = 2\alpha\sigma^2 TA + \alpha I A T \tag{2.4.25}$$

and

$$\mathrm{Var}[k] = 2\alpha\sigma^2 TA[1 + 2\alpha\sigma^2 TA] + \alpha I A T[1 + 4\alpha\sigma^2 TA] \tag{2.4.26}$$

If we determine the mean and variance of the level density in (2.4.12), we find

$$E_{m_V}[m] = 2\alpha\sigma^2 TA + \alpha I A T \tag{2.4.27}$$

$$\mathrm{Var}[m_V] = [2\alpha\sigma^2 TA]^2 + \alpha I A T[4\alpha\sigma^2 TA] \tag{2.4.28}$$

Notice again that the difference between the variance in (2.4.26) and that in (2.4.28) is

$$\mathrm{Var}[k] - \mathrm{Var}[m_V] = 2\alpha\sigma^2 TA + \alpha I A T \tag{2.4.29}$$

and can again be related to an increase in variance resulting from Poisson photodetection (i.e., $2\alpha\sigma^2 TA$ from detecting the noise field and $\alpha I A T$ from detecting the monochromatic field, both appearing as additive independent effects).

We can extend the result of this example to include the sum of M independent variables each having densities similar to (2.4.12), each with identical σ^2 but distinct values of intensity, $\hat{I}_i$. (This would correspond to summing the amplitudes of M independent, monochromatic signal fields, each with an additive noise field.) The characteristic function becomes [2]

$$\psi_{m_V}(\omega) = \prod_{i=1}^{M} \left\{ \frac{1}{1 - 2\alpha\sigma^2 TAj\omega} \exp\left[\frac{\alpha\hat{I}_i ATj\omega}{1 - 2\alpha\sigma^2 TAj\omega}\right]\right\}$$

$$= \left(\frac{1}{1 - 2\alpha\sigma^2 TAj\omega}\right)^M \exp\left[\frac{\sum_{i=1}^{M} \alpha\hat{I}_i ATj\omega}{1 - 2\alpha\sigma^2 TAj\omega}\right] \tag{2.4.30}$$

The probability density associated with this characteristic function is the noncentral chi-squared density of degree M, resulting from the sum of M independent Rician variables. This density has the form [2, p. 396]

$$p_{m_V}(m) = \frac{1}{2\alpha\sigma^2 TA} \left(\frac{m}{\alpha I A}\right)^{\frac{M-1}{}} \exp\left[-\frac{m + \alpha I A}{2\alpha\sigma^2 TA}\right] I_{M-1}\left(\frac{(m\alpha I AT)^{1/2}}{\alpha\sigma^2 TA}\right), m \geq 0$$

$$= 0 \qquad \text{elsewhere} \tag{2.4.31}$$

where $I = \sum_{i=1}^{M} \hat{I}_i$ and $I_{m-1}(\cdot)$ is the Bessel function of order $M - 1$. The probability density of k is obtained by using (2.4.30) in (2.3.5) and transforming. Equivalently, it can be obtained by convolving the densities (2.4.19), making use of the convolution properties of Laguerre functions in Appendix C. After some algebra this yields

$$P_k(k) = \frac{(2\alpha\sigma^2 TA)^k}{(1 + 2\alpha\sigma^2 TA)^{k+M}} \exp\left[\frac{-\alpha I AT}{1 + 2\alpha\sigma^2 TA}\right] L_k^{M-1}\left[\frac{I/2\sigma^2}{1 + 2\alpha\sigma^2 TA}\right] \tag{2.4.32}$$

where $L_k^D(x)$ is the generalized Laguerre polynomial of index k and order D, defined as

$$L_k^D(x) = \sum_{i=0}^{k} \binom{k + D}{i} \frac{(-x)^i}{i!} \tag{2.4.33}$$

where

$$\binom{k + D}{i} = \frac{(k + D)!}{(k + D - i)!i!} \tag{2.4.34}$$

(For further discussion, of this polynomial see Appendix C.) The mean and variance of the count are given by

$$E_k[k] = M(2\alpha\sigma^2 TA) + \alpha I AT \tag{2.4.35}$$

$$\text{Var}[k] = M(1 + 2\alpha\sigma^2 TA)(2\alpha\sigma^2 TA) + \alpha I AT(1 + 2\alpha\sigma^2 TA) \tag{2.4.36}$$

Similar results can be determined for other modifications of this example, as in Problem (2.4).

Example 2.4   Turbulent Field Plus Narrow Band Noise

Consider again Example 2.3, except we replace the monochromatic field by a turbulent field having amplitude scintillation. We model this as a monochromatic field with a random intensity. At each t, the intensity of the turbulent field is described by the probability density $p_I(I)$, $I \geq 0$. We again assume a point detector and an additive narrow band Gaussian noise field. The count probability can be determined by averaging the count for a given I over the density on I. When conditioned on I the count probability is given in (2.4.19). Thus

$$P_k(k) = \int_0^\infty \text{Lag}[k, \alpha I A T, 2\alpha\sigma^2 T A, 0]p_I(I)\,dI \qquad (2.4.37)$$

where Lag represents the Laguerre count probability in (2.4.20). Explicit results are hindered by the complexity of the integrand, and most often computer integration must be used. In certain cases, limiting forms of (2.4.37) can be determined (Problem 2.5). The interested reader is referred to the work of Diament and Teich [3–5], Lachs [6], and Teich and Rosenberg [7], who have obtained counting probabilities for stochastic radiation after transmission through lognormal turbulent channels.

2.5   INVERSION OF THE COUNT PROBABILITY

We have shown that the CP count probability can be generated from the probability density of the integrated random intensity m_V by averaging a Poisson function. In this section we show that the operation can be inverted. That is, if we are given the CP count probabilities, we can regenerate the probability density on m_V from which they were derived. To accomplish this we make use of the Laguerre wave function discussed in Appendix C:

$$l_n(2xy) = \sqrt{2x}\, e^{-xy}L_n(2xy) \qquad (2.5.1)$$

and its orthonormality over the interval $(0, \infty)$ in y:

$$\int_0^\infty l_n(2xy)l_m(2xy)\,dy = \begin{cases} 1, & n = m \\ 0, & n \neq m \end{cases} \qquad (2.5.2)$$

This means any function integrably square bounded can be uniquely expanded in an infinite series of such functions. In particular, any probability density on the random level variable m_V can be so expanded. Thus, we consider the expansion

$$p_{m_V}(m) = \sum_{n=0}^\infty a_n l_n(2\alpha A T m) \qquad (2.5.3)$$

where

$$a_n = \int_0^\infty p_{m_V}(m) l_n(2\alpha A T m) \, dm \qquad (2.5.4)$$

Now using (2.5.3) in (2.3.1), we have

$$P_k(k) = \int_0^T \frac{(\alpha A T m)^k}{k!} e^{-\alpha m T A} \left[\sum_{n=0}^\infty a_n l_n(2\alpha A T m) \right] dm \qquad (2.5.5)$$

If we multiply both sides by the factor $(2\alpha A T)^{1/2}(-1)^k\binom{n}{k}z^k$ and sum from 0 to n, we will generate the Laguerre polynomial on the right-hand side. Substitution of (2.5.1) and use of the orthonormality in (2.5.2) then yields an identity for the a_n coefficients in (2.5.3):

$$a_n = (2\alpha A T)^{1/2} \sum_{k=0}^n (-1)^k \binom{n}{k} 2^k P_k(k) \qquad (2.5.6)$$

Thus,

$$p_{m_V}(m) = \sum_{n=0}^\infty (2\alpha A T)^{1/2} \left[\sum_{k=0}^n (-1)^k\binom{n}{k}2^k P_k(k) \right] l_n(2\alpha A T m) \qquad (2.5.7)$$

This gives us the desired inversion rule. The count probabilities $P_k(k)$ must be substituted into (2.5.7), and the resulting sum is the density $p_{m_V}(m)$. Although (2.5.7) is mathematically valid, its use generally depends on the ability to obtain a closed form expression for the infinite summation.

Example 2.5

Consider the Poisson count probability

$$P_k(k) = \frac{(\alpha I A T)^k}{k!} e^{-\alpha I A T} \qquad (2.5.8)$$

Substituting this into (2.5.7) yields

$$p_{m_V}(m) = \sum_{n=0}^\infty l_n(2\alpha A T m) \left[\sum_{k=0}^n (-1)^k\binom{n}{k} \frac{(2\alpha I A T)^k}{k!} e^{-\alpha I A T} \right] \qquad (2.5.9)$$

Relating the bracket to (2.5.1) and (2.5.6) shows that (2.5.9) is the form

$$p_{m_V}(m) = \sum_{n=0}^\infty l_n(2\alpha A T m) l_n(2\alpha I A T) \qquad (2.5.10)$$

We recognize this as the expansion of a delta function [Appendix C, (C. 3.14)]. This yields (2.4.4) as the desired density.

Example 2.6

Consider the Bosé–Einstein density

$$P_k(k) = \left(\frac{1}{1 + 2\alpha\sigma^2 TA}\right)\left(\frac{2\alpha\sigma^2 TA}{1 + 2\alpha\sigma^2 TA}\right)^k \qquad (2.5.11)$$

Equation 2.5.7 becomes

$$p_{m_V}(m) = \frac{1}{1 + 2\alpha\sigma^2 TA} \sum_{m=0}^{\infty} l_n(2\alpha AmT)\left[\sum_{k=0}^{n} \binom{n}{k}\left[\frac{-4\alpha\sigma^2 TA}{1 + 2\alpha\sigma^2 TA}\right]^k\right] \qquad (2.5.12)$$

Use of the binomial expansion

$$\sum_{k=0}^{n} (-x)^k \binom{n}{k} = (1 - x)^n \qquad (2.5.13)$$

produces

$$p_{m_V}(m) = \frac{1}{1 + 2\alpha\sigma^2 TA} \sum_{n=0}^{\infty} l_n(2\alpha mTA)\left[1 - \frac{4\alpha\sigma^2 TA}{1 + 2\alpha\sigma^2 TA}\right]^n \qquad (2.5.14)$$

Use of (2.5.1) and the identity in (C.6.2) simplifies this to

$$p_{m_V}(m) = \frac{1}{2\sigma^2 \alpha TA} \exp\left(-\frac{m}{2\sigma^2 \alpha TA}\right) \qquad (2.5.15)$$

corresponding to (2.4.7). The inversion rule in (2.5.7) has been proposed as a method for determining actual field statistics. In this technique count probabilities are measured and inverted in (2.5.7) to determine level statistics, from which field envelope densities can be derived. Such techniques have been used to aid in the development of mathematical models for the output of certain types of noisy laser sources [8, 9].

REFERENCES

[1] Rice, S. O., "Mathematical Analysis of Random Noise," *Bell System Tech. J.*, **23**, 283–382 (1944); **24**, 46–156 (1945).

[2] Van Trees, H. L., *Detection, Estimation, and Modulation Theory*, Part I, John Wiley and Sons, New York, 1968.

[3] Diament, P. and Teich, M. C., "Photoelectron-Counting Distributions for Irradiance-Modulated Radiation," *J. Opt. Soc. Am.*, **60**, 682–689 (May 1970).

[4] ———, "Photodetection of Low-Level Radiation Through the Turbulent Atmosphere," *J. Opt. Soc. Am.*, **60**, 1489–1494 (November 1970).

[5] ———, "Optical Detection of Laser or Scattered Radiation Transmitted Through the Turbulent Atmopshere," *Appl. Opt.*, **10**, 1664–1667 (July 1971).

[6] Lachs, G., "Theoretical Aspects of Mixtures of Thermal and Coherent Radiation," *Phys. Rev.*, **138**, B1012–B1016 (May 1965); also G. Lachs, "Approximate Photocount Statistics for Coherent and Chaotic Radiation of Arbitrary Spectral Shape," *J. Appl. Phys.*, **42**, 602–609 (February 1971).

[7] Teich, M. C. and Rosenberg, S., "*N*-Fold Joint Photocounting Distribution for Modulated Laser Radiation: Transmission Through the Turbulent Atmosphere," *J. Opt. Electron*, **3**, 63–76 (May 1971).

[8] Bedard, G., "Photon Counting Statistics of Gaussian Light," *Phys. Rev.*, **151**, No. 4, 151 (November 1966).

[9] Arecchi, F., "Measurement of Distributions of Gaussian and Laser Sources," *Phys. Rev. Lett.*, **15–24**, 912 (December 1965).

PROBLEMS

1. Electrons are emitted from a photodetector surface according to a Poisson probability with fixed level m. However, the probability that a given electron will be collected at the output is η. Determine the count distribution collected at the output. Relate the probability η to an energy efficiency parameter.

2. Show that Taylor series expansion of a characteristic function $\psi(\omega)$ has moments as coefficients. That is,

$$\psi(\omega) = \sum_{i=0}^{\infty} \frac{m_i}{i!} (j\omega)^i$$

where m_i is the ith moment.

3. Determine the mean and variance of the probability density in (2.4.24) by (a) using the characteristic function, (b) using the probability density directly.

4. Evaluate the characteristic function of the count process generated by an intensity with envelope density:

$$p_{m_V}(m) = m \int_0^{\infty} J_0(Ax)J_0(Bx)J_0(mx) \exp\left[-\frac{\sigma^2 x^2}{2}\right] x \, dx.$$

Also compute the mean and variance. *Hint*: Use the following integrals

$$\int_0^{\infty} r \, dr \, J_0(xr)e^{-a^2 r^2} = \frac{\exp(-x^2/4a^2)}{2a^2}$$

$$\int_0^{\infty} J_0(Ay)J_0(By)e^{-b^2 y^2} y \, dy = \frac{\exp\{-[(A^2 + B^2)/4b^2]\}}{2b^2} I_0\left(\frac{AB}{2b^2}\right)$$

5. Compute the integral in (2.4.37) where $p_I(I)$ is log-normal. Repeat when the Laguerre probability is approximated by a Poisson probability.

6. Determine the probability transform of a Poisson random variable. Invert this using (2.3.7) to obtain the Poisson probability.

7. Consider a normal variable x, with mean μ and variance σ^2. Consider the transformation $y = \exp(x)$. The random variable y is lognormal.
(a) Show that

$$p_y(y) = \frac{1}{\sqrt{2\pi}\sigma}\exp\left\{-\left[\frac{(\log\mu - \mu)^2}{2\sigma^2}\right]\right\}\frac{dy}{y}$$

(b) Calculate the moments of y.
(c) Repeat (a) and (b) for the variable $z = y^2$.

8. Consider x, y as two identically distributed, independent Gaussian random variables with zero mean and variance σ^2.
(a) What is their joint distribution? If we define the envelope as

$$R = (x^2 + y^2)^{1/2}$$

what is the distribution of the envelope?
(b) If we add the mean value A_0 to one of the variables, what does the distribution of the envelope become? Does it matter which variable has mean? What if we add means to both variables? Does the envelope change?
(c) Repeat (a) and (b) with three independent Gaussian variables.

9. Set all the χ_{q_i} in (2.2.28) equal to a fixed value. Now calculate the coefficients for the leading three terms $(i = q; i = q - 1; i = q - 2)$ of $E_k[k^q]$.

10. Given the noncentral chi-squared distribution:

$$p_z(z) = \left[\begin{array}{l}\left(\dfrac{z}{z_0}\right)^{(M-1)/2}e^{-(z-z_0)}I_{m-1}(2\sqrt{zz_0}), \qquad z \geq 0 \\[2ex] 0, \qquad z < 0\end{array}\right]$$

with moments

$$E(z^u) = \frac{(u + m - 1)!}{(m - 1)!}\,{}_1F_1\,(-u_1 m_1 - z_0)$$

The log of the moments can be approximated by

$$\ln E(z^u) \approx \left[\psi(m) - \ln m + \ln(1 - z_0) + \frac{z_0}{m}\right] + \frac{\psi'(m)}{2}u^2$$

where $\psi(m)$ is the digamma function and $|z_0|/m \ll 1$.
(a) Show that all the moments of z approximate those of a lognormal variable. (Hence, a chi-squared density approximates a lognormal density.)
(b) Let N be a Poisson variable with level z, where z has a lognormal density. Use the result of (a) to show that

$$P_N(N) \cong 2^{-N-m}e^{-z_0/2}[L_N^{m-1}(\tfrac{1}{2}z_0)]$$

11. Define a Poisson branching process as a stochastic amplification whereby an event is transformed into a Poisson event with parameter δ. That is, the Poisson event can take on values from 0 to ∞ whereas the original event was unity with probability one. Consider an n-stage Poisson branching process with δ_i associated with the ith stage.

(a) Given an event at the input to the first stage, calculate the mean and variance of the output number from the nth stage. Assume that the events produced at each stage are independent.

(b) Show that after n stages the ratio of mean squared value to variance is given by

$$\frac{E_k(k^2)}{\text{Var } k} = \frac{\delta_1}{1 + 1/\delta_2 + 1/\delta_2 \delta_3 + \cdots}$$

12. Given two conditional Poisson counting processes, show that the correlation between the counts is related to the integral of the correlation of the field intensities generating them.

3 Conditional Poisson Counting

In the preceding chapter we derived a general equation that determines the conditionally Poisson count probabilities resulting from the photodetection of stochastic fields. This equation required the determination of the probability density of the level of the counts, which is integrally related to the detected field intensity. We avoided the mathematical complexity of determining this level density rigorously by invoking a short-term counting assumption, which allowed us to relate the level density to the field statistics in a rather simple way. In this chapter we return to the general equation and develop a more rigorous approach to computing conditional Poisson (CP) counting probabilities. We are eventually able to relate the results to the short-term counting results, and somewhat formalize their range of validity and physical interpretation.

3.1 FIELD MODE DECOMPOSITION

In CP counting the count probability is related to the probability density of the integrated field intensity by

$$P_k(k) = \int_0^\infty \frac{m^k}{k!} e^{-m} p_{m_V}(m) \, dm \qquad (3.1.1)$$

where $p_{m_V}(m)$ is the probability density of m_V. The latter is in turn related to the received field envelope $a(t, \mathbf{r})$ on the detector surface over a time interval $(t, t + T)$ by the integral†

$$m_V = \alpha \int_{\mathscr{A}} \int_t^{t+T} |a(\rho, \mathbf{r})|^2 \, d\rho \, d\mathbf{r} \qquad (3.1.2)$$

Here the subscript V denotes the specific observation volume defined by the area $\mathscr{A}$ and interval T. In the following we treat V as a fixed volume, and

---

† The integral in (3.1.2) should be normalized by the $1/2Z_W$ factor in (1.3.2). For convenience we incorporate this factor into the envelope notation $a(t, \mathbf{r})$.

attempt to evaluate (3.1.1) for various types of stochastic fields. Although we initially allow some degree of generality we will eventually find it necessary to constrain our study to random Gaussian fields in order to attain some degree of tractability.

The key to solving (3.1.1), using (3.1.2), depends on the ability to expand the random field envelope $a(t, \mathbf{r})$ into a orthonormal Fourier series over V as described in Section 1.3. That is, we write

$$a(\mathbf{v}) = \sum_{i=0}^{\infty} a_i \phi_i(\mathbf{v}) \tag{3.1.3}$$

where $\mathbf{v} = (t, \mathbf{r})$ and

$$a_i = \int_V a(\rho, \mathbf{r}) \phi_i^*(\rho, \mathbf{r}) \, d\rho \, d\mathbf{r} \tag{3.1.4}$$

are the random Fourier coefficients and $\{\phi_i(\mathbf{v})\}$ represents a complete set of complex orthonormal basis functions over V. With (3.1.3) we can now substitute with (3.1.4) into (3.1.2) and derive

$$m_V = \int_V \alpha |a(\rho, \mathbf{r})|^2 \, d\rho \, d\mathbf{r}$$

$$= \alpha \int_{\mathscr{A}} \int_t^{t+T} \sum_{i=0}^{\infty} a_i \phi_i(\mathbf{v}) \sum_{j=0}^{\infty} a_j^* \phi_j^*(\mathbf{v}) \, d\mathbf{v}$$

$$= \alpha \sum_{i=0}^{\infty} \sum_{j=0}^{\infty} a_i a_j^* \int_V \phi_i(\mathbf{v}) \phi_j^*(\mathbf{v}) \, d\mathbf{v} \tag{3.1.5}$$

The orthonormality of the basis functions then yields

$$m_V = \alpha \sum_{i=0}^{\infty} |a_i|^2 \tag{3.1.6}$$

Thus we have expressed the counting level as the sum of the squared magnitudes of the coefficients. In essence we have replaced a volume integration of a random intensity by an infinite sum of random coefficient variables. If $R_a(t_1, \mathbf{r}_1; t_2, \mathbf{r}_2)$ is the correlation function of the field envelope at volume points $(t_1, \mathbf{r}_1)$ and $(t_2, \mathbf{r}_2)$, and if each member of the function set satisfies the integral equation

$$\int_V R_a(\mathbf{v}_1, \mathbf{v}_2) \phi_i(\mathbf{v}_2) \, d\mathbf{v}_2 = \gamma_i \phi_i(\mathbf{v}_1) \tag{3.1.7}$$

for some positive constants $\{\gamma_i\}$, the expansion in (3.1.3) will be a Karhunen–Loeve expansion. This means that the random coefficients in the sum in

(3.1.3) are uncorrelated random variables, and the constant γ_i is the mean squared value of a_i.

To apply these equations, let us first examine the case of a deterministic field. This is a special case where the random coefficients are simply complex numbers, and m_V is a scalar for a fixed V. We see from (3.1.5) that the infinite sum of these coefficients must add up to the total energy over V, which guarantees the convergence of the sum if the field energy is bounded. Mathematically, we can consider m_V as a discrete random variable with a density containing a unit delta function at this scalar value. This means we use

$$p_{m_V}(m) = \delta\left(m - \alpha \sum_{i=0}^{\infty} |a_i|^2\right) \qquad (3.1.8)$$

and the resulting evaluation of (3.1.1) yields

$$
\begin{aligned}
P_k(k) &= \mathrm{Pos}\left(k, \alpha \sum_{i=0}^{\infty} |a_i|^2\right) \\
&= \frac{(\alpha \sum_{i=0}^{\infty} |a_i|^2)^k}{k!} \exp\left[-\alpha \sum_{i=0}^{\infty} |a_i|^2\right]
\end{aligned} \qquad (3.1.9)
$$

This is the usual Poisson counting associated with deterministic fields. Note that the mean count is the count energy, and therefore is proportional to the detected field intensity over V, substantiating the interpretation of a photodetector as an energy detector. In addition, (3.1.9) has an alternative interpretation. Recall from Section 2.2 that the sum of independent Poisson variables is itself a Poisson variable with level given by the sum of the individual levels. In this light, (3.1.9) can be interpreted as the probability density of a sum of independent Poisson variates, say $k_1, k_2, \ldots$, in which k_i has level $\alpha|a_i|^2$. Thus the count variable k can be interpreted as

$$k = \sum_{i=0}^{\infty} k_i \qquad (3.1.10)$$

where each k_i has count probability

$$P_{k_i}(k_i) = \mathrm{Pos}[k_i, \alpha|a_i|^2] \qquad (3.1.11)$$

That is, the total count k appears as if it were produced by the summation of independent Poisson counts, one associated with each component of the envelope expansion. If we think of these components as being associated with the modes of the radiation field over time and space, then each mode essentially contributes a random Poisson count to the total count. Since the sum of the $|a_i|^2$ adds up to the total field energy, each $|a_i|^2$ can be considered the energy of the individual mode. Therefore, the photodetector energy detects each mode independently, and produces a count equal to the sum of the individual mode counts. Note that this particular interpretation does not depend on the orthonormal basis set chosen for the expansion, which makes the physical meaning of a mode somewhat arbitrary. Different bases redistribute the energy differently over the modes, but it is always the total energy that affects the count statistic.

A physical interpretation of the notion of a detector mode can be made if we assume the counting time T is fairly long and the detector area fairly large, so that the set of sine waves in time and plane waves in space represents the orthonormal basis set over V. Then the expansion in (3.1.3) represents the expansion of the envelope into its frequency content and plane wave arrival directions, and the components $|a_i|^2$ represent energy at each frequency and at each direction of arrival. In this case the independent mode functions are the usual cissoids, $\exp j[\omega_i t - \mathbf{z} \cdot \mathbf{r}]$ describing electromagnetic fields, and the individual Poisson mode counts are the detector's response to each frequency arriving from each direction.

It would be extremely convenient if a similar interpretation is possible for stochastic fields. However, in order to develop this, we must determine the probability from (3.1.1). For stochastic fields, (3.1.6) corresponds to an infinite sum of random variables, and the density of this sum must first be determined. (The convergence of the sum in probability to a unique random variable is now guaranteed by the integral boundedness of the envelope correlation function.) The use of m_V as a sum of random variables, as opposed to the integral of a random process in (3.1.2), affords us several convenient systematic approaches to determining the general count probabilities. We can either attempt to compute the probability density of the sum in (3.1.6) and evaluate (3.1.1) or, alternatively, compute the characteristic function of the sum, $\psi_{m_V}(\omega)$ and evaluate (2.3.6) as

$$P_k(k) = \frac{1}{2\pi} \int_{-\infty}^{\infty} \psi_{m_V}[-j(e^{j\omega} - 1)]e^{-j\omega k}\, d\omega \qquad (3.1.12)$$

We hasten to point out that although we may have a clear-cut procedure for the stochastic field case, the computation of the density, or characteristic function, of m_V requires joint statistics of all the $\{a_i\}$. The KL expansion produces uncorrelated coefficients, but the joint density is still needed. However, by confining our attention to Gaussian fields, we can, to some extent, overcome this obstacle. In the next section we develop this approach and attempt to expand our detector mode concept.

3.2 MODAL DECOMPOSITION WITH GAUSSIAN FIELDS

When the stochastic field $f(t)$ of the detector is Gaussian, the KL expansion has the added feature of producing independent random coefficients. This follows from the fact that the $\{a_i\}$ in (3.1.5) are now complex Gaussian variables, and the uncorrelatedness produced by the KL expansion implies independence. Thus for Gaussian fields the level m_V in (3.1.6) is the sum of squared magnitudes of independent complex Gaussian coefficients. The density of m_V can now be developed by properly manipulating these Gaussian summations.

Let us consider the random field at the detector to be composed of the sum of a deterministic field $\{s(t, \mathbf{r}) \exp(j\omega_0 t)\}$ and a zero mean, Gaussian noise field $\{b(t, \mathbf{r}) \exp(j\omega_0 t)\}$, where ω_0 is the wave frequency, and $s(t, \mathbf{r})$ and $b(t, \mathbf{r})$ are complex, narrow band envelopes about ω_0. The total field is then

$$f(t, \mathbf{r}) = \{s(t, \mathbf{r}) + b(t, \mathbf{r})\} \exp[j\omega_0 t] \qquad (3.2.1)$$

We assume $b(t, \mathbf{r})$ has a known covariance function $R_b(t_1, \mathbf{r}_1; t_2, \mathbf{r}_2)$. The combined field is a Gaussian field and, in typical application, $s(t, \mathbf{r})$ is a signal envelope while $b(t, \mathbf{r})$ represents the envelope of additive background noise. The Karhunen–Loeve expansion of the field envelope into the basis set $\{\phi_i(t, \mathbf{r})\}$ produces the random coefficients

$$a_i = s_i + b_i \qquad (3.2.2)$$

where

$$s_i = \int_V s(\mathbf{v})\phi_i^*(\mathbf{v}) \, d\mathbf{v}$$

$$b_i = \int_V b(\mathbf{v})\phi_i^*(\mathbf{v}) \, d\mathbf{v}$$

Note that a_i is the sum of a complex constant s_i and a zero mean complex Gaussian random variable b_i. For narrow band Gaussian envelope processes, the real and imaginary parts of b_i are independent, each with variance $\gamma_i/2$, where γ_i is the mean square value of $|b_i|$. Since $|b_i|^2$ corresponds to an energy component, its mean squared value γ_i is the average energy of the noise field in this component. The probability density of $|a_i|$ is then the magnitude of a complex Gaussian variate, which is known to have the Rice density

$$p_{|a_i|}(a) = \left(\frac{2a}{\gamma_i}\right) \exp\left[-\frac{|s_i|^2 + a^2}{\gamma_i}\right] I_0\left[\frac{2|s_i|a}{\gamma_i}\right], \qquad a \geq 0$$

$$= 0 \qquad \text{elsewhere} \qquad (3.2.3)$$

This means that $|a_i|^2$ has the probability density

$$p_{|a_i|^2}(m) = \left(\frac{1}{\gamma_i}\right) \exp\left[-\frac{|s_i|^2 + m}{\gamma_i}\right] I_0\left(\frac{2|s_i|\sqrt{m}}{\gamma_i}\right), \qquad m \geq 0$$

$$= 0 \qquad \text{elsewhere} \qquad (3.2.4)$$

and characteristic function

$$\psi_{|a_i|^2}(\omega) = E[e^{j\omega|a_i|^2}]$$

$$= \frac{1}{1 - \gamma_i j\omega} \exp\left[\frac{|s_i|^2 j\omega}{1 - \gamma_i j\omega}\right] \quad (3.2.5)$$

Since the $\{a_i\}$ are independent, the random variables $|a_i|^2$ are independent, and m_V in (3.1.6) is the sum of independent random variables. Its characteristic function is therefore

$$\psi_{m_V}(\omega) = \prod_{i=0}^{\infty} \frac{1}{1 - \gamma_i j\omega} \exp\left[\frac{\alpha|s_i|^2 j\omega}{1 - \alpha\gamma_i j\omega}\right] \quad (3.2.6)$$

The probability of the counts can now be obtained by using (3.1.12), and inverse Fourier transforming

$$\psi_k(\omega) = \psi_{m_V}[-j(e^{j\omega} - 1)]$$

$$= \prod_{i=0}^{\infty} \left\{\left[\frac{1}{1 + \alpha\gamma_i(1 - e^{j\omega})}\right] \exp\left[\frac{\alpha|s_i|^2(e^{j\omega} - 1)}{1 + \alpha\gamma_i(1 - e^{j\omega})}\right]\right\} \quad (3.2.7)$$

Note that this characteristic function of k appears as a product of individual characteristic functions. This means it too can be interpreted as the characteristic function of a sum of random variables $\{k_i\}$, where each k_i has a characteristic function given in the brackets. Furthermore, the quantity in the braces can be recognized from (2.4.24) as the characteristic function of a Laguerre counting variable, which had probability

$$P_{k_i}(k) = \frac{(\alpha\gamma_i)^k}{(1 + \alpha\gamma_i)^{k+1}} \exp\left[\frac{-\alpha|s_i|^2}{1 + \alpha\gamma_i}\right] L_k\left[-\frac{|s_i|^2}{\gamma_i(1 + \alpha\gamma_i)}\right] \quad (3.2.8)$$

Thus, for the case of Gaussian fields, the count statistic k can indeed be interpreted as the sum

$$k = \sum_{i=0}^{\infty} k_i \quad (3.2.9)$$

of independent counts k_i, where each k_i has the Laguerre count probability given by (3.2.8). Again, each k_i can be considered the random count associated with a mode of the Gaussian field. However, the modes have specific meaning, since they must be associated with the particular orthonormal set of the KL expansion. Each such mode of a Gaussian field therefore contributes an independent Laguerre count variable to the total count. If the noise field is not Gaussian, this interpretation cannot be used; that is, there does not exist an orthonormal basis set into which the stochastic envelopes can be expanded such that k can be represented as the sum of independent mode counts.

Note that if $s_i = 0$, for some i, implying no deterministic component in that particular mode, (3.2.8) becomes

$$P_{k_i}(k) = \frac{(\alpha\gamma_i)^k}{(1 + \alpha\gamma_i)^{k+1}} \qquad (3.2.10)$$

which is the Bosé–Einstein probability. Hence, any mode of the Gaussian field that does not contain a deterministic field component contributes a Bosé–Einstein counting variable to k. Similarly, from our discussion of the preceding section, we know that a mode with no noise, $b_i = 0$, contributes a Poisson counting variable to k. All other components contribute the Laguerre count of (3.2.8). We point out that the Laguerre densities in (3.2.8) require knowledge of the eigenvalues $\{\gamma_i\}$, which in turn require solution of the integral equation (3.1.7) associated with the KL expansion.

3.3 COMPUTATIONAL PROCEDURES FOR COUNT STATISTICS WITH GAUSSIAN FIELDS

Although we have shown that the count statistic with narrow band Gaussian envelopes can be decomposed into the sum of independent Laguerre modal counts, as with deterministic fields, the actual probability of the combined count statistic may still be desired. There are two basic procedures we may use, both of which are difficult to carry out in a general manner. The first procedure is the formal approach in which we recognize k in (3.2.9) to be a sum of discrete random variables, whose probabilities are known. The desired probability $P_k(k)$ is then the infinite discrete convolution of $\{P_{k_i}(k)\}$ over all i. This is equivalent to finding a closed form for the characteristic function in (3.2.7) and inverse transforming. This formal procedure requires first solving for all the eigenvalues of the KL expansion by using the integral equation (3.1.7), with the correlation of the noise field inserted. Integral equations are usually quite difficult to solve, and even when solutions are available, the eventual inverse transformation of (3.2.7) is an equally formidable task. Nevertheless, the procedure is quite straightforward.

The second method makes use of a moment expansion and an iterated kernel property derived from integral equations of the type in (3.1.7). If we expand the exponential the desired probability can be expanded as

$$
\begin{aligned}
P_k(k) &= \int_0^\infty \frac{m^k}{k!} e^{-m} p_{m_V}(m)\, dm \\
&= \int_0^\infty \left[\sum_{j=0}^\infty \frac{m^{k+j}}{k!\,j!} \right] p_{m_V}(m)\, dm \\
&= \frac{1}{k!} \sum_{j=0}^\infty \frac{E_{m_V}[m^{k+j}]}{j!} \qquad (3.3.1)
\end{aligned}
$$

which is an infinite series expansion involving only the moments of the random variable m_V. Thus the alternative procedure is to determine all the moments of m_V and reconstruct (3.3.1). These moments can often be determined by first computing the semiinvariants of m_V, and then converting to moments through the identities in (2.2.28). The semiinvariants of m_V can be obtained from (3.2.6) by taking its logarithm, and expanding $\log(1 - \alpha\gamma_i j\omega)$ and $(1 - \alpha\gamma_i j\omega)^{-1}$ into a power series. This results in an expansion of the form

$$\log \psi_{m_V}(\omega) = \sum_{i=0}^{\infty} \frac{\chi_i(j\omega)^i}{i!} \tag{3.3.2}$$

where the semiinvariants $\{\chi_i\}$ are

$$\chi_i = \alpha^i \left\{ (i-1)! \left[\sum_{j=0}^{\infty} (\gamma_j)^i \right] + i! \left[\sum_{j=0}^{\infty} |s_i|^2 \gamma_j^{i-1} \right] \right\} \tag{3.3.3}$$

Formally, these semiinvariants still depend on $\{\gamma_i\}$, but their computation can be avoided by using the identities (Problem 3.1)

$$\sum_{i=0}^{\infty} \gamma_i^q \triangleq \int_V R_b^{(q)}(\mathbf{v}, \mathbf{v}) \, d\mathbf{v} \tag{3.3.4a}$$

$$\sum_{i=0}^{\infty} |s_i|^2 \gamma_i^q \triangleq \int_V s(\mathbf{v}) y_q^*(\mathbf{v}) \, d\mathbf{v} \tag{3.3.4b}$$

where

$$y_q(\mathbf{v}) = \int_V R_b^{(q)}(\mathbf{v}, \mathbf{v}_1) s(\mathbf{v}_1) \, d\mathbf{v}_1$$

$$R_b^{(0)}(\mathbf{v}, \mathbf{v}_1) \triangleq \delta(\mathbf{v} - \mathbf{v}_1)$$

$$R_b^{(1)}(\mathbf{v}, \mathbf{v}_1) \triangleq R_b(\mathbf{v}, \mathbf{v}_1)$$

$$R_b^{(q)}(\mathbf{v}, \mathbf{v}_1) \triangleq \int_V R_b^{(q-1)}(\mathbf{v}, \mathbf{v}') R_b(\mathbf{v}', \mathbf{v}_1) \, d\mathbf{v}'$$

The integrals just given are called *iterated kernels* [1]. Evaluation of the succession of integrals in (3.3.4) determines the summations needed in (3.3.3). Thus iterated kernels can be used to determine directly the semi-invariants χ_i from which the moments $E_{m_V}[m^j]$ can be derived and the series in (3.3.1) can be constructed. A rigorous application of this procedure is obviously quite difficult, but on the other hand it provides an elegant means of bypassing the problem of solving the KL integral equation (3.1.7), and arrives directly at a solution in terms of the fundamental quantities; that is, the signal envelope $s(t, \mathbf{r})$ and the noise correlation function. We may remark that although it may be a lengthy calculation to evaluate the iterated kernels beyond the first few orders, they are often all that is needed for a good approximation, particularly for analyses requiring only the count mean and variance. For example, we see from (2.3.4) that

$$E_k[k] = E_{m_V}[m] = \chi_1$$

$$= \alpha \int_V R_b(\mathbf{v}, \mathbf{v}) \, d\mathbf{v} + \alpha \int_V |s(\mathbf{v})|^2 \, d\mathbf{v} \qquad (3.3.5)$$

and

$$\text{Var}(k) = \alpha \int_V R_b(\mathbf{v}, \mathbf{v}) \, d\mathbf{v} + \alpha \int_V |s(\mathbf{v})|^2 \, d\mathbf{v} + \alpha^2 \int_V R_b(\mathbf{v}, \mathbf{v}')R_b(\mathbf{v}', \mathbf{v}) \, d\mathbf{v}'$$

$$+ 2\alpha^2 \int_V \int_V s(\mathbf{v})R_b(\mathbf{v}, \mathbf{v}')s(\mathbf{v}') \, d\mathbf{v} \, d\mathbf{v}' \qquad (3.3.6)$$

Note from (3.1.2) and (3.1.6) that

$$\int_V R_b(\mathbf{v}, \mathbf{v}) \, d\mathbf{v} = \int_V E[b(\mathbf{v})b^*(\mathbf{v})] \, d\mathbf{v}$$

$$= E \int_{\mathscr{A}} \int_t^{t+T} b(\rho, \mathbf{r})b^*(\rho, \mathbf{r}) \, d\rho \, d\mathbf{r} \qquad (3.3.7)$$

and is actually the average noise energy over the counting interval. Thus, even with stochastic fields, the mean count is directly related to the average total energy collected during the counting time.

When comparing the formal procedure to the iterated kernel procedure, we see from the discussion of this section that neither method can be considered best in general for computing $P_k(k)$. Whichever method may be more successful will often depend on the particular covariance function involved. However, it should be emphasized that these types of eigenvalue approaches to noise theory have been analyzed historically in the past in the communication theory literature. As a result there has been a great deal of engineering intuition that has developed in attacking these noise problems which, as we shall see later, can lead to approximate solutions suitable for practical applications.

3.4 COHERENCE-SEPARABLE FIELDS

The solution of the KL equation in (3.1.7) leads to a set of eigenfunctions that define the field modes and a set of eigenvalues that generate the count probabilities, using the procedures of the preceding section. Unfortunately, the KL expansion functions are difficult to determine in general, hindering an exact solution to the count statistics. However, some degree of simplification occurs when the noise field envelope is, or is assumed to be, coherence-separable. In this case the coherence function factors as

$$R_b(t_1, \mathbf{r}_1; t_2, \mathbf{r}_2) = R_s(\mathbf{r}_1, \mathbf{r}_2)R_t(t_1, t_2) \tag{3.4.1}$$

where $R_s(\mathbf{r}_1, \mathbf{r}_2)$ is the spatial coherence function and $R_t(t_1, t_2)$ is the temporal correlation function. For coherence-separable fields, the KL expansion has the property that the left side of (3.1.7) becomes

$$\int_{\mathscr{A}} \int_t^{t+T} R_t(t_1, t_2)R_s(\mathbf{r}_1, \mathbf{r}_2)\phi_j(t_2, \mathbf{r}_2)\, dt_2\, d\mathbf{r}_2$$

$$= \int_{\mathscr{A}} R_s(\mathbf{r}_1, \mathbf{r}_2) \int_t^{t+T} R_t(t_1, t_2)\phi_j(t_2, \mathbf{r}_2)\, dt_2\, d\mathbf{r}_2 \tag{3.4.2}$$

The form of this shows that an eigenfunction solution always occurs as

$$\phi_j(t, \mathbf{r}) = g_j(t)w_j(\mathbf{r}) \tag{3.4.3}$$

$$\gamma_j = \gamma_{jt}\gamma_{js} \tag{3.4.4}$$

where the terms above satisfy

$$\int_t^{t+T} R_t(t_1, t_2)g_j(t_2)\, dt_2 = \gamma_{jt}g_j(t_1) \tag{3.4.5}$$

$$\int_{\mathscr{A}} R_s(\mathbf{r}_1, \mathbf{r}_2)w_j(\mathbf{r}_2)\, d\mathbf{r}_2 = \gamma_{js}w_j(\mathbf{r}_1) \tag{3.4.6}$$

That is, for coherence-separable stochastic fields the eigenfunctions and eigenvalues factor into a product of time and spatial components. This means the eigenfunctions in (3.1.7) can be determined by separately determining the family of eigenfunctions for both the time and space coherence kernels in (3.4.5) and (3.4.6). Since the product of any two such time and space eigenfunctions satisfies (3.1.7), the set of all eigenfunctions must involve all possible pairwise products of the $\{g_i(t)\}$ and $\{w_j(\mathbf{r})\}$. The KL expansion of coherence-separable fields therefore takes the form

$$b(t, \mathbf{r}) = \sum_i \sum_j b_{ij}g_i(t)w_j(\mathbf{r}) \tag{3.4.7}$$

where

$$b_{ij} = \int_{\mathscr{A}} \int_t^{t+T} b(t, \mathbf{r})g_i^*(t)h_j^*(\mathbf{r})\, dt\, d\mathbf{r} \tag{3.4.8}$$

The functions $\{g_i(t)\}$ are said to define the temporal modes of the received noise field, while the functions $\{w_j(\mathbf{r})\}$ designate its spatial modes. The corresponding $\{\gamma_{js}\gamma_{it}\}$ eigenvalues are the mean square energy in each of these modes. (i.e., the mean squared value of $|b_{ij}|$). By collecting all the modal energy, we have the total average energy over the volume. That is, by substituting with (3.4.7),

$$\sum_{i=0}^{\infty}\sum_{j=0}^{\infty}\gamma_{js}\gamma_{it} = E\int_{\mathscr{A}}\int_{t}^{t+T}|b(\rho,\mathbf{r})|^2\,d\rho\,d\mathbf{r} \qquad (3.4.9)$$

The summations $\gamma_{it}\sum_{j=0}^{\infty}\gamma_{js}$ represent the total energy over the area during time mode i, and $\gamma_{js}\sum_{i=0}^{\infty}\gamma_{it}$ is the average energy in $(t, t+T)$ in spatial mode j. If the field is coherence-separable and spatially homogeneous, then $R_s(\mathbf{r}_1,\mathbf{r}_1) = R_s(0)$ and

$$\sum_{j=0}^{\infty}\gamma_{js} = R_s(0)A \qquad (3.4.10)$$

where A is the area of $\mathscr{A}$. This suggests dealing instead with the normalized mutual coherence function $\tilde{R}_s(\mathbf{r}_1,\mathbf{r}_2) \triangleq R_s(\mathbf{r}_1 - \mathbf{r}_2)/R_s(0)$, whose eigenvalues therefore sum to the area A. If the field is completely spatially coherent over $\mathscr{A}$, such that $\tilde{R}_s(\mathbf{r}_1,\mathbf{r}_2) = 1$ over all $\mathbf{r}_1, \mathbf{r}_2$ in $\mathscr{A}$, then (3.4.6) is satisfied with the single eigenfunction $w_0(\mathbf{r}) = 1/\sqrt{A}$ and eigenvalue $\gamma_{0s} = A$. Thus the field will have only one spatial mode in $\mathscr{A}$. Thus spatial coherence over an area is equivalent to the existence of a single spatial mode over that area, when dealing with stochastic fields. Substitution into (2.1.21) also shows that $n(t) = \alpha A|f(t)|^2$, as is assumed in the definition of the point detector in Section 2.4. The point detector assumption therefore is valid for stochastic fields whenever the field is spatially coherent over the aperture area, or equivalently, when only one spatial mode exists.

In deriving a Karhunen–Loeve expansion associated with a received field in an optical receiver, we may choose to deal with either the field correlation function at the receiver lens (aperture plane) or with the transformed correlation function in the focal plane. Since the orthonormal function set of the aperture plane Fourier transforms to an orthonormal function set in the focal plane, field expansions in either plane have the same coefficients (Problem 3.3). Thus a Karhunen–Loeve expansion in either plane has the same eigenvalues and expansions can therefore be made in either plane. Generally, field correlations are most easily described at the receiver lens and analysis is usually confined to the aperture plane.

3.5 SPATIAL MODES

Computation of the count statistics for coherence-separable fields requires the solution to the integral equations in (3.4.5) and (3.4.6). Both equations are basically identical in form, with one involving the scalar parameter t, while the other involves the space vector **r**. In this section we derive some eigenfunction solutions for the spatial equation in (3.4.6), for several examples of typical receiver operations. We avoid unnecessary mathematical complication by resorting to engineering approximations whenever possible. Our objective is to develop some insight into the notion of a spatial mode for stochastic fields.

Rectangular Receiver Areas. Consider an optical receiver composed of a rectangular receiving area of length a and width b, as shown in Figure 3.1. The received field is a random, coherence-separable field produced from a source at distance L. We assume the source is described by an irradiance

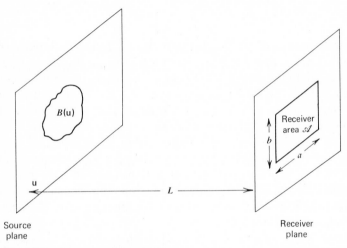

Source
plane

Receiver
plane

Figure 3.1. Source-receiver geometry.

function $B(\mathbf{u})$, where $\mathbf{u}$ is a spatial vector in a plane parallel to the receiver area plane. The normalized spatial coherence function over the receiver surface due to this source is given by the Fresnel–Kirckoff approximation [2, see also Appendix D],

$$\tilde{R}_s(\mathbf{r}_1, \mathbf{r}_2) = \left[\frac{\beta(\mathbf{r}_2 - \mathbf{r}_1)}{\beta(0)}\right] \exp\left[j\frac{\pi}{L\lambda}(|\mathbf{r}_1|^2 - |\mathbf{r}_2|^2)\right] \qquad (3.5.1)$$

where $\mathbf{r}_1, \mathbf{r}_2$ are vector points on the receiver area, λ is the optical wavelength, A is the receiver area ($A = ab$), and $\beta(\mathbf{r})$ is the two-dimensional Fourier transform of the source irradiance

$$\beta(\mathbf{r}) = \int_{\mathcal{U}} B(\mathbf{u}) \exp\left[\frac{j2\pi(\mathbf{u} \cdot \mathbf{r})}{\lambda L}\right] d\mathbf{u} \qquad (3.5.2)$$

Here $\mathcal{U}$ represents the two-dimensional space of $\mathbf{u}$ vectors. By writing the desired eigenfunctions as

$$w_j(\mathbf{r}) = \hat{w}_j(\mathbf{r}) \exp\left[\frac{-j\pi r^2}{\lambda L}\right] \tag{3.5.3}$$

we can rewrite the spatial integration equation in (3.4.6) as

$$\gamma_{js} \hat{w}(\mathbf{r}_2) = [\beta(0)]^{-1} \int_{\mathcal{A}} \beta(\mathbf{r}_2 - \mathbf{r}_1) \hat{w}_j(\mathbf{r}_1) \, d\mathbf{r}_1 \tag{3.5.4}$$

where $\mathcal{A}$ is the spatial receiver area. Now, if the irradiance $B(\mathbf{u})$ subtends a "small" solid angle, $\beta(\mathbf{r}) \approx \beta(0)$ over the whole receiver area, and (3.5.4) is is satisfied with the single term $w_0(\mathbf{r}) = 1/\sqrt{A}$, where A is the area of $\mathcal{A}$. In this case the detected field appears coherent and the source and receiver can be considered to exist at single points. When the source irradiance subtends a "large" solid angle, the correlation function over the receiver area is smaller than the receiver area itself, and (3.5.4) must be evaluated. For the rectangular receiver, the eigenfunctions can be approximated by two-dimensional spatial plane waves

$$\hat{w}_j(\mathbf{r}) = \frac{\exp[j2\pi q x/a + j2\pi l y/b]}{\sqrt{A}} \tag{3.5.5}$$

where x and y are the components of $\mathbf{r}$, and q, l are integers. That is, the eigenfunctions can be taken as spatial harmonics, with spatial frequencies that are multiples of the periods a and b. If we substitute with (3.5.2) and (3.5.5), the right side of (3.5.4) becomes

$$[\beta(0)]^{-1} \int_{\mathcal{A}} \int_{\mathcal{U}} B(\mathbf{u}) \exp\left[\frac{j2\pi \mathbf{u} \cdot (\mathbf{r}_2 - \mathbf{r}_1)}{\lambda L}\right] \frac{\exp[j2\pi q x_1/a + j2\pi l y_1/b] \, d\mathbf{r}_1 \, d\mathbf{u}}{\sqrt{A}}$$

$$= [\beta(0)]^{-1} \int_{\mathcal{U}} B(\mathbf{u}) \exp\left[j\frac{2\pi(\mathbf{u} \cdot \mathbf{r}_2)}{\lambda L}\right] d\mathbf{u}$$

$$\cdot \frac{1}{\sqrt{A}} \int_{\mathcal{A}} \exp\left[j2\pi x\left(\frac{q}{a} - \frac{u_1}{L\lambda}\right) + j2\pi y\left(\frac{l}{b} - \frac{u_2}{L\lambda}\right)\right] dx \, dy \tag{3.5.6}$$

where u_1 and u_2 are the components $\mathbf{u}$. The right-hand integral over $\mathcal{A}$ integrates to approximately

$$\frac{1}{\sqrt{A}} \delta\left(\frac{q}{a} - \frac{\mu_1}{\lambda L}\right) \delta\left(\frac{l}{b} - \frac{\mu_2}{\lambda L}\right) \tag{3.5.7}$$

so that (3.5.6) becomes approximately

$$\left[\frac{\lambda^2 L^2}{\beta(0)}\right] B\left(\frac{q\lambda L}{a}, \frac{l\lambda L}{b}\right) \frac{\exp[j(2\pi q/a)x_2 + j(2\pi l/b)y_2]}{\sqrt{A}} \qquad (3.5.8)$$

where $B(u_1, u_2) = B(\mathbf{u})$. Now substituting (3.5.8) and (3.5.3) into (3.5.4), we have

$$\gamma(q, l) = \left[\frac{\lambda^2 L^2}{\beta(0)}\right] B\left(\frac{q\lambda L}{a}, \frac{l\lambda L}{b}\right) \qquad (3.5.9)$$

Therefore, the spatial eigenvalue (average energy) associated with each of the two-dimensional harmonic spatial frequencies $(q/a, l/b)$ is given by the evaluation of the source irradiance function $B(u_1, u_2)$ at the value $(u_1 = qa_0, u_2 = lb_0)$ where $a_0 = \lambda L/a$, $b_0 = \lambda L/b$. Thus a significant spatial frequency harmonic, or spatial mode, will exist for all (q, l) combinations such that

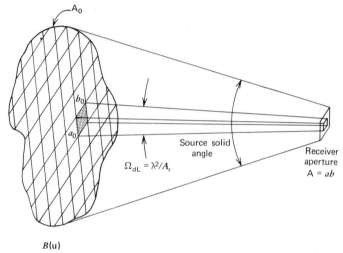

$\Omega_{dL} = \lambda^2/A_r$

Source solid angle

Receiver aperture
$A = ab$

$B(\mathbf{u})$

Figure 3.2. Object sampling with square aperture.

(3.5.9) is nonnegligible. Note the above is equivalent to effectively sampling $B(\mathbf{u})$ at points separated by $a_0 = \lambda L/a$ in the u_1 dimension and by $b_0 = \lambda L/b$ in the u_2 dimension. This means that if A_0 is interpreted as the spatial area subtended by the source irradiance function $B(\mathbf{u})$, then the number of significant spatial modes (harmonics) can be determined by partitioning A_0 into grids, or disjoint rectangles, of width a_0 and b_0 and determining the number of such rectangles needed to cover A_0, as shown in Figure 3.2. This means

$$[\text{number of spatial modes}] \approx \frac{A_0}{a_0 b_0} + 1 \qquad (3.5.10)$$

where the 1 factor is needed to include the case where $A_0 \ll a_0 b_0$. Note that the area $a_0 b_0$, when viewed from the receiver, subtends at the detector a solid angle $a_0 b_0/L^2 = (\lambda L/a)(\lambda L/b)/L^2 = \lambda^2/ab = \lambda^2/A$, which is the diffraction limited field of view of the receiver Ω_{dL}. Therefore, the number of significant spatial modes over a rectangular receiver surface can alternatively be viewed as the number of solid angles needed to cover the solid angle subtended by the irradiance function $B(\mathbf{u})$ located at the source (Figure 3.2). Thus we can also write

$$[\text{number of spatial modes}] = \frac{A_0/L^2}{\lambda^2/A} + 1 = \frac{AA_0}{\lambda^2 L^2} + 1 \qquad (3.5.11)$$

When A_0 is smaller than $a_0 b_0$ only one mode exists, and we have a point source. Note carefully the manner in which this mode model has developed. Given any $\beta(\mathbf{r})$ on the receiver surface, we have interpreted its Fourier transform [the corresponding irradiance function $B(\mathbf{u})$] as the effective source area, located in the plane of the source, a distance L away. Thus no matter how the field coherence function is in fact generated, an equivalent source area can still be defined. In particular, a turbulent medium producing a stochastic field has an equivalent source area relative to a receiver. This area is all we need to determine the number of significant spatial modes. Note that the eigenfunctions in (3.5.5) correspond to plane waves arriving at different arrival angles. Hence, spatial modes can be equivalently considered as different directions of arrival, with all arrival angles within a diffraction limited field of view grouped together to effectively represent one direction. Thus a large source (large irradiance area) transmits to the receiver from many different directions, the combination of such arriving fields determining the spatial coherence function $\beta(\mathbf{r})$ at the receiver surface.

In deriving (3.5.10) and (3.5.11) we have implicitly assumed that the angle subtended by the source irradiance area A_0 is always entirely within the field of view of the receiver. If the source angle is larger than (i.e., fills) the receiver field of view, the number of modes is limited to that observed within this angle. In this case

$$D_s \cong \frac{\Omega_{fv}}{\lambda^2/A} = \frac{\Omega_{fv}}{\Omega_{dL}} \geq 1 \qquad (3.5.12)$$

where Ω_{fv} is the receiver field of view. Note that D_s no longer depends on the distance to the source, L, and depends only on the design properties of the receiver. Diffuse, background radiation noise is always present from the whole background hemisphere, and therefore always fills the receiver field of view, so that (3.5.12) describes the background noise case. (Compare this with our discussion in Section 1-7.) Thus diffuse background noise always appears as a source in all spatial modes, and effectively "transmits" to the receiver from all directions. For a typical value of D_s, consider a 0.1 meter receiver with a $1° \times 1°$ field of view (2.5×10^{-4} sr), operating at frequency 6×10^{14} Hz. For this case, $D_s = 2.5 \times 10^{-4}/2.5 \times 10^{-11} = 10^7$. We see therefore that in typical optical receivers the number of spatial modes is quite large.

Circular Receivers. Consider now a circular receiver area of radius b, and again assume a coherence-separable received field from a circular source of

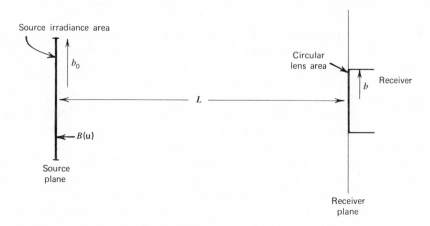

Figure 3.3. Circular irradiance source with circular receiver lens.

radius b_0 at distance L, as in Figure 3.3. We assume the source irradiance function $B(\mathbf{u})$ is homogeneous with circular symmetry, so that we can write

$$B(\mathbf{u}) = B(|\mathbf{u}|) \qquad (3.5.13)$$

This immediately implies that the coherence function $\beta(\mathbf{r})$ in (3.5.2) is also circular symmetric. This is evident if we convert to polar coordinates and note that (3.5.2) becomes

$$\beta(\mathbf{r}) = \int_0^{b_0} \int_0^{2\pi} B(\mathbf{u}) \exp\left[j \frac{2\pi}{\lambda L} ur \cos(\theta_u - \theta_r)\right] u \, d\theta_u \, du$$

$$= \int_0^{b_0} B(u) J_0\left(\frac{2\pi}{\lambda L} ur\right) u \, du$$

$$= \beta(r) \qquad (3.5.14)$$

where $r = |\mathbf{r}|$. The eigenvalue solution in (3.5.4) requires substitution for $\beta(\mathbf{r}_2 - \mathbf{r}_1)$. Using (3.5.14) and some Bessel function identities, we can expand

$$\beta(|\mathbf{r}_2 - \mathbf{r}_1|) = \int_0^{b_0} B(u) \sum_{q=-\infty}^{\infty} J_q\left(\frac{2\pi}{\lambda L} ur_1\right) J_q\left(\frac{2\pi}{\lambda L} ur_2\right) \exp[jq(\theta_2 - \theta_1)]u\, du$$

(3.5.15)

Using (3.5.15), and converting (3.5.4) to polar coordinates, yields the eigenfunction equation

$$\gamma_{js}\hat{w}_j(r_2, \theta_2) = \sum_{q=-\infty}^{\infty} e^{jq\theta_2} \int_0^b r_1\, dr_1 \int_0^{b_0} B(u)$$

$$\cdot J_q\left(\frac{2\pi}{\lambda L} ur_1\right) J_q\left(\frac{2\pi}{\lambda L} ur_2\right) u\, du \cdot \int_0^{2\pi} \hat{w}_j(r_1, \theta_1)e^{-jq\theta_1}\, d\theta_1 \quad (3.5.16)$$

The eigensolutions for $\hat{w}_j(r, \theta)$ for an arbitrary irradiance function $B(u)$ requires solution of the above. No general solution corresponding to the sampling interpretation for the rectangular case is known. However, if $B(u)$ is uniform [i.e., $B(u) = C_0, u \leq b_0$], then the integral equation (3.5.16) is of the type studied by Slepian [3]. He has shown that the eigenfunctions are proportional to generalized prolate spheroidal wave functions, and the eigenvalues γ_{js} depend on the parameter

$$\beta \triangleq \frac{2\pi b_0 b}{\lambda L}$$

(3.5.17)

The use of spheroidal expansion functions in optical imaging has been discussed by Toraldo di Francia [4]. For our interest here, Slepian has shown that if the source is large so that $\beta \gg 1$, then the number of significant eigenvalues (i.e., the number of field modes) is approximately

$$D_s \approx \frac{\beta^2}{4} = \frac{\pi^2 b_0^2 b^2}{(\lambda L)^2} = \frac{A_0 A}{\lambda^2 L^2} = \frac{A_0/L^2}{\lambda^2/A}$$

(3.5.18)

if we interpret $A_0 = \pi b_0^2$, $A = \pi b^2$. The result is the same as (3.5.11) with the rectangular areas replaced by the equivalent circular areas. In summary, then, we see that for either type receiver, the number of significant spatial modes is obtained by effectively dividing the field of view subtended by the diffraction limited field of view of the receiver. Here the source area is the effective area of the source irradiance $B(u)$ which can be obtained by transforming the field coherence function at the receiver, as indicated by (3.5.2). When the source irradiance area and receiver area is neither rectangular nor circular, Helstrom [5] has suggested that the number of modes

D_s be estimated from (3.5.18) with A_0 taken as

$$A_0 = \frac{[\int_{\mathscr{A}} B(\mathbf{u})\, d\mathbf{u}]^2}{\int_{\mathscr{A}} B^2(\mathbf{u})\, d\mathbf{u}} \tag{3.5.19}$$

for an arbitrary irradiance function $B(\mathbf{u})$. This approximation depends on the integrated source irradiance function and the area A of the aperture, but not on their shape.

3.6 TEMPORAL MODES

We now examine the temporal portion of the coherence-separable field. Recall that within each spatial mode there exists a random time variation describing the stochastic field. This becomes apparent if we rewrite (3.4.7) as

$$f(t, \mathbf{r}) = \sum_{j=0}^{\infty} f_j(t) w_j(\mathbf{r}) \tag{3.6.1}$$

where

$$f_j(t) = \sum_{i=0}^{\infty} f_{ij} g_i(t) \tag{3.6.2}$$

The random modal processes $\{f_j(t)\}$ can therefore be viewed as independent Gaussian processes, each having the Karhunen–Loeve expansion in (3.6.2). The function set $\{g_i(t)\}$ now describes the temporal modes of each space mode process. As in Section 3.5 we again seek a physical interpretation to these temporal modes for typical operating conditions. Without loss of generality, we can drop the spatial subscript j and investigate an arbitrary Gaussian random process $b(t)$. We consider several types of such processes. In each case we assume the process is being observed over the interval $(t, t + T)$ and over a fixed receiver area.

Integrated White Noise Process. Consider a Gaussian process $b(t)$ generated by the integration of purely white Gaussian noise. (The process is often called a Wiener process.) This particular example illustrates a case in which a straightforward method of obtaining $P_k(k)$ directly can be applied. The properties associated with this Gaussian process are known to be

$$b(0) = 0, \qquad E[b(t)] = 0, \quad \text{all } t$$
$$E[b^2(t)] = \rho^2 t$$
$$R_t(t_1, t_2) = \rho^2 \min(t_1, t_2)$$
$$= \begin{cases} \rho^2 t_1, & t_1 \leq t_2 \\ \rho^2 t_2, & t_2 \leq t_1 \end{cases} \tag{3.6.3}$$

When this correlation function is used in (3.4.5) as the integral kernel, the eigenfunctions become the sinusoidal set

$$g_i(u) = \left(\frac{2}{T}\right)^{1/2} \sin\left[\left(i - \frac{1}{2}\right)\frac{\pi u}{T}\right], \qquad t \le u \le t + T \qquad (3.6.4)$$

with eigenvalues

$$\gamma_i = \frac{4\rho^2 T^2}{\pi^2(2i - 1)^2}, \qquad i = 1, 2, \ldots \qquad (3.6.5)$$

Thus the temporal modes become the harmonic frequencies of the process over the observation interval. Inserting these values of γ_i into (3.2.6) yields the closed form

$$\psi_{m_T}(\omega) = \sec(\rho T \sqrt{\alpha j \omega}) \qquad (3.6.6)$$

where m_T replaces m_V in our temporal model. The transform inversion of the above is not obvious. However, if we expand it in a Taylor series, the coefficients in the expansion become precisely the moments of m_T (Problem 2.2). Since

$$\sec x = \sum_{i=0}^{\infty} \frac{|\varepsilon_{2i}|}{(2i)!} x^{2i}, \qquad x^2 < \frac{\pi^2}{4}$$

where $\{\varepsilon_i\}$ are the Euler numbers [13], we therefore have

$$E_{m_T}(m^q) = \frac{q!}{(2q)!} |\varepsilon_{2q}|(\alpha \rho^2 T^2)^q \qquad (3.6.7)$$

Equation 3.3.1 now gives the $P_k(k)$ generated within the particular spatial mode as [6]

$$P_k(k) = \sum_{i=0}^{\infty} \frac{(-1)^i}{(2k + 2i)!} \binom{k + i}{i} |\varepsilon_{2k + 2i}|(\alpha \rho^2 T^2)^{k+i} \qquad (3.6.8)$$

Note that because of the relatively simple form for $\psi_{m_T}(\omega)$, we are able to find the counting distribution without first having to compute the semiinvariants $\{\chi_i\}$. The mean and variance of the counts follow as (Problem 3.4)

$$E_k(k) = \frac{\alpha}{2} \rho^2 T^2 \qquad (3.6.9)$$

$$\text{Var}[k] = \frac{\alpha}{2} \rho^2 T^2 + \frac{\alpha^2}{6} \rho^4 T^4 \qquad (3.6.10)$$

Markov Noise Process. Consider a Markovian random process, obtained by the first-order filtering of a Gaussian white noise process. (Such processes generate a Lorenzian power spectral density, and are usually used to model a thermally broadened (noisy) laser source radiating in the optical region.) Its correlation function has the form

$$R_t(t_1, t_2) = Q \exp[-\beta|t_1 - t_2|] \tag{3.6.11}$$

The eigenfunction solutions to (3.4.5) become nonharmonically related sinusoids at frequencies $\{q_i\}$, and the corresponding eigenvalues become (Problem 3.5)

$$\gamma_i = \frac{2Q\beta}{\beta^2 + q_i^2} \tag{3.6.12}$$

where $\{q_i\}$ are the solutions to the transcendental equation

$$\tan\left[q_i\frac{T}{2} + \frac{q_i}{\beta}\right]\tan\left[\frac{q_i T}{2} - \frac{\beta}{q_i}\right] = 0 \tag{3.6.13}$$

Since the power spectral density corresponding to (3.6.11) is

$$S(\omega) = \frac{2Q\beta}{\beta^2 + \omega^2} \tag{3.6.14}$$

we see that the eigenvalues in (3.6.12) are obtained by sampling the power spectrum at the eigenfunction frequencies $\{q_i\}$. Thus the significant eigenfunctions are those whose frequencies occur at the significant values of (3.6.14).

For this example the characteristic function becomes

$$\psi_{m_T}(\omega) = \prod_i \left(\frac{1}{1 - \alpha\gamma_i j\omega}\right) \tag{3.6.15}$$

$$= e^{\beta T}\left\{\cosh\left[\beta T\left(1 - \frac{2Q\alpha}{\beta}j\omega\right)\right]\right.$$

$$\left. + \frac{1 - (Q\alpha/\beta)j\omega}{1 - (2Q\alpha/\beta)j\omega}\sinh\left[\beta T\left(1 - \frac{2Q\alpha}{\beta}j\omega\right)\right]\right\}^{-1} \tag{3.6.16}$$

Unfortunately, this characteristic function does not easily yield a useful expression for $P_k(k)$. However, Bedard [7] has been able to obtain recurrence relations for the counting distribution and its factorial moments, while Helstrom [8] has explored certain limiting conditions and obtained approximate counting distributions. We omit these results here. The mean and variance of k in a spatial mode are easily found to be

$$E_k(k) = \alpha Q T$$

$$\text{Var}[k] = \alpha Q T + \frac{(\alpha Q)^2}{2\beta^2}[2\beta T + e^{-\beta T} - 1] \tag{3.6.17}$$

If we let βT and $\beta/4\alpha Q$ become large, then

$$\lim_{\substack{\beta T \to \infty \\ (\beta/4\alpha Q) \to \infty}} \psi_k(\omega) = \exp[2\alpha Q T(e^{j\omega} - 1)] \tag{3.6.18}$$

which is the characteristic function of a Poisson density with level equal to $2\alpha Q T$. Thus we see that in this limit the photocount generated by a Gaussian first-order Markov source is Poisson distributed.

Band Limited Gaussian White Processes. Consider a Gaussian white process, with two-sided level N_{0b} and band limited to $\pm B$ Hz. The correlation function is then

$$R_t(t_1, t_2) = N_{0b} \frac{\sin 2\pi B(t_1 - t_2)}{2\pi B(t_1 - t_2)} \tag{3.6.19}$$

The process is a familiar one in communication theory, and extensive discussions have been devoted to its Karhunen–Loeve expansion [9–11]. The eigenfunctions again involve angular and radial prolate spheroidal functions, and we need not repeat the results here. It has been shown [12] that a good approximation for the eigenvalues is to take the first $(2BT + 1)$ equal to N_{0b} and set the remaining ones equal to zero, when considering the interval $(t, t + T)$. With this approximation, (3.2.7) becomes

$$\psi_k(\omega) \cong \left[\frac{1}{1 - \alpha N_{0b}(e^{j\psi} - 1)} \right]^{2BT+1} \tag{3.6.20}$$

This is the characteristic function of a negative binomial random variate. Hence,

$$P_k(k) = \binom{2BT + k}{k} \left(\frac{1}{1 + \alpha N_{0b}} \right)^{2BT+1} \left(\frac{\alpha N_{0b}}{1 + \alpha N_{0b}} \right)^k \tag{3.6.21}$$

This is then an accurate approximation for the count probability associated with the photodetection over a single spatial mode of a Gaussian process whose spectrum is flat and band limited to $\pm B$ Hz. The result is interesting to examine further. Consider first the condition $2BT \ll 1$. We immediately see that (3.6.20) becomes

$$\psi_k(\omega) = \left[\frac{1}{1 - \alpha N_{0b}(e^{j\omega} - 1)} \right] \tag{3.6.22}$$

and $P_k(k)$ is then

$$P_k(k) = \left(\frac{1}{1 + \alpha N_{0b}} \right) \left(\frac{\alpha N_{0b}}{1 + \alpha N_{0b}} \right)^k \tag{3.6.23}$$

which is the Bosé–Einstein probability. Hence, the count becomes Bosé–Einstein distributed, when only one eigenvalue is considered. However, this is not surprising from our earlier work since one temporal eigenvalue implies the count is due to short-term counting of the Gaussian field, and we have already shown that such counting contributes individually a Bosé–Einstein count. The interesting point is that this occurs if $2BT \ll 1$, or if $T \ll 1/2B$. That is, when the counting time is much smaller than the reciprocal of the noise bandwidth. This is precisely the short-term counting condition defined in Chapter 2. Thus we can now formally redefine short-term counting as, equivalently, counting over a single temporal mode, or counting under the condition $T \ll 1/2B$. This means we have now a somewhat rigorous rule we can apply to determine if the short-term counting results are valid. More specifically, if $2BT \ll 1$, the short-term counting model can be used. If this is not true, the procedure of Section 3.3 must be used.

On the other hand, consider $2BT \gg 1$, and simultaneously let $\alpha N_{0b} \ll 1$. If we note that

$$\left(\frac{1}{1 + \alpha N_{0b}}\right)^{2BT} \to \exp[-2BT\alpha N_{0b}]$$

$$\binom{2BT + k}{k} \to \frac{(2BT)^k}{k!} \qquad (3.6.24)$$

$$\left(\frac{\alpha N_{0b}}{1 + \alpha N_{0b}}\right)^k \to (\alpha N_{0b})^k$$

we see that (3.6.21) becomes

$$P_k(k) \to \frac{(2BT\alpha N_{0b})^k}{k!} \exp[-2BT\alpha N_{0b}] \qquad (3.6.25)$$

which is the Poisson probability with level $(2BT\alpha N_{0b})$. Recall that we earlier associated Poisson counting with the detection of deterministic (nonrandom) envelopes. This means the detector is acting as if it is observing a nonrandom envelope. This is precisely what the condition $2BT \gg 1$ implies, since the counting time is now long compared to the time variations in the envelope, and the detector "smoothes" out the intensity variations. Roughly speaking, we can conclude that when observing band limited noise, short-term counting approaches Bosé–Einstein statistics, long-term counting approaches Poisson statistics, while the true counting is negative binomial. Note that since $2BN_{0b}$ is the noise power (area under the power spectrum) the level of the Poisson probability in (3.6.25) is the average noise energy over the T sec counting interval. That is, the noise acts as if it were producing counts at a rate $2B\alpha N_{0b}$ per second.

Signal Plus a Gaussian White Band Limited Process. We have been considering only a random envelope process arising from an effective received noise. We now extend to the case where we add to the random noise a deterministic signal envelope $s(t)$, as in (3.2.1). If we generate the Karhunen–Loeve expansion, we add to each envelope coordinate a deterministic, complex number as in (3.2.2). This is equivalent to adding a mean value to each Gaussian field coordinate. If the random process is a white band limited process of level N_{0b} two sided, and bandwidth B Hz, and if the signal is also band limited to B Hz, the count probability in each temporal mode has the Laguerre density of (3.2.8). The overall count during the T sec observation time in a single spatial mode is then the $(2BT + 1)$-fold convolution of this Laguerre density with itself. We noted earlier in (2.4.32) that this results in

$$P_k(k) = \frac{(\alpha N_{0b})^k}{(1 + \alpha N_{0b})^{k + 2BT + 1}} \exp\left[\frac{-\alpha P_s T}{1 + \alpha N_{0b}}\right] L_k^{2BT}\left[\frac{-\alpha P_s T}{\alpha N_{0b}(1 + \alpha N_{0b})}\right]$$

(3.6.26)

where P_{si} is the signal power in the ith time mode and

$$P_s = \sum_{i=0}^{2BT} P_{si},$$

(3.6.27)

and $L_k^{2BT}(x)$ is the $2BT$th-order Laguerre polynomial. Note that $\alpha P_s T$ is the total signal field count energy occurring in T sec within the spatial mode involved. The mean and variance of k are shown to be (equations 2.4.35 and 2.4.36)

$$E_k(k) = (2BT + 1)\alpha N_{0b} + \alpha P_s T$$
$$\text{Var}[k] = (2BT + 1)(1 + \alpha N_{0b})\alpha N_{0b} + (1 + \alpha N_{0b})\alpha P_s T$$

(3.6.28)

Furthermore, we recall that for this case $p_{m_t}(m)$ corresponds to a noncentral chi-square density (i.e., the density of the sum of $2BT + 1$ independent Rician variables). As in the case for $s(t) = 0$, the condition $2BT \ll 1$ is equivalent to one significant temporal eigenvalue, and (3.6.26) reduces to

$$P_k(k) = \frac{(\alpha N_{0b})^k}{(1 + \alpha N_{0b})^{1 + k}} \exp\left[\frac{-\alpha P_s T}{1 + \alpha N_{0b}}\right] L_k\left[\frac{-\alpha P_s T}{\alpha N_{0b}(1 + \alpha N_{0b})}\right]$$

(3.6.29)

This is now the single mode (one spatial mode and one temporal mode) counting probability when signal is included. For $2BT \gg 1$, however, $P_k(k)$ again approaches a Poisson distribution when αN_{0b} is small. This results from applying the asymptotic forms in (3.6.24), with the addition of the limits

$$\frac{\alpha P_s T}{1 + \alpha N_{0b}} \to \alpha P_s T$$

$$L_k^{2BT}\left[\frac{-\alpha P_s T}{\alpha N_{0b}(1 + \alpha N_{0b})}\right] \to \frac{1}{k!}\left(2BT + \frac{P_s T}{N_{0b}}\right)^k$$

(3.6.30)

for $\alpha N_{0b}' \ll 1, 2BT \gg 1$. These equations, together with (3.6.24), combine to yield

$$P_k(k) \approx \frac{[\alpha(N_{0b}2BT + P_sT)]^k}{k!} \exp[-\alpha(N_{0b}2BT + P_sT)] \quad (3.6.31)$$

Thus for small αN_{0b} and large $2BT$, the counting distribution is again Poisson, with the rate parameter as the sum of the signal intensity and noise intensity. Alternatively, we can state that in the limiting form, k appears as the sum of two independent Poisson variables, one associated with the signal field level P_sT, and one associated with the noise field level $2BTN_{0b}$. This interpretation is useful in subsequent system analyses.

3.7 OPTICAL COMMUNICATION SYSTEM FIELD MODELS

The results of our discussion can now be used to derive a basic field model for a typical optical communication system. We consider the system shown in Figure 3.4. A signal transmitter transmits a deterministic signal intensity by

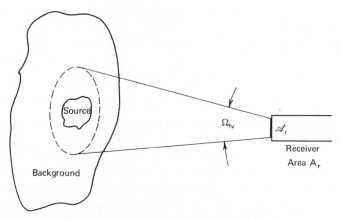

Figure 3.4. Basic optical communication system model.

modulating an optical carrier of wavelength λ. This modulated signal field is received by a receiver having optical bandwidth B_0, area A_r, field of view Ω_{fv}, and diffraction limit $\Omega_{dL} = \lambda^2/A_r$. A coherence-separable Gaussian background noise field produces background radiation at the receiver. We assume the noise field has a flat intensity spectrum of level N_0 watts/hz-area over the bandwidth B_0, and a constant irradiance function over the receiver field of view. This is equivalent to assuming that the noise field has a mutual coherence function

$$R_b(t_1, \mathbf{r}_1; t_2, \mathbf{r}_2) = N_0 \delta(t_1 - t_2) \delta(\mathbf{r}_1 - \mathbf{r}_2) \qquad (3.7.1)$$

at the receiver aperture. If the receiver observes for T sec the combined fields arising from both the signal and background noise, the number of temporal modes will be

$$D_t = 2B_0 T + 1 \qquad (3.7.2)$$

where all temporal modes have identical eigenvalues $\gamma_t = N_0$. The number of spatial modes at the receiver is then given by (3.5.12):

$$D_s = \frac{\Omega_{fv}}{\Omega_{dL}} + 1 \qquad (3.7.3)$$

where each spatial mode has identical eigenvalue $\gamma_s = \lambda^2$. (We show in Chapter 6 that λ^2 is actually the coherence area of the noise field at the receiver.) Thus, the observed radiation field has a total of

$$D = D_t D_s \qquad (3.7.4)$$

independent time-space modes, each of equal eigenvalue $\gamma_s \gamma_t$. This eigenvalue is the average noise field energy per mode, and therefore

$$\gamma_s \gamma_t = N_0 \lambda^2 = N_{0b} \qquad (3.7.5)$$

where N_{0b} was introduced in Section 1.7. For the blackbody noise model, N_{0b} is given in (1.7.8). The resulting characteristic function of the photo-detected count level is given in (3.2.7), which is now a product of a finite number of similar terms, and therefore has the form

$$\psi_{m_V}(\omega) = \left[\frac{1}{1 + \mu_{b0}(1 - e^{j\omega})} \right]^D \exp\left[\frac{\mu_s(e^{j\omega} - 1)}{1 + \mu_{b0}(1 - e^{j\omega})} \right] \qquad (3.7.6)$$

where $\mu_{b0} = \alpha N_{0b}$ is the background average count energy per mode, and μ_s is the total signal count energy collected over the receiver field of view and observation time T. If s_i is the signal field envelope coordinate of the ith mode, then

$$\mu_s = \alpha A_r \sum_{i=1}^{D} |s_i|^2 \qquad (3.7.7)$$

Substituting (3.7.6) into (3.1.12) and integrating yields the resulting count probability

$$P_k(k) = \frac{(\mu_{b0})^k}{(1 + \mu_{b0})^{k+D}} \exp\left[\frac{-\mu_s}{1 + \mu_{b0}} \right] L_k^{D-1} \left[\frac{-\mu_s}{\mu_{b0}(1 + \mu_{b0})} \right] \qquad (3.7.8)$$

Equation 3.7.8 is the extension of (3.6.26) into many spatial modes. The limiting conditions in (3.6.30) can again be applied, with $D \gg 1$ replacing the temporal condition $2BT \gg 1$. The limiting form in (3.6.31) is again the Poisson probability

$$P_k(k) \rightarrow \frac{[\mu_s + D\mu_{b0}]^k}{k!} \exp[(\mu_s + D\mu_{b0})] \qquad (3.7.9)$$

when $D \gg 1$ and $\mu_{b0} \ll 1$. Note that the count probability in (3.7.8), or its approximation in (3.7.9), depends on the total number of observed modes, and not on the manner in which they are divided among time and space modes. That is, one may observe D modes by counting for a short period of time over a wide field of view, or by observing relatively few spatial modes for a long period of time. The signal count energy μ_s, however, may depend on the specific time-space modes, since signal energy may be distributed differently in time and space. The condition $D = D_t$ implies optical detection over a single spatial mode, or a receiver with a diffraction limited field of view. The condition $D = D_s$ implies detection in one temporal mode, or in a time period much less than the optical bandwidth. We emphasize that in typical optical systems, the parameter D is extremely large, generally in the range 10^3 to 10^7, so that the Poisson approximation in (3.7.9) is generally valid. In subsequent communication analysis we shall find it convenient to invoke the system model presented in this section. Count probabilities are then governed by the Laguerre probability in (3.7.8), or by its limiting Poisson form in (3.7.9).

It is important to note that the above discussion is valid regardless of the physical meaning of a mode. However the coefficients s_i in (3.7.7) depend specifically on the mode interpretation (i.e., on the orthonormal set of time-space functions used in the field expansion). Since the noise field is assumed to have the coherence function in (3.7.1), it can be shown (Problem 1.7) that any orthonormal expansion set leads to independent noise modes. It is therefore advantageous to select an orthonormal set that is particularly convenient for both system analysis and interpretation. This is especially important in the temporal domain where receiver processing is usually described. If the field envelope is bandlimited to a bandwidth B_0, it is convenient to use orthonormal sampling functions that produce time samples as signal coordinates. That is, we write

$$\mu_s = \alpha A_r \sum_{j=1}^{D_s} \sum_{i=1}^{D_t} |s_j(t_i)|^2 \, \Delta t \qquad (3.7.10)$$

where $s_j(t_i)$ is a time sample of the signal field in spatial mode j at time $i \, \Delta t$, where Δt is equal to $1/2B_0$. We then decompose (3.7.6) into a product of D functions, and interpret k in (3.7.8) as being accumulated from a sum of D individual counts, each having the form

$$P_{k_{ij}}(k_{ij}) = \frac{(\mu_{b0})^{k_{ij}}}{(1 + \mu_{b0})^{k_{ij}+1}} \exp\left[\frac{-\alpha A_r |s_j(t_i)|^2 \, \Delta t}{1 + \mu_{b0}}\right] L_k\left[\frac{-A_r |s_j(t_i)|^2 \, \Delta t}{N_{0b}(1 + \mu_{b0})}\right] \quad (3.7.11)$$

Notice that k_{ij} is well defined as the count over the ith time interval in the jth spatial mode. Summing k_{ij} over i yields the total space mode count, while summing over j yields the total interval count in the interval $(i \, \Delta t, (i + 1) \, \Delta t)$.

In the typical application, the signal intensity field will have a bandwidth much smaller than the detector bandwidth B_0, the latter defining the bandwidth of the noise field. In this case, it is convenient to deal with sampling intervals τ_s on the order of the reciprocal of the signal bandwidth. This means that $\Delta t \ll \tau_s$, so that each interval τ_s contains many noise mode intervals. The count in the ith τ_s interval of the jth spatial mode can therefore be written using the Poisson assumption of (3.7.9) with $\mu_s = \alpha A_r \tau_s |s_j(i\tau_s)|^2$ and $D = \tau_s/\Delta t$. This interpretation of collecting counts that have signal intensity time samples as coordinates is fundamental to much of our analysis in later chapters.

We can extend to the case where the signal field is itself a Gaussian stochastic process received in the presence of a background noise field. (This occurs, for example, if the signal field is passed through a random channel having scattering, turbulence, or time and space dispersion.) In general, the problem is complicated by the necessity to define field modes that are independent modes for both the noise and signal fields simultaneously. However by modeling the noise with the coherence in (3.7.1) we avoid this problem, since we are free to expand in the signal field eigenfunctions and be guaranteed simultaneous independent noise field modes. Thus we expand the received Gaussian signal field as

$$s(t, \mathbf{r}) = \sum_{i=1}^{D_s} \sum_{j=1}^{D_t} s_{ij} g_j(t) w_i(\mathbf{r}) \quad (3.7.12)$$

where $\{g_i(t)\}$ and $\{w_j(\mathbf{r})\}$ are its KL eigenfunctions, and $|s_{ij}|^2$ has mean squared value $\gamma_{tj}\gamma_{si}$. When the combined signal and noise Gaussian fields are expanded into the same function set, we generate random field coordinates that are independent, with eigenvalues $(\gamma_{tj}\gamma_{si} + N_{0b})$. This allows us to describe a stochastic field completely in terms of the signal field modes. We emphasize however that the modes now have a specific interpretation, and the convenience of a time sampling model cannot be automatically applied.

REFERENCES

[1] Middleton, D., *Introduction to Statistical Communication Theory*, McGraw-Hill Book Co., New York, 1960, Chap. 17.

[2] Born, M. and Wolf, E., *Principles of Optics*, 4th ed., Pergamon Press, London, 1970.

[3] Slepian, D., "Prolate Spheroidal Wave Functions," *Bell Systems Tech. J.*, No. 43, 3009 (1964).

[4] Toraldo di Francia, G., "Degrees of Freedom of an Image," *J. Opt. Soc. Am.*, **59**, No. 7, 799 (July 1969).

[5] Helstrom, C. W., "Modal Decomposition of Aperture Fields in Detection and Estimation of Incoherent Objects," *J. Opt. Soc. Am.*, **60**, No. 4, 521 (April 1970).

[6] Karp, S. and Clark, J. R., "Photon Counting: A Problem in Classical Noise Theory," *IEEE Trans. Inf. Theory*, **IT-21**, January 1970.

[7] Bedard, G., "Photon Counting Statistics of Gaussian Light," *Phys. Rev.*, **151**, 1038 (November 1966).

[8] Helstrom, C. W., "The Distribution of Photoelectron Counts from Partially Polarized Gaussian Light," *Proc. Phys. Soc.*, **83**, 777 (1964).

[9] Slepian, D. and Pollak, H. O., "Prolate Spheroidal Wave Function, Fourier Analysis, and Uncertainty, I," *Bell Systems Tech. J.*, **40**, 43–64 (1961).

[10] Landau, H. J. and Pollak, H. O., "Prolate Spheroidal Wave Functions, Fourier Analysis and Uncertainty, II," *Bell System Tech. J.*, **40**, 65–84 (1961).

[11] Landau, H. J. and Pollak, H. O., "Prolate Spheroidal Wave Functions, Fourier Analysis and Uncertainty, III," *Bell System Tech. J.*, **41**, 1295–1336 (1962).

[12] Van Trees, H. L., *Detection Estimation and Modulation Theory*, Part I, John Wiley and Sons, New York, 1968, Chap. 3.

[13] Abramowitz, M. and Stegum, I., *Handbook of Mathematical Functions*, National Bureau of Standards Appl. Math. Series, Washington, D.C., June 1964, p. 810.

PROBLEMS

1. Let $R_b(\mathbf{r}_1, \mathbf{r}_2)$ be a covariance function defined over V. Using the KL expansion in (3.1.3) and (3.1.7) prove the following identities:

(a)
$$\int_V R_b(\mathbf{v}, \mathbf{v}) \, d\mathbf{v} = \sum_{i=0}^{\infty} \gamma_i$$

(b)
$$\sum_{i=0}^{\infty} \gamma_i^q = \int_V R_b^{(q)}(\mathbf{v}, \mathbf{v}) \, d\mathbf{v}$$

$$R_b^{(q)}(\mathbf{v}_1, \mathbf{v}_2) = \int_V R_b^{(q-1)}(\mathbf{v}_1, \mathbf{v}) R_b(\mathbf{v}, \mathbf{v}_2) \, d\mathbf{v}$$

(c)
$$\sum_{i=0}^{\infty} |s_i|^2 \gamma_i^q = \int_V s(\mathbf{v}) y_q^*(\mathbf{v}) \, d\mathbf{v}$$

$$y_q(\mathbf{v}) = \int_V R_b^{(q)}(\mathbf{v}, \mathbf{v}_1) s(\mathbf{v}_1) \, d\mathbf{v}_1$$

2. Consider the eigenfunction problem for the circular aperture, (3.5.16). Determine the eigenfunctions and eigenvalues under the assumption that $w_i(\mathbf{r})$ is circular symmetric.

3. We have shown (Problem 1.4) that the orthonormal eigenfunctions over the aperture transform to orthonormal eigenfunctions in the focal plane. Now show that the Fourier coefficients remain the same.

4. Compute the mean and variance of the count density in (3.6.8).

5. Consider the covariance function $Qe^{-\beta|\tau|}$ with the associated integral equation

$$\int_{-T}^{T} Q \exp[-\beta|t - u|]\phi_i(u)\, du = \gamma_i \phi_i(t)$$

(a) By differentiation find the corresponding differential equation.
(b) Show that the equation can only be satisfied with the eigenvalues $0 < \gamma < 2Q/\beta$.
(c) Find the solution in (b) and discuss the results.

6. Show that the mean and variance of the negative binomial distribution is given by

$$E_k(k) = (2BT + 1)\alpha N_{ob}$$

$$\mathrm{Var}[k] = (2BT + 1)(1 + \alpha N_{ob})\alpha N_{ob}$$

7. Consider the integral equation

$$\gamma_i(T)\phi_i(t, T) = \int_{0}^{T} R_b(t, t_1)\phi_i(t_1, T)\, dt_1, \qquad 0 \le t \le T$$

where we have emphasized the dependence on T. $R_b(t, t_1)$ is a square integrable correlation function. Show that $\gamma_i(T)$ is a monotone increasing function of T. [Hint: Differentiate with respect to T and show that $\gamma_i(T)$ is positive. Notice that $\int_0^T \phi_i^2(t, T)\, dt = 1$.]

8. Prove that the largest eigenvalue of the integral equation

$$\gamma\phi_i(t) = \int_{-T}^{T} R_b(t, t_1)\phi(t_1)\, dt_1, \qquad -T \le t \le T$$

satisfies the inequality

$$\gamma_{max} \ge \int_{-T}^{T}\int_{-T}^{T} f(t)R_b(t, t_1)f(t_1)\, dt\, dt_1$$

where $f(t)$ is any function with unit energy on $(-T, T)$. [*Hint*: Assume $R_b(t, t_1)$ is positive definite, and hence satisfies

$$\int_{-T}^{T} \int_{-T}^{T} |R_b(t, t_1)|^2 \, dt \, dt_1 < \infty$$

and the $\{\phi_i(t)\}$ are a complete orthonormal set.]

9. Consider the Wiener process defined in (3.6.3).
(a) Show that the KL equation becomes

$$\gamma_i \phi_i(t) = \rho^2 \int_0^t t_1 \phi_i(t_1) \, dt_1 + \rho^2 t \int_t^T \phi_i(t_1) \, dt_1$$

(b) Differentiate (a) twice with respect to t. Discuss the three ranges for γ_i:

(i) $\qquad\qquad\qquad\qquad\qquad\quad \gamma_i < 0$
(ii) $\qquad\qquad\qquad\qquad\qquad\quad \gamma_i = 0$
(iii) $\qquad\qquad\qquad\qquad\qquad\quad \gamma_i > 0$

(c) Compute the eigenvalue and eigenfunctions.

10. Consider the eigenvalues in (3.5.9):

$$\gamma(q, l) = \frac{\lambda^2 L^2}{\beta(0)} B\left[\frac{q\lambda L}{a}, \frac{l\lambda L}{b}\right]$$

Assume that the source has a constant brightness C_0 over the area A_0. Evaluate $\gamma(q, l)$ in terms of D_s, a, and b and discuss the result.

4 The Optical
Detector Response Process

In the system model of Chapter 1, the optical receiver contains a photodetector, which responds to incident radiation by releasing electrons at random from its surface. Preceding chapters discuss the relationship of the number of released electrons (which we call the electron count) to the incident radiation. From a design point of view, however, we observe at the detector output only the temporal effect of these electrons as they are produced over the entire detector area. In this chapter we mathematically examine the resulting current that appears at the photodetector output because of the random occurrences of these electrons. This output current is the time process physically observed during radiation reception, and therefore represents the time function eventually processed in the subsequent receiver circuitry. In Chapter 1 we labeled the class of processes so produced as shot noise processes. Hence, a study of this process and its inherent statistical properties is mandatory before discussing communication limitations and design.

4.1 SHOT NOISE PROCESSES

The mechanism that produces the detector output current is discussed in Section 1.8, and is summarized in Figure 4.1. As the radiation impinges upon the detector surface, electrons are released and travel to the cathode surface. The electron motion produces a current pulse effect to an external observer, or sensor, at the detector output. Thus the detector output is caused by the superposition of the pulse effects from the total collection of released electrons. As such, it is generated intrinsically as a current process, which can be converted to voltage by multiplication by the detector load impedance. We can therefore represent this output process as

$$x(t) = \sum_{j=1}^{k(V)} h(t - z_j) \qquad (4.1.1)$$

where $h(t)$ is the current or voltage response due to a single electron, z_j is the time of release of the jth electron, and $k(V)$ again is the count variable. This

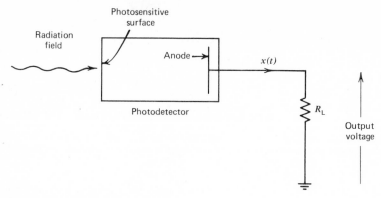

Figure 4.1. Photodetector signal model.

count is the number of such electrons occurring in the volume V composed of the detector area $\mathscr{A}$ and time interval up to t. The count variable indicates the number of terms considered in the summation of (4.1.1). Such processes are called shot noise processes [1–3]. Since we are primarily interested in the time evolution of $x(t)$ and its inherent statistics it is convenient to suppress the spatial variation in the count variable, and to express only its temporal dependence. Thus we write k(V) as k(0, t) to represent the total count over V as it evolves in time, over the interval 0 to t. (Here we assume the detection is initiated at a particular time $t = 0$, otherwise the 0 is replaced by $-\infty$.) Thus k(0, t) becomes a random process in t, describing the electron count over the total detector area. It is also convenient to suppress the spatial variation in the count intensity $n(t, \mathbf{r})$ and deal entirely with the spatially integrated count intensity

$$n(t) = \int_{\mathscr{A}} n(t, \mathbf{r}) \, d\mathbf{r} \tag{4.1.2}$$

When the counting statistic is pure Poisson, we refer to the resulting shot noise process in (4.1.1) as a *Poisson shot noise* (PSN) process. When the count is conditionally Poisson, the resulting shot noise is called a *conditional Poisson shot noise* (CPSN). The difference between the two is only in the statistics associated with the electron occurrences.

Although (4.1.1) is a general representation of shot noise, the physical applications in which we are interested allow us to impose constraints that simplify the formulation. Since $h(t)$ is the detector current response of a single electron moving in one direction, it must be everywhere nonnegative, and its area must be equal to the electron charge e. Thus $h(t)$ is an integrably bounded, nonnegative function. Furthermore, if we neglect space charge effects in the photodetector model, the travel time of each electron is finite, which means that the function $h(t)$ must be time limited to some interval τ_h. That is, $h(t) = 0$ for $t < 0$ and $t > \tau_h$ as shown in Figure 4.2a. This travel time τ_h is inversely related to the detector bandwidth, and is relatively short (10^{-7}–10^{-9} sec) with respect to the time variations in most information bearing waveforms. For an infinite bandwidth detector, $\tau_h \to 0$ and the electron functions can be regarded as delta functions of fixed area e (Figure 4.2b). The shot noise in (4.1.1) is then the sum of randomly occurring delta functions, a process that is mathematically convenient, but physically unrealizable, as we shall see. The time limitation on $h(t)$ also means that only the electron functions that have occurred over the past τ_h sec will affect the process in (4.1.1) at any t. Thus

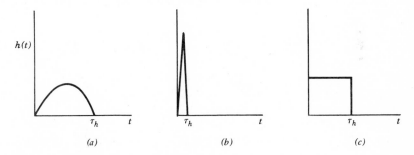

Figure 4.2. Photodetector response functions.

the count k(0, t) in the equation can be replaced by k($t - \tau_h$, t), the count over the finite interval ($t - \tau_h$, t). If $h(t)$ is simplified to a flat, rectangular function over (0, τ_h) (Figure 4.2c), then each such electron function that occurs in the past τ_h sec contributes a constant e/τ_h to the shot noise sum. Thus $x(t)$ in (4.1.1) becomes

$$x(t) = \frac{e}{\tau_h} k(t - \tau_h, t) \tag{4.1.3}$$

and the output process is proportional to the counting process governing the electron occurrences. Thus the statement that the detector output current produces the counting process tacitly assumes the presence of rectangular electron functions. When a nonrectangular impulse shape is to be accounted for, however, one is forced into a closer examination of the shot noise process. Clearly, when bandwidth considerations are significant, and ideal rectangular shapes cannot be assumed, further examination must be made.

4.2 OCCURRENCE TIMES STATISTICS WITH POISSON SHOT NOISE

We see from (4.1.1) that the output process depends not only on the count statistic but also on the occurrence times $\{z_j\}$. These occurrence times of the individual electrons in the shot noise process represent a collection of random locations in time inherently linked to the counting process. In this section we derive the joint probability density of a given number of such locations in a specified time interval (T_1, T_2) for a PSN process. Specifically, we derive the joint density of the random sequence $\mathbf{z} = (z_1, z_2, \ldots, z_k)$ over (T_1, T_2), given k Poisson occurrences in that interval. We write this conditional joint density as $p_z(z_1, z_2, \ldots, z_k | k)$. This can be derived as follows. Consider the interval (T_1, T_2) to contain the infinitesimal slots $(t_1, t_1 + \Delta t)$, $(t_2, t_2 + \Delta t)$, ..., $(t_k, t_k + \Delta t)$, as shown in Figure 4.3. The probability of getting a Poisson occurrence during each of the slots above and simultaneously getting exactly k

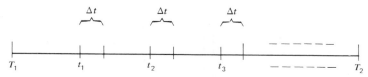

Figure 4.3. Time axis model.

occurrences is simply the probability of getting one occurrence in each slot and not getting any occurrence outside the slots. Since the slots and their exterior regions are disjoint, the independent interval property of Poisson processes means this joint probability is

Probability$[z_1 \in (t_1, t_1 + \Delta t), z_2 \in (t_2, t_2 + \Delta t), \ldots, z_k \in (t_k, t_k + \Delta t)]$

$$= \prod_{j=1}^{k} \left[\int_{t_j}^{t_j + \Delta t} n(t)\, dt \right] \left\{ \exp\left[- \int_{T_1}^{T_2} n(t)\, dt \right] \right\} \quad (4.2.1)$$

where $n(t)$ is the count intensity of the received field, and indicates the rate of the count occurrences. Now as $\Delta t \to 0$, (4.2.1) approaches

$$\prod_{j=1}^{k} n(t_j)\, \Delta t \left\{ \exp\left[- \int_{T_1}^{T_2} n(t)\, dt \right] \right\} \quad (4.2.2)$$

obtained by again applying (2.4.1). On the other hand, for small Δt, (4.2.1) could have been written

Prob$[z_1 \in (t_1, t_1 + \Delta t), \ldots, z_k \in (t_k, t_k + \Delta t)]$

$$= k!\ \text{Prob}[z_1 = t_1, \ldots, z_k = t_k](\Delta t)^k \quad (4.2.3)$$

where the $k!$ term takes into account all the ways in which k events can occur in k slots. Equating (4.2.3) and (4.2.2) identifies the joint density as

$$p(z_1, z_2, \ldots, z_k, k) = \frac{1}{k!} \prod_{j=1}^{k} n(z_j) \exp\left[- \int_{T_1}^{T_2} n(t) \, dt \right] \qquad z_j \in (T_1, T_2) \quad (4.2.4)$$

Now the probability of k Poisson occurrences in an arbitrary interval (T_1, T_2) given the count intensity $n(t)$ is, from (2.1.12),

$$P_k(k) = \text{Pos}\left[k, \int_{T_1}^{T_2} n(t) \, dt \right]$$

$$= \frac{[\int_{T_1}^{T_2} n(t) \, dt]^k}{k!} \exp\left[- \int_{T_1}^{T_2} n(t) \, dt \right] \qquad (4.2.5)$$

The conditional density of the k occurrence times $\{z_j\}$ is then obtained by dividing (4.2.4) by (4.2.5), yielding

$$p_z[z_1, z_2, \ldots, z_k | k] = \frac{\prod_{j=1}^{k} n(z_j)}{[\int_{T_1}^{T_2} n(t) \, dt]^k}$$

$$= \prod_{j=1}^{k} \frac{n(z_j)}{(m)^k} \qquad (4.2.6)$$

where

$$m = \int_{T_1}^{T_2} n(t) \, dt \qquad (4.2.7)$$

Here m is again the level of the Poisson density in (4.2.5), or, equivalently, the average number of counts occurring over (T_1, T_2). Note that the desired conditional density for the occurrence times factors into the product of the individual densities

$$p_{z_j}(z) = \frac{n(z)}{m}, \qquad z \in (T_1, T_2) \qquad (4.2.8)$$

for each j. This immediately implies that Poisson occurrence times are independent, each having a location density over t given by the normalized count intensity process $n(t)/m$. Note that this latter probability density is non-negative and of course integrates to unity. Furthermore, the result is valid for any interval (T_1, T_2). The result also serves as a further interpretation of the meaning of the intensity process $n(t)$—during any time interval it indicates the probability of releasing an electron at any particular time. Thus times at which the intensity level is high, physically implies times at which electrons are most likely to be released, while low intensities can be interpreted as times when electrons are least likely to be released. In particular times at which $n(t) = 0$ have a zero probability of releasing electrons.

For CPSN processes, governed by conditional Poisson counting statistics, the results above correspond to the density of occurrence times for any particular sample function of the random intensity process, defined over (T_1, T_2). The counting is Poisson and the location density is given by (4.2.6). To emphasize this inherent conditioning, were write this density as

$$p_z[z_1, z_2, \ldots, z_k | k, n(t)] = \prod_{j=1}^{k} \left(\frac{n(z_j)}{m} \right), \qquad z_j \in (T_1, T_2) \qquad (4.2.9)$$

To get the marginal density of occurrences with CPSN, we must average over the ensemble of all sample functions of $n(t)$. Denoting this average by the operator $E_n[\cdot]$, we have

$$p_z[z_1, z_2, \ldots, z_k | k] = E_n[p_z[z_1, \ldots, z_k | k, n(t)]]$$

$$= E_n\left[\prod_{j=1}^{k} \frac{n(z_j)}{m} \right]$$

$$= \frac{E_n[n(z_1) \cdots n(z_j)]}{m^k} \qquad (4.2.10)$$

The numerator is now the joint kth-order moment of the intensity process $n(t)$ at time $(z_1, z_2, \ldots, z_k)$. Note in particular that it does not necessarily factor into a product of terms. Hence, for CPSN the occurrence times are not necessarily independent. That is, given a simple function $n(t)$ the location times are generated independently, with the density in (4.2.9). However, the additional averaging over the randomness of $n(t)$ produces nonindependent location times. In subsequent work, we shall find it convenient to first deal with conditioned statistics (moments, correlation, spectral density, etc.) to take advantage of this independence, and follow it with the necessary averaging for CPSN.

4.3 MEAN AND VARIANCE OF SHOT NOISE PROCESSES

The results of the preceding section can now be used to evaluate statistical moments of the shot noise process. We calculate first conditional moments for a given number of counts k in a specified interval, and a given intensity function $n(t)$, then average over k for PSN, and average over k and $n(t)$ for CPSN.

Mean Value. The mean of the shot noise is formally defined as

$$E[x(t)] \triangleq E_k[E[x(t)|k]] \qquad (4.3.1)$$

where k represents the random count $k(0, t)$ up to time t. Using subscripts to denote the averaging variable, this is then

$$E[x(t)] = E_k\left\{E_z\left[\sum_{j=1}^{k} h(t - z_j)|k\right]\right\}$$

$$= E_k\left\{\int_{-\infty}^{t} \cdots \int_{-\infty}^{t} \sum_{j=1}^{k} h(t - z_j)p_z[z_1,\ldots,z_k|k]\, dz_1 \ldots dz_k\right\} \quad (4.3.2)$$

The density in the integrand is given by (4.2.9) so that (4.3.2) becomes

$$E[x(t)] = E_k\left\{\sum_{j=1}^{k} \int_{-\infty}^{t} h(t - z)p_{z_j}(z)\, dz\right\}$$

$$= E_k\left\{k \int_{-\infty}^{t} h(t - z)\left(\frac{n(z)}{m}\right)\, dz\right\}$$

$$= E_k[k]\left[\int_{-\infty}^{t} h(t - z)\left(\frac{n(z)}{m}\right)\, dz\right] \quad (4.3.3)$$

However, since $E_k[k]$ = the level of Poisson probability = m, this reduces to

$$E[x(t)] = \int_{-\infty}^{t} h(t - z)n(z)\, dz \quad (4.3.4)$$

Thus the mean of the shot noise appears as a filtering (convolution) of the shot noise intensity by the electron functions $h(t)$. In particular, we note that the mean depends on t and therefore evolves as a function of time. For infinite bandwidth detectors $h(t) = e\,\delta(t)$ and (4.3.4) becomes

$$E[x(t)] = \int_{-\infty}^{t} e\,\delta(t - z)n(z)\, dz$$

$$= en(t) \quad (4.3.5)$$

and the mean function is directly related to the instantaneous count intensity. Note that the dependence on t in (4.3.4) and (4.3.5) indicates that the general detector process is nonstationary in time. We may compare this ensemble mean to the time averaged mean, defined as

$$\overline{x(t)} \triangleq \lim_{T \to \infty} \frac{1}{2T} \int_{-T}^{T} x(t)\, dt \quad (4.3.6)$$

For the shot noise in (4.1.1), we have

$$\overline{x(t)} = \lim_{T \to \infty} \frac{1}{2T} \int_{-T}^{T} \sum_{j=1}^{k} h(t - z_j)\, dt$$

$$= \lim_{T \to \infty} \left(\frac{1}{2T}\right) ek[-T, T]$$

$$= e \lim_{T \to \infty} \left[\frac{k[-T, T]}{2T}\right]$$

$$= e\bar{n} \qquad (4.3.7)$$

where we have used $\bar{n}$ to represent the time averaged electron count rate over the infinite interval. This result is of course independent of time, and therefore differs from the ensemble average previously computed. However, if the intensity is constant, $n(t) = n_0$, as for monochromatic fields in (2.4.4), $\bar{n} = \lim_{T \to \infty} (n_0 2T/2T) = n_0$ and (4.3.5) and (4.3.7) become identical.

It is also convenient to define a "short term" time average as

$$\overline{x(t)}^T \triangleq \frac{1}{T} \int_{t-T}^{t} x(t)\, dt \qquad (4.3.8)$$

which becomes

$$\overline{x(t)}^T = \frac{e}{T} k(t - T, t) \qquad (4.3.9)$$

where we have assumed T is much greater than τ_h, the time width of the electron functions. Thus the short-term time average "follows" the number of occurrences over the past T sec, at each t. Note that (4.3.9) would be approximately the output of a filter of bandwidth $1/T$ operating upon the shot noise. If T is small compared to the time variations in $n(t)$, then $k(t - T, t) \approx T n(t)$ and (4.3.9) is identical to (4.3.5). Thus the ensemble mean of a shot noise process can be interpreted roughly as the output of a filter narrow band compared to the detector bandwidth, but wide band in comparison to the intensity bandwidth.

Mean Square Value. The second moment, or mean squared value of the shot noise, is related to the power in the process, and can be determined similarly:

$$E[x^2(t)] = E_k \left\{ \int_{-\infty}^{t} \cdots \int_{-\infty}^{t} \left[\sum_{j=1}^{k} h(t - z_j) \right]^2 p_z(z_1, \ldots, z_k | k)\, dz_1 \cdots dz_k \right\}$$

$$(4.3.10)$$

Expanding the square, and making use of the independence of the location times, we obtain

$$E[x^2(t)] = E_k \left\{ k \int_{-\infty}^t h^2(t - z_j) p_{z_j}(z_j) \, dz_j \right.$$

$$\left. + (k^2 - k) \left[\int_{-\infty}^t h(t - z_j) p_{z_j}(z_j) \, dz_j \right]^2 \right\} \quad (4.3.11)$$

To evaluate the averages over the k parameters, we must invoke the Poisson counting moments in (2.2.7). Therefore, (4.3.11) becomes

$$E[x^2(t)] = m \int_{-\infty}^t h^2(t - z) p_{z_j}(z) \, dz$$

$$+ m^2 \left[\int_{-\infty}^t h(t - z) p_{z_j}(z) \, dz \right]^2$$

$$= \int_{-\infty}^t h^2(t - z) n(z) \, dz + \{E[x(t)]\}^2 \quad (4.3.12)$$

The variance of the shot noise about the mean is precisely the first term above:

$$\text{Var}[x(t)] = \int_{-\infty}^t h^2(t - z) n(z) \, dz \quad (4.3.13)$$

For rectangular electron functions, $h(t) = e/\tau_h, \, 0 \le t \le \tau_h$, and the variance reduces to

$$\text{Var}[x(t)] = \int_{t-\tau_h}^t \left(\frac{e}{\tau_h} \right)^2 n(z) \, dz$$

$$= \left(\frac{e}{\tau_h} \right)^2 k(t - \tau_h, t) \quad (4.3.14)$$

In this case, the variance of the shot noise also follows the counting process as a function of t. Equation 4.3.13 also shows that for infinite bandwidth detectors,

$$\text{Var}[x(t)] = \frac{e^2 \tau_h n(t)}{\tau_h^2} = \frac{e^2 n(t)}{\tau_h} \quad (4.3.15)$$

and the variance becomes unbounded, as $\tau_h \to 0$. Thus infinite bandwidth PSN processes do not correspond to bounded mean square processes. The mean squared time average and the mean square short time average, can be calculated as in (4.3.6) and (4.3.8) and shown to be

$$\overline{[x(t)]^2} = \delta_2 \bar{n} + [\overline{x(t)}]^2$$

$$\overline{[x(t)]^2}^T = \left(\frac{\delta_2}{T}\right) k(t - \tau_h, t) + \{\overline{x(t)}^T\}^2 \qquad (4.3.16)$$

where

$$\delta_2 \triangleq \int_{-\infty}^{\infty} h^2(t)\, dt \qquad (4.3.17)$$

is the area under the square of $h(t)$. Thus time averaged values and ensemble averaged values lead to different results, a fact attributed to the nonstationarity of the process.

The first two moments of the Poisson shot noise process are inherently linked to the received intensity function $n(t)$ of the radiation field. We might examine this relation in more detail. Consider the ensemble mean and variance in (4.3.4) and (4.3.13) when the electron function width τ_h is much smaller than the time variations in $n(t)$ (i.e., the detector bandwidth is suitably larger than the bandwidth of the intensity being received). In this case, the mean and variance are approximately $en(t)$ and $\delta_2 n(t)$, respectively, where δ_2 is given in (4.3.17). Since the square root of the variance of any random variable is an indication of the average spread about the mean, we can depict the PSN process as in Figure 4.4. Here we have chosen an arbitrary intensity function $n(t)$, and shown the mean variation and corresponding average spread region of the random process $x(t)$. Because of these results, the spread varies in accordance with the mean. At any time t, we can consider the PSN

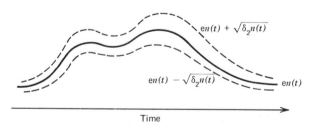

Figure 4.4. Plot of shot noise process mean value and spread due to standard deviation.

to be close, in an average sense, to the intensity function (its mean value) if the spread is small with respect to the mean. Specifically, if we require the squared ratio of the spread to be small, we would require

$$\frac{\delta_2 n(t)}{[en(t)]^2} \ll 1 \tag{4.3.18}$$

or equivalently,

$$n(t) \gg \frac{\delta_2}{e^2} \tag{4.3.19}$$

This means that the intensity function, which we recall from (1.3.3) is related to the received power, must be suitably large. Hence, accurate representation of the intensity by the photodetector output shot noise will occur if it receives enough power. This is even more meaningful if we assume the electron functions $h(t)$ are rectangular over τ_h with area e. Then $\delta_2 = e^2/\tau_h$ and the requirement above becomes $n(t) \gg e^2/e^2\tau_h = 1/\tau_h$ or

$$n(t)\tau_h \gg 1 \tag{4.3.20}$$

Since the intensity indicates the average number of electrons per second, the quantity on the left can be interpreted as the average number of electrons occurring in the time width of an electron function. Hence, (4.3.20) requires that at any time the electron "density," that is, the amount of function overlap, be large. Roughly speaking, we conclude that if the PSN is generated so that there is a large electron overlap (many produced in a short time), the PSN will strongly resemble the intensity function. As we have seen, this occurs in a high power situation. If the overlap is slight, the PSN spread is large relative to the mean, and no longer resembles the intensity. This latter condition appears during low power operation, and essentially generates a "threshold" condition on our detector. The PSN spread (the average variation from the true intensity) can therefore be attributed to the "discreteness" of our detector process. At low power operation, the discrete structure of the PSN is evident and affects the detector output process. At high power operation, the discreteness is no longer evident, and the detector approaches an ideal detector of the intensity function. Since the count intensity is proportional to received field power, the optical detector behaves as an instantaneous power detector. It is evident that if an optical communication system is to be operated at or near threshold condition, further statistical properties of the output shot noise process must be developed in order to carry out standard analysis and design procedures.

4.4 CORRELATION FUNCTIONS

The correlation function of the PSN process is formally defined as

$$R_x(t, t + \tau) = E[x(t)x(t + \tau)]$$

$$= E\left[\sum_{j=1}^{k} h(t - z_j) \sum_{q=1}^{k} h(t + \tau - z_q) \right] \qquad (4.4.1)$$

Expanding the sum and carrying out the expectation operation as in (4.3.11) yields

$$R_x(t, t + \tau) = \int_{-\infty}^{t} h(t - z)h(t + \tau - z)n(z)\, dz$$

$$+ \int\int_{-\infty}^{t} h(t - z_j)n(z_j)h(t + \tau - z_q)n(z_q)\, dz_j\, dz_q$$

$$= \int_{-\infty}^{t} h(t - z)h(t + \tau - z)n(z)\, dz + E[x(t)]E[x(t + \tau)] \qquad (4.4.2)$$

The result is again nonstationary, depending explicitly on the time t. For this reason it is often more convenient to deal with the time averaged autocorrelation function:

$$\bar{R}_x(\tau) = \overline{x(t)x(t + \tau)}$$

$$= \lim_{T \to \infty} \frac{1}{2T} \int_{-T}^{T} \sum_{j=1}^{k} h(t - z_j) \sum_{q=1}^{k} h(t + \tau - z_q)\, dt$$

$$= \lim_{T \to \infty} \frac{k(-T, T)}{2T} \int_{-T}^{T} h(t - z)h(t + \tau - z)\, dt + [\overline{x(t)}]^2$$

$$= \bar{n} \int_{-\infty}^{\infty} h(u)h(u + \tau)\, du + (e\bar{n})^2 \qquad (4.4.3)$$

The last term arises from the time-averaged rate of the process. The integral is similar to the correlation of the electron functions $h(t)$. Hence, the time averaged autocorrelation function is specifically related to the correlation of the electron functions $h(t)$. In particular, note that $\bar{R}_x(\tau)$ has a correlation time only over the correlation time of these functions. Thus PSN is a pure white process (infinite bandwidth process) only if the $h(t)$ are themselves delta functions, that is, the detectors have infinite bandwidth.

4.5 SPECTRAL DENSITY OF PSN

The power spectral density of a random process is important in specifying the power content of the process as a function of frequency. The usual procedure for determining the power spectrum of a stationary process is by Fourier transforming its autocorrelation function. The PSN process, however, is nonstationary as evidenced by the correlation function in (4.4.2). If the time averaged correlation in (4.4.3) is used, it transforms to

$$\bar{S}_x(\omega) \triangleq \text{Fourier transform of } \bar{R}_x(\tau)$$
$$= \bar{n}|H(\omega)|^2 + (e\bar{n})^2 2\pi \, \delta(\omega) \tag{4.5.1}$$

when $H(\omega)$ is the Fourier transform of the electron function $h(t)$. Note that the spectrum always appears as the sum of a delta function (because of the time averaged mean of the PSN) and a continuous portion, given by the transformed shape of the electron functions. Unfortunately, this is true no matter what the form of the intensity process $n(t)$. That is, the frequency content of the intensity is not exhibited in the spectral density when defined in this way. Basically, the frequency variations have been averaged out. Hence, this definition of spectral density is meaningful only for constant intensity processes. To avoid this problem, we derive the power spectral density by computing the time averaged power at any radian frequency ω. That is, we consider the spectrum defined by

$$\hat{S}_x(\omega) \triangleq \lim_{T \to \infty} \frac{1}{2T} E[|X_T(\omega)|^2] \tag{4.5.2}$$

where $X_T(\omega)$ is the Fourier transform of a sample function of $x(t)$ restricted to the interval $(-T, T)$. (Equation 4.5.2 is the classical definition of a power spectrum, specializing to a correlation transform only for wide sense stationary processes.) The function $|X_T(\omega)|^2$ is the energy density at each ω for the sample function, and therefore is a random variable at each ω. It must therefore be averaged over the statistics of the process. Thus $\hat{S}_x(\omega)$ is the time average of the ensemble average of the energy density at each ω. Now if we condition upon $\{z_j\}$ and k, we have

$$X_T(\omega) = \int_{-T}^{T} \sum_{j=1}^{k} h(t - z_j)e^{-j\omega t} \, dt$$
$$= \sum_{j=1}^{k} \int_{-T}^{T} h(t - z_j)e^{-j\omega t} \, dt$$
$$= \sum_{j=1}^{k} e^{-j\omega z_j} H_T(\omega) \tag{4.5.3}$$

where $H_T(\omega)$ is the Fourier transform of $h(t)$, $-T < t < T$. Thus,

$$|X_T(\omega)|^2 = X_T(\omega)X_T^*(\omega)$$

$$= \sum_{j=1}^{k} e^{-j\omega z_j} H_T(\omega) \sum_{q=1}^{k} e^{j\omega z_q} H_T^*(\omega)$$

$$= |H_T(\omega)|^2 \sum_{j=1}^{k} \sum_{q=1}^{k} \exp[j\omega(z_q - z_j)] \qquad (4.5.4)$$

The ensemble average over $\{z_j\}$, conditioned upon k, can now be taken. Noting that the summand is unity when $q = j$, we have

$$E_z[|X_T(\omega)|^2 | k] = |H_T(\omega)|^2 \left[k + (k^2 - k) \int_{-T}^{T} e^{j\omega z} p_{z_q}(z) \, dz \right.$$

$$\left. \cdot \int_{-T}^{T} e^{-j\omega z} p_{z_j}(z) \, dz \right] \qquad (4.5.5)$$

The subsequent averaging over k finally yields

$$E[|X_T(\omega)|^2] = |H_T(\omega)|^2 [m_T + N_T^*(\omega)N_T(\omega)] \qquad (4.5.6)$$

where

$$m_T = \int_{-T}^{T} n(t) \, dt \qquad (4.5.7)$$

$$N_T(\omega) = \int_{-T}^{T} e^{-j\omega u} n(u) \, du \qquad (4.5.8)$$

Here $N_T(\omega)$ is the Fourier transform of the intensity function $n(t)$, $-T \leq t \leq T$. Finally, we have

$$\hat{S}_x(\omega) = \lim_{T \to \infty} \frac{1}{2T} E[|X_T(\omega)|^2]$$

$$= |H(\omega)|^2 [\bar{n} + |N(\omega)|^2] \qquad (4.5.9)$$

where $\bar{n}$ is the time average rate in (4.3.7) and $N(\omega) = \lim_{T \to \infty} |N_T(\omega)|^2/2T$. Comparison with (4.5.1) shows that the spectral density derived in this way now exhibits the frequency content of the intensity.

For conditional Poisson shot noise (CPSN), the intensity $n(t)$ is a sample function of a random process and the calculation of the spectrum in (4.5.9) requires an additional averaging over the ensemble of intensity processes. Hence,

$$S_x(\omega) = E_n[\hat{S}_x(\omega)]$$

$$= |H(\omega)|^2 [\overline{E(n)} + S_n(\omega)] \qquad (4.5.10)$$

where now

$$\overline{E[n]} = \lim_{T\to\infty} \left[\frac{1}{2T} \int_{-T}^{T} E_n[n(t)] \, dt \right] \tag{4.5.11}$$

$$S_n(\omega) = \lim_{T\to\infty} \left[\frac{1}{2T} E_n[|N_T(\omega)|^2] \right] \tag{4.5.12}$$

The spectrum is sketched in Figure 4.5. The term $\overline{E(n)}$ is now the time averaged mean count rate, averaged over the intensity statistics. The term $S_n(\omega)$ is the spectral density of the stochastic intensity process $n(t)$, according to a formal definition in (4.5.2). Thus, in going from PSN to CPSN power spectra, the rate $\bar{n}$ is replaced by $\overline{E(n)}$, and the transform spectrum $|N(\omega)|^2$ is replaced by the power spectral density $S_n(\omega)$. The spectrum in (4.5.10) is valid for all CPSN processes, and therefore is valid for a shot noise process having any of the counting probabilities developed in Chapter 3.

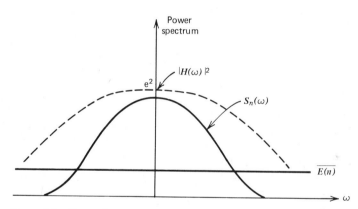

Figure 4.5. Shot noise power spectrum.

Note that the spectrum in (4.5.10) always takes the form of a filtered intensity spectrum $|H(\omega)|^2 S_n(\omega)$ with an additive spectrum $\overline{E(n)}|H(\omega)|^2$. For infinite bandwidth detectors, $H(\omega) = e$ for all ω, and the second term appears as a spectrally white additive process of height $e^2 \overline{E[n]}$. Since this appears as an added "noise" to the desired intensity spectrum, it is called the *shot noise level* of the detector. Note that this noise always appears when photodetecting an optical field, and is due to the discrete nature of the detector model. If the detector has a finite bandwidth, then $|H(\omega)|^2$ effectively "shapes" the signal and shot noise spectrum. If the desired frequency content of $S_n(\omega)$ lies within the frequency region where $H(\omega) \approx e$, its spectrum is not distorted. This simply means the detector bandwidth should exceed the desired signal frequency band of $S_n(\omega)$. The intensity spectrum $S_n(\omega)$ in general contains portions due to desired intensity, portions due to background noise effects or local oscillator injections (as in heterodyning), and possible cross-spectral terms. An examination of these spectra, with specific applications to communication receivers, is considered in Chapter 5.

Let us relate the shot noise spectrum to the variance of the detector output. For PSN, this variance was derived in (4.3.13). For CPSN the variance is obtained by averaging over the intensity in this latter result. Hence, for stationary mean intensities,

$$\text{Var}[x(t)] = \int_{-\infty}^{\infty} h^2(t - z)E_n[n(z)]\,dz$$

$$= \overline{E[n]} \int_{-\infty}^{\infty} h^2(t)\,dt \tag{4.5.13}$$

Now by application of Parseval's theorem, we rewrite this as

$$\text{Var}[x(t)] = \overline{E[n]}\left(\frac{1}{2\pi} \int_{-\infty}^{\infty} |H(\omega)|^2\,d\omega\right)$$

$$= \frac{1}{2\pi} \int_{-\infty}^{\infty} \overline{E[n]}|H(\omega)|^2\,d\omega \tag{4.5.14}$$

The integrand is recognized as the spectral density of the shot noise portion of the detector spectrum. Hence, the power in the shot noise spectrum (area under the shot noise portion) is precisely the variance of the detector output process, for stationary intensities.

Dark current in photodetectors corresponds to the random emission of electrons at a fixed rate, when no field is being detected. As such, dark current electrons add directly to the average rate at the detector output, and therefore add directly to the shot noise level in (4.5.10). Thus, if the dark current produces an average electron rate $\bar{n}_d$, the detector output spectrum is modified to

$$S_x(\omega) = |H(\omega)|^2[\overline{E[n]} + \bar{n}_d + S_n(\omega)] \tag{4.5.15}$$

In typical operation, values for $\bar{n}_d$ are much less than those for $\overline{E[n]}$, because of the received field, and the dark current level can often be neglected.

Since the spectrum in (4.5.10) has the form of a "signal in noise," there is a tendency to view the photodetected output as a signal plus noise. Although such an interpretation can be applied when determining detector output signal-to-noise ratios, the reader is cautioned against a liberal application of this model. The difficulty is that the signal and noise are not independent, and usual "signal plus noise" interpretations familiar to communication engineers, sometimes lead to false conclusions. This point is stressed further in Chapter 5.

4.6 LINEAR FILTERING OF SHOT NOISE PROCESSES

Consider the system in Figure 4.6 where a shot noise process $x(t)$ is filtered by a linear system. We describe the linear filter by its impulse response $u(t)$, or by its system transfer function $U(\omega)$, that is, the Fourier transform of $u(t)$. We assume the system is causally realizable in the sense that $u(t) = 0$ for $t < 0$.

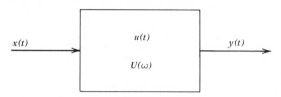

Figure 4.6. Linear filter model.

The output of the linear system can immediately be written as

$$y(t) = \int_{-\infty}^{\infty} u(t - \xi)x(\xi)\, d\xi$$

$$= \int_{-\infty}^{\infty} u(t - \xi)\left[\sum_{j=1}^{\mathbf{k}(-\infty, t)} h(\xi - z_j)\right] d\xi$$

$$= \sum_{j=1}^{\mathbf{k}(-\infty, t)} \hat{h}(t - z_j)$$

where

$$\hat{h}(t) = \int_{-\infty}^{\infty} u(t - \xi)h(\xi)\, d\xi \qquad (4.6.1)$$

The function $\hat{h}(t)$ is the response of the system to a single electron function $h(t)$. Hence, the output of the linear system is itself a shot noise process with counting statistics identical to the input, and with component functions given by the filtered version of the input functions. These latter functions have transform

$$\hat{H}(\omega) = H(\omega)U(\omega) \qquad (4.6.2)$$

The preceding statistical characteristics derived in the earlier sections are now applicable, with an appropriate replacement of $h(t)$ by $\hat{h}(t)$. In particular, the mean value of the output is, from (4.3.4),

$$E[y(t)] = \int_{-\infty}^{\infty} \hat{h}(t - \rho)n(\rho)\,d\rho$$

which can be rewritten as

$$E[y(t)] = \int_{-\infty}^{\infty} \left[\int_{-\infty}^{\infty} u(t - \rho - \xi)h(\xi)\,d\xi\right]n(\rho)\,d\rho$$

$$= \int_{-\infty}^{\infty} h(\xi)\left[\int_{-\infty}^{\infty} u(t - \rho - \xi)n(\rho)\,d\rho\right]d\xi \qquad (4.6.3)$$

The inner integral can be denoted as $\hat{n}(t - \rho)$ where

$$\hat{n}(t) = \int_{-\infty}^{\infty} u(t - \xi)n(\xi)\,d\xi \qquad (4.6.4)$$

and appears as a filtered version of the input intensity. When written in this way, the mean value appears as if the linear system has directly filtered the input intensity function without affecting the electron functions. Hence, one may consider the mean output in either of two points of view—the linear system filters the functions, with the output intensity the same as that of the input, or the linear system filters the intensity, without affecting the output functions. Note that the mean square value of the output shot noise $y(t)$, obtained by substitution of (4.6.1) into (4.3.12), does not have a similar interpretation.

The spectral density for $y(t)$ is obtained directly from (4.5.10):

$$S_y(\omega) = |H(\omega)|^2|U(\omega)|^2[\overline{E[n]} + S_n(\omega)] \qquad (4.6.5)$$

Thus the filter function $U(\omega)$ shapes the shot noise spectral level, while creating the new signal spectrum $|U(\omega)|^2|H(\omega)|^2 S_n(\omega)$. We see that the frequency components of the input intensity $S_n(\omega)$ that are outside the spectral width of $U(\omega)$, do not appear at the filter output, as shown in Figure 4.7a. In fact, if the intensity spectrum $S_n(\omega)$ lies wholly outside the filter band [$U(\omega) = 0$ over the band of $S_n(\omega)$], the signal portion of the spectrum may be completely removed by the filter. A baseband filter centered about zero frequency with bandwidth B passes only the frequency components within B. On the other hand, a bandpass filter of spectral width $2B$ centered at ω_0 passes only the intensity components within B of the band center frequency

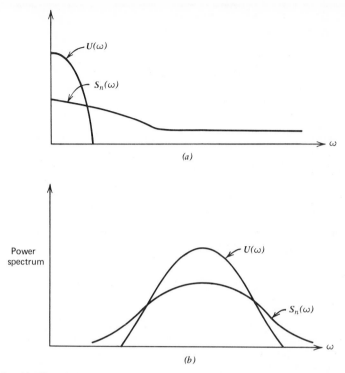

Figure 4.7. (*a*) Filtered shot noise process. (*b*) Bandpass filtered shot noise process.

ω_0 (Figure 4.7*b*). It should be noted, however, that even if the signal portion is removed, the shot noise process itself still exists at the filter output. That is, the output still appears as a summation of randomly located component functions.

One particular type of linear filtering often of interest in communication system is the short-term integrator. That is, a filter whose impulse response is specifically

$$u(t) = 1, \qquad 0 \le t \le T$$
$$= 0, \qquad \text{elsewhere} \tag{4.6.6}$$

The output of such a device due to any input $x(t)$ is

$$y(t) = \int_{t-T}^{t} x(u)\, du \tag{4.6.7}$$

Thus the device effectively integrates over the finite past of the input process. For shot noise inputs, this becomes

$$y(t) = \int_{t-T}^{t} \sum_{j=1}^{k(-\infty,\,t)} h(t - z_j)\, dt$$

$$= ek(t - T, t) \tag{4.6.8}$$

where the approximation assumes the function time width τ_h is much less than the integration time T. Thus the short-term integrator produces at its output at each t the count over the past T sec of the input shot noise. If the input is PSN, then $y(t)$ is a Poisson process, and is a conditional Poisson process if the input is CPSN.

The short-term integrator can be viewed in an alternative way, using earlier results. We have shown earlier that $y(t)$ in (4.6.8) can be considered a shot noise process with rectangular (T sec wide) functions. Thus it has spectrum

$$S_y(\omega) = e^2 \left| \frac{\sin(\omega T/2)}{\omega T/2} \right|^2 [\overline{E[n]} + S_n(\omega)] \tag{4.6.9}$$

where the magnitude term is the filter function $U(\omega)$ for the integrator. If the intensity process has bandwidth B, and if $BT \ll 1$, then

$$S_y(\omega) \underset{BT \ll 1}{\approx} e^2 \left| \frac{\sin(\omega T/2)}{(\omega T/2)} \right|^2 \overline{E[n]} + e^2 S_n(\omega) \tag{4.6.10}$$

and the integrator reproduces the intensity spectrum without distortion, with the added shot noise given by the first term. The condition $BT \ll 1$ is equivalent to the condition $T \ll 1/B$, or that of the integration time being small compared to the period of the highest frequency component in the intensity. Hence, the integrator output follows the time variations of the intensity, with the effective additive noise. This, of course, could have been predicted directly from (4.6.8) by noting that $k(t - T, t)$ is a random process whose average value becomes

$$E[k(t - T, t)] = \int_{t-T}^{t} n(\rho)\, d\rho$$

$$\underset{T \ll 1/B}{\longrightarrow} n(t)T \tag{4.6.11}$$

which is the level associated with short-term counting. Therefore, the average value of the count process is proportional to the intensity process when using short-term integration. Similarly, if $T \gg 1/B$, then

$$\frac{\sin(\omega T/2)}{\omega T/2} \to 0, \qquad \omega \neq 0$$

$$\to 1, \qquad \omega = 0$$

and

$$S_y(\omega) \underset{BT \gg 1}{\to} e^2 S_n(0) = e^2 2\pi \, \delta(\omega) \qquad (4.6.12)$$

In this case the integrator has completely filtered the intensity variations, and its output is the time averaged count over the T sec integration time. We previously labeled this as long-term counting.

4.7 FIRST-ORDER STATISTICS OF PSN

The preceding discussions are concerned with basically mean values and second-order statistics of the shot noise process. In many situations, complete first-order probability densities of the shot noise process, at any t, are desired. As we shall see, an explicit expression for this density is difficult to write, except in the integral form. The analytical procedure is to develop the characteristic function first, then attempt to inverse transform into the desired density. When dealing with PSN processes (deterministic intensities) the analysis is somewhat straightforward, but the extension to CPSN processes (random intensities) becomes complicated rather quickly.

The characteristic function of the PSN process at any time t is, formally,

$$\Psi_x(\omega, t) = E[e^{j\omega x(t)}]$$

$$= E\left[\exp\left\{j\omega \sum_{j=1}^{k(-\infty, t)} h(t - z_j)\right\}\right] \qquad (4.7.1)$$

Conditioning on k and averaging over the densities of $\{z_j\}$ first, yields

$$\Psi_x(\omega, t) = E_k\left[\prod_{j=1}^{k} E_z\{\exp[j\omega h(t - z_j)]|k\}\right]$$

$$= E_k[\Psi_n(\omega, t)]^k$$

where

$$\Psi_n(\omega, t) = \int_{-\infty}^{t} \exp[j\omega h(t - \rho)]\left(\frac{n(\rho)}{m}\right) d\rho \qquad (4.7.2)$$

is the characteristic function of the random variable $h(t - u)$. For PSN, the ensuing average over k yields

$$\Psi_x(\omega, t) = E_k[\Psi_n(\omega, t)]^k$$

$$= \sum_{k=0}^{\infty} \left[\frac{m^k}{k!} e^{-m}\right][\Psi_n(\omega, t)]^k$$

$$= \exp\{[\Psi_n(\omega, t) - 1]m\} \qquad (4.7.3)$$

where again m is the average number of counts in $(-\infty, t)$. Using (4.7.2), this characteristic function can be rewritten as

$$\Psi_x(\omega, t) = \exp\left\{\int_{-\infty}^t [\exp[j\omega h(t - \rho)] - 1]n(\rho) \, d\rho\right\} \qquad (4.7.4)$$

This equation represents the characteristic function of the shot noise at any t. We emphasize the dependence on t, which exhibits the nonstationary property that we have observed all along. The probability density, $p(x, t)$, at corresponding times t is obtained in theory by inverse transforming (4.7.4); that is,

$$p(x, t) = \frac{1}{2\pi} \int_{-\infty}^{\infty} e^{-j\omega x}\Psi_x(\omega, t) \, d\omega \qquad (4.7.5)$$

This transform is difficult to evaluate in all but certain special cases. For example, with rectangular electron functions $h(t) = e/\tau_h, 0 \leq t \leq \tau_h$,

$$\Psi_x(\omega, t) = \exp\left\{\left[\exp\left(\frac{j\omega e}{\tau_h}\right) - 1\right]\int_{t-\tau_h}^t n(\rho) \, d\rho\right\}$$

$$= \exp\left\{\tilde{m}_t\left[\exp\left(\frac{j\omega e}{\tau_h}\right) - 1\right]\right\} \qquad (4.7.6)$$

where $\tilde{m}_t$ is the average count over $(t - \tau_h, t)$. The probability density is then

$$p(x, t) = \frac{1}{2\pi} \int_{-\infty}^{\infty} e^{-j\omega x}\Psi_x(\omega, t) \, d\omega$$

$$= \frac{e^{-\tilde{m}_t}}{2\pi} \int_{-\infty}^{\infty} e^{-j\omega x}\left\{\sum_{q=0}^{\infty} \frac{(\tilde{m}_t)^q}{q!}\left[\exp\left(\frac{j\omega e}{\tau_h}\right)\right]^q\right\} d\omega$$

$$= \frac{e^{-\tilde{m}_t}}{2\pi} \sum_{q=0}^{\infty} \frac{(\tilde{m}_t)^q}{q!}\left\{\int_{-\infty}^{\infty} \exp\left[-j\omega\left(x - \frac{qe}{\tau_h}\right)\right] d\omega\right\}$$

$$= \sum_{q=0}^{\infty} \left[\frac{(\tilde{m}_t)^q}{q!} e^{-\tilde{m}_t}\right]\delta\left(x - \frac{qe}{\tau_h}\right) \qquad (4.7.7)$$

which is a modified Poisson density with level $\tilde{m}_t$. Note that the density at any t takes on values at the discrete points $\{qe/\tau_h\}$ with a Poisson probability on q. Hence, the probability density of the shot noise is governed by the density of the counting process itself, when rectangular electron functions are used. This of course could have been predicted from our result in (4.1.3).

Even though the density in (4.7.5) in general may not be transformed in closed form, all the moments of the PSN process can be derived directly from (4.7.4) by differentiation, using

$$E[x^q(t)] = \frac{1}{(j)^q} \left[\frac{\partial^q}{\partial \omega^q} \Psi_x(\omega, t) \right]_{\omega = 0} \tag{4.7.8}$$

In particular,

$$E[x(t)] = \left[\left\{ \frac{1}{j} \int_{-\infty}^{t} jh(t - u)n(u) [\exp(j\omega h(t - u)) - 1] \, du \right\} \Psi_x(\omega, t) \right]_{\omega = 0}$$

$$= \int_{-\infty}^{t} h(t - \rho)n(\rho) \, d\rho \tag{4.7.9}$$

$$E[x^2(t)] = \frac{1}{j^2} \left\{ \int_{-\infty}^{t} j^2 h^2(t - u)n(u) \, du + j^2 \left[\int_{-\infty}^{t} h(t - u)n(u) \, du \right]^2 \right\}$$

$$= \int_{-\infty}^{t} h^2(t - u)n(u) \, du + \left[\int_{-\infty}^{t} h(t - u)n(u) \, du \right]^2 \tag{4.7.10}$$

as in (4.3.4) and (4.3.12). Alternatively, the moments can be generated by noting that the log of the characteristic function in (4.7.4) is

$$\log \Psi_x(\omega, t) = \int_{-\infty}^{t} \{\exp[j\omega h(t - u)] - 1\}n(u) \, du$$

$$= \sum_{q=1}^{\infty} \frac{(j\omega)^q}{q!} \int_{-\infty}^{t} h^q(t - u)n(u) \, du \tag{4.7.11}$$

Therefore the PSN has semiinvariants

$$\chi_q = \int_{-\infty}^{t} h^q(t - u)n(u) \, du \tag{4.7.12}$$

The moments can then be derived by their relation to these semiinvariants. The first few moments take the specific form

$$E[x(t)] = \chi_1$$
$$E[x^2(t)] = \chi_2 + \chi_1^2 \tag{4.7.13}$$
$$E[x^3(t)] = \chi_3 + 3\chi_2\chi_1 + \chi_1^3$$
$$\vdots$$

Some interesting limiting forms for the probability density $p(x, t)$, as the intensity $n(t)$ is increased at any t, can be obtained from the semiinvariants. Since both the mean and variance increase as the intensity $n(t)$ increases at any t, we consider the normalized shot noise process

$$\hat{x}(t) = \frac{x(t) - \chi_1}{\sqrt{\chi_2}} = \frac{x(t)}{\sqrt{\chi_2}} - \frac{\chi_1}{\sqrt{\chi_2}} \tag{4.7.14}$$

Since $\hat{x}(t)$ is linearly related to $x(t)$, its characteristic function is

$$\Psi_{\hat{x}}(\omega, t) = \Psi_x\left(\frac{\omega}{\sqrt{\lambda_2}}, t\right) \exp\left(\frac{-j\omega\chi_1}{\sqrt{\chi_2}}\right)$$

$$= \exp\left\{\int_{-\infty}^t \left[\exp\left(\frac{j\omega h(t - u)}{\sqrt{\chi_2}}\right) - 1\right] n(u) \, du - j\frac{\chi_1}{\sqrt{\chi_2}}\omega\right\}$$

$$= \exp\left\{-\frac{\omega^2}{2} + \sum_{q=1}^\infty \frac{(j\omega)^q}{q!} \frac{\chi_q}{\chi_2^{q/2}}\right\} \tag{4.7.15}$$

The semiinvariants of this normalized shot noise are then seen to be

$$\hat{\lambda}_1 = 0$$
$$\hat{\lambda}_2 = 1$$
$$\vdots \tag{4.7.16}$$
$$\hat{\lambda}_q = \frac{\int_{-\infty}^t h^q(t - u)n(u) \, du}{[\int_{-\infty}^t h^2(t - u)n(u) \, du]^{q/2}}, \qquad q > 2$$

Thus the normalized shot noise process $\hat{x}(t)$ has zero mean and unit variance. If the functions $h(t)$ are time limited to $(0, \tau_h)$ and τ_h is much smaller than the time variations in $n(t)$, the last term above is approximately

$$\hat{\lambda}_q \approx \frac{n(t) \delta_q}{[n(t) \delta_2]^{q/2}}$$

$$= \frac{\delta_q}{\delta_2^{q/2}[n(t)]^{(q/2)-1}}, \qquad q > 2 \tag{4.7.17}$$

where

$$\delta_q = \int_{-\infty}^\infty h^q(t) \, dt \tag{4.7.18}$$

If δ_q is finite for all q, then $\lambda_q \to 0$, $q > 2$, for all t for which $n(t) \to \infty$. Hence, only the first two semiinvariants are nonzero, and $x(t)$ approaches a Gaussian density with zero mean and unit variance.

By inverting (4.7.4) the original shot noise $x(t)$, which is in turn linearly related to $\hat{x}(t)$, must also be a Gaussian random process, with mean $\chi_1 = en(t)$ and variance $\chi_2 = \delta_2 n(t)$. This establishes

$$p(x, t) \cong \frac{1}{[2\pi \, \delta_2 n(t)]^{1/2}} \exp\left[-\frac{(x - en(t))^2}{2\delta_2 n(t)} \right] \qquad (4.7.19)$$

for $n(t) \gg 1$. Thus the first-order density of PSN approaches a Gaussian density with mean $en(t)$ and variance $\delta_2 n(t)$, provided δ_q is finite for all q. We conclude that at any instant of time at which the instantaneous intensity is significantly "large," the instantaneous probability density of the shot noise appears approximately Gaussian with density given by (4.7.19).

For CPSN processes, the characteristic function in (4.7.4) becomes a conditional characteristic function, conditioned upon a sample function $n(t)$ of the random intensity process. The subsequent manipulations, including the calculation of probability density and semiinvariants, are valid for a given sample function. In order to determine the total statistics of the shot noise, however, subsequent averaging over the statistics of $n(t)$ must be used. Thus the characteristic functions become

$$\Psi_x(\omega, t) = E_n[\Psi_x(\omega, t) | n(t)]$$

$$= E_n\left[\exp \int_{-\infty}^{t} [\exp(j\omega h(t - \rho)) - 1]n(\rho)\, d\rho \right]$$

$$= E_n\left[\sum_{q=0}^{\infty} \frac{1}{q!} \left[\int_{-\infty}^{t} [\exp(j\omega h(t - \rho)) - 1]n(\rho)\, d\rho \right]^q \right] \qquad (4.7.20)$$

When the multiplicity of the integral is written as a multiple integral, and the subsequent average is taken, it is clear that the summation involves integration of joint moments $E_n[n(u_1)n(u_2) \cdots n(u_q)]$ of the process $n(t)$ for all q. Thus the first-order statistics of CPSN depend on the higher order joint moments of the intensity process. The moments of $x(t)$ can be obtained by averaging (4.7.12):

$$E[x(t)] = E_n[\chi_1]$$
$$E[x^2(t)] = E_n[\chi_2] + E_n[\chi_1^2]$$
$$E[x^3(t)] = E_n[\chi_3] + 3E_n[\chi_2\,\chi_1] + E_n[\chi_1^3]$$
$$\vdots \qquad\qquad (4.7.21)$$

We again note that the averaging in the above involves joint moments of the process $n(t)$. In particular, we see

$$E[x(t)] = \int_{-\infty}^{\infty} h(t - \rho)E_n[n(\rho)] \, d\rho$$

$$E[x^2(t)] = \int_{-\infty}^{\infty} h^2(t - \rho)E_n[n(\rho)] \, d\rho \qquad (4.7.22)$$

$$+ \int\int_{-\infty}^{\infty} h(t - \rho)h(t - \xi)E_n[n(\rho)n(\xi)] \, d\rho \, d\xi$$

Note that the mean square value of the shot noise depends on the correlation of the intensity process.

4.8 SECOND-ORDER STATISTICS OF PSN

The procedures for determining the second-order, or even higher order, statistics of the shot noise process can be developed as straightforward extensions of the first-order case. For PSN the second-order characteristic function is

$$\Psi_x(\omega_1, \omega_2; t_1, t_2) = E[\exp[j\omega_1 x(t_1) + j\omega_2 x(t_2)]]$$

$$= E[\exp[j\omega_1 \sum_{q=1}^{k_1} h(t_1 - z_q) + j\omega_2 \sum_{j=1}^{k_2} h(t_2 - z_j)]] \qquad (4.8.1)$$

where $k_1 = k(-\infty, t_1)$ and $k_2 = k(-\infty, t_2)$. The random location times $\{z_q\}$ in the summations above can be categorized into those common to $x(t_1)$ and $x(t_2)$, and those occurring in the interval (t_1, t_2) or (t_2, t_1), whichever is positive. For the location points in the latter intervals, however, either $h(t_1 - z_q)$ or $h(t_2 - z_q)$ is zero. This means the exponent in (4.8.1) can be written under a single summation

$$\Psi_x(\omega_1, \omega_2; t_1, t_2) = E\left\{\exp\left[\sum_{q=1}^{k} j\omega_1 h(t_1 - z_q) + j\omega_2 h(t_2 - z_q)\right]\right\} \qquad (4.8.2)$$

where $k = k_1$ if $z_1 \geq z_2$, or $k = k_2$ otherwise. The resulting averaging over the $\{z_q\}$ and k yields

$$\Psi_x(\omega_1, \omega_2; t_1, t_2) = \exp\{[\psi_h(t_1, t_2) - 1](m_{t_{max}})\} \qquad (4.8.3)$$

where

$$\Psi_h(t_1, t_2) = \int_{-\infty}^{t_{max}} \exp[j\omega_1 h(t_1 - \rho) + j\omega_2 h(t_2 - \rho)]n(\rho) \, d\rho$$

and t_{max} is the maximum of t_1 and t_2. The second-order probability joint density at time $t = t_1$ and $t = t_2$ follows as the two-dimensional transform:

$$p(x_1, x_2; t_1, t_2) = \frac{1}{(2\pi)^2} \int\int_{-\infty}^{\infty} \exp(-j\omega_1 x_1 - j\omega_2 x_2)$$

$$\cdot \Psi_x(\omega_1, \omega_2; t_1, t_2)\, d\omega_1\, d\omega_2 \qquad (4.8.4)$$

The two-dimensional moments of the shot noise, at times t_1 and t_2, can be obtained by differentiation of (4.8.3). Thus

$$E[x^{q_1}(t_1)x^{q_2}(t_2)] = \frac{1}{j^{q_1+q_2}} \left\{ \frac{\partial^{q_1}}{\partial\omega_1^{q_1}} \frac{\partial^{q_2}}{\partial\omega_2^{q_2}} [\Psi_x(\omega_1, \omega_2; t_1, t_2)] \right\}\Bigg|_{\omega_1=\omega_2=0}$$

4.9 RELATION OF CPSN AND INTENSITY STATISTICS

In Section 4.7 the relation between the PSN process and its inherent intensity function is discussed. Basically it is shown that if the electron pulse rate is sufficiently high, the PSN process accurately resembles the intensity function. When dealing with CPSN, a similar interpretation can be made for a specific sample function of the random intensity process. However, the relation between the statistics of the shot noise and the intensity process is not at all clear. If a photodetector detects the incident radiation, then the statistics at the detector output should, in some way, represent those of the radiation. In this section we examine the previously derived moment relations, and attempt to make this representation more definitive. To accomplish this, we inject some degree of rigor by initially placing some basic constraints on the intensity process $n(t)$.

Let $n(t)$ be a sample function from a random process that we assume is continuous, bounded, and nonnegative for all t. If the electron function width τ_h is less than the time variations in $n(t)$, we can write the conditional semiinvariants in (4.7.12) as

$$\chi_q | n(t) = n(\hat{t}) \int_{t-\tau_h}^{t} h^q(t - x)\, dx, \qquad t - \tau_h \leq \hat{t} \leq t$$

$$= n(t)\, \delta_q \qquad (4.9.1)$$

The first-order moments of the CPSN process $x(t)$ can now be written directly from (4.7.21) as

$$E[x(t)] = \delta_1 E_n[n(t)]$$

$$E[x^2(t)] = \delta_1^2 E_n[n^2(t)] + \delta_2 E_n[n(t)] \qquad (4.9.2)$$

$$E[x^3(t)] = \delta_1^3 E_n[n^3(t)] + 3\delta_1\delta_2 E_n[n^2(t)] + \delta_3 E_n[n(t)]$$

where $E_n[n^j(t)]$ is the jth moment of the intensity process at time t. It is evident that the jth moment of $x(t)$ can be cast in the form

$$E[x^j(t)] = \delta_1{}^j E_n[n^j(t)] + F_j \qquad (4.9.3)$$

where F_j represents a summation of terms involving lower moments of $n(t)$. Each term of the sum takes the general form

$$\frac{1}{q!} \sum_{I(j)} \frac{j!}{i_1! i_2! \cdots i_q!} \delta_{i_1} \cdots \delta_{i_q} E_n[n^q(t)] \qquad (4.9.4)$$

and $I(j)$ is the set of q positive integers $\{i_q\}$ that sum to j. Equation 4.9.3 represents a general expression relating the moments of the CPSN to the moments of the process $n(t)$ and is basically a rewritten version of (4.7.21). The short time assumption on $h(t)$ has allowed us the freedom to write these moments directly in terms of the moments of $n(t)$ rather than in terms of integrals as in (4.7.12). Note that the electron functions $h(t)$ enter the equation through the $\{\delta_q\}$ terms. When written as in (4.9.3), F_j represents the difference between the moments of the output CPSN process $x(t)$ and the moments of the normalized intensity $\delta_1 n(t) = en(t)$. If the CPSN is to represent this intensity current statistically in the jth moment, then F_j should be small compared to that moment. We may first inquire then if there is one function $h(t), 0 \le t \le \tau_h$, that will minimize δ_q, for a fixed $\delta_1 = e$. By straightforward application of calculus of variations using Lagrange multipliers [4], we obtain the solution $h(t) = e/\tau_h, 0 \le t \le \tau_h$, that is, a rectangular function over $(0, \tau_h)$. The rectangular function also simplifies (4.9.4), allowing further insight. Let $d = e/\tau_h$, the function height. Then $\delta_q = d^q \tau_h$, and terms in (4.9.4) now take the form $E_n[n^q(t)]\tau_h{}^q C(j,q) d^j$, where

$$C(j,q) = \frac{1}{q!} \sum_{I(j)} \left(\frac{j!}{i_1! \cdots i_q!} \right) \qquad (4.9.5)$$

Equation 4.9.3 is then

$$E[x^j(t)] = E_n\{[en(t)]^j\} + \sum_{q=1}^{j-1} C(j, q) E_n[n^q(t)] d^j \tau_h{}^q$$

$$= E_n\{[en(t)]^j\} \left[1 + \sum_{q=1}^{j-1} \left\{ \frac{C(j,q) E_n[n^q(t)]}{\tau_h{}^{j-q} E_n[n^j(t)]} \right\} \right]$$

$$= E_n\{[en(t)]^j\} [1 + \hat{D}(j)] \qquad (4.9.6)$$

where $\hat{D}(j)$ denotes the right-hand summation term. Equation 4.9.6 allows us to conclude that the jth moment of the CPSN, with short rectangular electron functions, will be approximately equal to the jth moment of the intensity current process $en(t)$ only if $\hat{D}(j) \ll 1$. The term $\hat{D}(j)$ can be evaluated knowing just the first j moments of $n(t)$. Since the intensity process $n(t)$ is nonnegative, we can now make use of well-known properties of absolute moments [5] to establish:

$$E_n[n^j] \geq [E_n(n^q)]^{j/q} = E_n[n^q][E_n\{n^q\}]^{j-q/q}$$
$$\geq E_n[n^q][E_n[n]]^{j-q} \qquad (4.9.7)$$

for all t. Substitution into (4.9.6) then implies

$$\hat{D}(j) \leq \sum_{q=1}^{j-1} \frac{C(j,q)}{\tau_h^{j-q} E_n[n(t)]^{j-q}} \qquad (4.9.8)$$

and an upper bound on $\hat{D}(j)$ is seen to be inversely related to the average value of $n(t)$. For $n(t)\tau_h$ large, the leading term in (4.9.8) can be used, yielding

$$\hat{D}(j) \leq \frac{j(j-1)}{2E_n[n(t)]\tau_h} \qquad (4.9.9)$$

where we have also used the fact that $C(j, j-1) = j(j-1)/2$. This result implies that the jth moment of $x(t)$ is approximately equal to the jth moment of $en(t)$ whenever

$$\tau_h E_n[n(t)] \gg \frac{j(j-1)}{2} \qquad (4.9.10)$$

Note that $\tau_h E_n[n(t)]$ is the average number of electron occurrences in τ_h sec. Thus (4.9.10) essentially states that the "denseness" of the electron occurrences (i.e., the average number of electron functions occurring in the time interval of one function) must be sufficiently large for moment representation. The right side of (4.9.10) serves as a rough rule of thumb for determining how large this "denseness" must be for approximate equality of the jth moment. It may be recalled from Section 4.3 that for PSN with deterministic intensities, a condition of a large number of occurrences is required before the PSN process loses its "discrete" nature. Equation 4.9.10 can therefore be interpreted as the statistical equivalent of this statement, that is, the condition under which the CPSN begins to take on the statistics of its intensity.

Similar results can be obtained with general functions $h(t)$ (i.e., not necessarily rectangular). It can be shown [6] that it is only necessary that δ_q be bounded for a similar type of convergence. However, larger values of $\tau_h E_n[n(t)]$ are required to guarantee the approximation.

4.10 PHOTODETECTOR MODELS

Throughout the development of the spectral characteristics of the shot noise processes in this chapter, we assume that the release of an electron is characterized by the component function $h(t)$, the latter satisfying the fundamental condition

$$\int_{-\infty}^{\infty} h(t) \, dt = e \qquad (4.10.1)$$

However, in practical photodetectors (and even at the same instant in time) two released electrons may yield vastly differing forms for $h(t)$, while still subject to the constraint of (4.10.1). To allow for these events in our analysis, we can generalize the form of $h(t)$ by embedding within it a random variable, or a set of random variables, ζ. We therefore write the component function as $h(t, \zeta)$, where each ζ effectively specifies a different component function. The Fourier transform, for a particular ζ, is then $H(\omega, \zeta)$. Using (4.10.1) we can now repeat the calculations in (4.5.4) to (4.5.5), with an additional averaging over ζ. The modified version of (4.5.10) is then

$$S_x(\omega) = E_\zeta[|H(\omega, \zeta)|^2]\overline{E_n(n)} + E_{\zeta, \zeta'}[H(\omega, \zeta)H^*(\omega, \zeta')]S_n(\omega) \qquad (4.10.2)$$

Here the first average is over ζ, while the second is over the joint statistics of ζ, ζ' from two different component functions. It is interesting to note that the signal portion (second term) is modified in a somewhat different manner than the shot noise term, when added randomness is present in the component functions. Depending on the specific device model, this difference may be significant. Let us examine some important photodetector models of this class.

Random Gain Photomultipliers. We point out in Section 1.8 that an ideal photomultiplier with gain G effectively multiplies $h(t)$ by G in the shot noise model. If the gain of the multiplier is itself a random variable, independent from electron to electron, the component functions are multiplied by independent random variables, producing a random amplitude from one $h(t)$ to the next. Hence, we let $\zeta = G$ and write $h(t, \zeta) = Gh(t)$. Equation 4.10.2 then becomes

$$S_x(\omega) = |H(\omega)|^2\{E[G^2]\overline{E_n(n)} + [E(G)]^2 S_n(\omega)\} \qquad (4.10.3)$$

Thus the shot noise is modified by the mean squared gain value, while the signal is modified by the square of the mean gain. Thus the difference between the two components now becomes a question of the magnitude of the gain "spread" (variance) in the random variable G. This is the usual method of modeling a photomultiplier and is used in later chapters.

Delay Dispersion. In practical photodetectors, the time between the release of an electron (the shot noise occurrence times) and the generation of current effect at the output can take on a random character. This can be accounted for by adding an additional random delay to the release time of the electron. This additional delay can in fact be multiple in cause, and actually appear as a sum of random delays. Hence, to model this electron delay dispersion we interpret $\zeta = \sum_i^M \zeta_i$ and write the component function as

$$h(t, \zeta) = h\left(t - \sum_{i=1}^{M} \zeta_i \right) \tag{4.10.4}$$

where M is generally taken as the number of dynode surfaces. The Fourier transform of (4.10.4) is then

$$\dot{H}(\omega) \prod_{i=1}^{M} e^{-j\omega\zeta_i} \tag{4.10.5}$$

If the delays $\{\zeta_i\}$ are pairwise independent, then (4.10.2) becomes

$$S_x(\omega) = |H(\omega)|^2 \left\{ \overline{E_n(n)} + \prod_{i=1}^{M} Q_i(\omega)S_n(\omega) \right\} \tag{4.10.6}$$

with

$$Q_i(\omega) = E_{\zeta_i, \zeta_i'}\{\exp[-j\omega(\zeta_i - \zeta_i')]\} \tag{4.10.7}$$

Here the average is taken over the joint statistics between the delay associated with two different electrons. [$Q_i(\omega)$ is actually the joint characteristic function of the ith delay.] Note that the shot noise (first term) is not affected by the additional delays. If the delays are also independent from electron to electron, (4.10.7) simplifies to

$$Q_i(\omega) = |E_{\zeta_i}[e^{j\omega\zeta_i}]|^2$$

$$= \left| \int_{-\infty}^{\infty} p_{\zeta_i}(\zeta)e^{-j\omega\zeta}\,d\zeta \right|^2 \tag{4.10.8}$$

where $p_{\zeta_i}(\zeta)$ is the probability density of ζ_i. The spectrum becomes

$$S_x(\omega) = |H(\omega)|^2\{\overline{E_n(n)} + Q(\omega)S_n(\omega)\} \tag{4.10.9}$$

where $Q(\omega) = \prod_{i=1}^{M} Q_i(\omega)$. The spectrum appears as if the characteristic function of the individual delays is linearly filtering the signal portion of the shot noise density. This interpretation in shot noise models has been noted previously in radiation [7] and scattering [8] models. For minimal distortion, it is clear that $Q(\omega)$ must be wide compared to $S_n(\omega)$, or equivalently, the highly probable delays must be small compared to the time variations in the intensity modulation. This becomes an important aspect of device design. Note that since the filtering is predictable, for a specific delay model there exists the possibility of using subsequent electronic filtering to compensate for the delay effect. Such filters, referred to as *inverse* filters, afford a tradeoff of signal restoration and increased noise interference.

REFERENCES

[1] Rice, S. O., "Mathematical Analysis of Random Noise—Part I," in *Selected Papers on Noise and Stochastic Processes*, (N. Way, ed.), Dover Publications, New York, 1964, p. 133.

[2] Middleton, D., *Introduction to Statistical Communication*, McGraw-Hill Book Co., New York, 1960, pp. 498–506.

[3] Parzen, M., *Stochastic Processes*, Holden-Day, San Francisco, California, 1962, Chap. 4.

[4] Gelfand, I. and Fomin, S., *Calculus of Variations*, Prentice-Hall, Englewood Cliffs, New Jersey, 1963.

[5] Cramer, H., *Mathematical Methods of Statistics*, Princeton University Press, Princeton, New Jersey, 1945, p. 176.

[6] Karp, S. and Gagliardi, R. M., "On the Representation of a Continuous Stochastic Intensity by Poisson Shot Noise," *IEEE Trans. Inf. Theory*, March 1970.

[7] Karp, S., Gagliardi, R. M., and Reed, I. S., "Radiation Models Using Discrete Radiation Ensembles," *Proc. IEEE*, **56**, No. 10, 1704–1711 (October 1968).

[8] Middleton, D., "A Statistical Theory of Reverberation and First Order Scattered Fields, Part I and Part II," *IEEE Trans. Inf. Theory*, **IT-13**, 372–414 (July 1967).

PROBLEMS

1. Two people observe a nonnegative random process over $(0, T)$. Each takes the observed sample function (the same for each) to a separate location and uses the normalized function as a probability density for selecting a number in $(0, T)$. That is, they each use the same function to induce a probability over $(0, T)$ and select randomly according to this density.

(a) Are the selected values independent? Explain.

(b) Relate this example to the discussion of occurrence times in Section 4.2.

2. Let W_n be the time for n events to occur in a counting process $k(t)$. (W_n is called the *waiting time* to the nth event.) Show that if the events are Poisson distributed with rate parameter v, then

(a)
$$\text{Prob}(W_n > t) = \text{Prob}[k(t) < n] = \sum_{j=0}^{n-1} e^{-vt} \frac{(vt)^j}{j!}$$

(b)
$$E[W_n] = \frac{n}{v}$$

(c)
$$\text{Var}[W_n] = \frac{n}{v^2}$$

(d) The time T between events is exponentially distributed in T.

3. Two separate shot noise processes are generated from two counting processes whose intensities have correlation $R_{n_1 n_2}(t_1, t_2)$. Assume each uses identical component functions $h(t)$. Determine the correlation of the two shot noise processes.

4. Consider a shot noise $x(t)$ with random intensity

$$n(t) = \sum_{s=1}^{R} c_s \delta(t - t_s)$$

with the times t_s distributed as $p(t_s)$, $E_{c_s}[c_s] = C_1 \geq 0$, $Var[c_s] = C_2$, and $\lim_{2T \to \infty} E_R[R/2T] = S$. If $E_n[n]/S = G$, compute the resulting power spectrum $S_x(\omega)$ and compare it to (4.10.3).

5. Consider the radiation ensemble [7]

$$r(t) = \sum_{i=1}^{k(0,t)} a_i h(t - t_i - x_i) \exp[j\omega_0 t + \lambda_i]$$

where $(a_i, x_i, \lambda_i, t_i, k)$ are random variables with k Poisson distributed.
(a) Compute the power spectral density and discuss the results.
(b) If $p_x(x) = \frac{1}{2}\delta(x) + \frac{1}{2}\delta(x - T)$, compute $\Psi_x(\omega)$. What are the conditions for a zero to occur?

6. Consider the shot noise process $x(t)$

$$x(t) = \sum_{i=0}^{k} h(t - t_i), \qquad 0 \leq t \leq 2T$$

where $\quad h(t) = pe^{-pt}, \quad (0, \infty)$
$\qquad\qquad\quad = 0 \qquad\quad$ elsewhere

(a) Compute the first and second moments of $x(t)$. Solve for $n(t) = n_0$.
(b) In the limit at $T \to \infty$, compute the power spectrum and the correlation function.
(c) Compute $E_x^2[x]/Var[x]$.

7. Consider a photodetector illuminated by a constant light source starting at $t = 0$. Ignore background and circuit noises. Assume that the detector response has a single pole at $p = 2\pi B$. Further assume that we can model the shot noise as Gaussian.
(a) Calculate the probability density as a function of t.
(b) What is the probability that the current output will be above some arbitrary threshold m? Below it?
(c) Suppose we look in N independent intervals of length T. Assume that the light source is on only in the ith interval. What is the probability of seeing the light source in the correct interval as a function of the threshold value m?

In the wrong interval? Not at all? Discuss the effect of the threshold m. (Assume that in the intervals with no light we have a Gaussian process with zero mean but the same variance.)

8. Repeat Problem 4.7 only assume that the detector response for each released electron is

$$h(t) = \frac{e}{\tau_h}, \qquad 0 \le t \le \tau_h$$

$$= 0, \qquad \text{elsewhere}$$

with Poisson distributed dark current.

9. Using the second characteristic function, compute the output spectrum and correlation of a square law device when the input is a Poisson shot noise process.

10. Consider a shot noise process with

$$h(t) = \text{Real}\{R(t)e^{j(\omega t + \phi)}\}$$

with ϕ uniformly distributed in the interval $(0, \pi)$. Show that the characteristic function of the envelope of the shot noise is

$$\Psi_R(\lambda) = \int_0^\infty \lambda J_0(\lambda R) f(\lambda, T) \, d\lambda$$

where

$$f(\lambda, T) = \exp\left[\int_0^T n(x)[J_0[\lambda h(t - x)] - 1] \, dx\right]$$

11. Consider the process $y(t) = \exp[j\omega t + jx(t)]$, where $x(t)$ is a shot noise process as in (4.1.1). (a) Using (4.8.1), determine the correlation function of $y(t)$, $R_y(t_1, t_2) = E[y(t_1)y(t_2)]$. (b) Expand the exponent and evaluate the first three terms when $h(t) = 1$ over $(0, T)$ and zero elsewhere.

12. Compute the two leading terms in the series in (4.9.8), and determine the conditions under which (4.9.9) is the dominant term.

5 Noncoherent (Direct) Detection

In the preceding chapters we concentrated on developing a statistical model for an optical receiver. We are now interested in applying this model to the analysis and synthesis of optical communication systems. We begin in this chapter by devoting attention to a communication system employing noncoherent or direct detection. Noncoherent detection occurs when no use is made of the spatial coherence of the optical field, and the detector responds only to the power in the received field. We consider as a prime objective the accurate transmission of a desired data waveform from the transmitter to the receiver output. The emphasis on accurate waveform transmission, and the corresponding analysis and design techniques, is referred to as *analog* communications. In later chapters, we examine *digital* systems, in which accurate detection of symbols, rather than accurate waveform transmission, is of prime importance. Analog communication design principally involves the preservation of waveforms during system transmission. As we have seen, optical detectors inherently introduce a degree of randomness into the transmission operation, and the optical communicator is always faced with an intrinsic distortion or quantization effect. In addition, additive noise, because of both inadvertent reception of background noise and internal generation of circuit noise, accompanies the transmission of signal waveforms. In this chapter we attempt to assess the degree of this distortion, and to examine design techniques when noncoherent reception is used in an analog communication system.

5.1 WAVEFORM DISTORTION AND SIGNAL-TO-NOISE RATIOS

We have stated that in the design of an analog communication system we are primarily interested in preserving the waveshape of the modulating baseband signal after optical transmission. It is standard practice in analog communication analysis to measure the degree by which the information has been successfully communicated by determining the output signal-to-noise ratio (SNR). Historically, there are various definitions of signal-to-noise ratios that are used and, unfortunately, often mistakenly interchanged, or used out of context. When the output waveform is composed of a deterministic signal, $s(t)$, plus zero mean random noise, the definitions of SNR are fairly evident. We can denote the *instantaneous SNR* at time t as

$$\text{SNR}_t = \frac{s^2(t)}{\left[\begin{array}{l}\text{mean square noise} \\ \text{value at time } t\end{array}\right]} \qquad (5.1.1)$$

That is, SNR_t is the ratio of the square of signal value to the mean square noise value, at time t. Similarly, we can denote the *power* SNR as

$$\text{SNR}_p = \frac{\text{power in } s(t)}{\text{noise power}} \qquad (5.1.2)$$

The noise power above can be either a time or statistical averaged noise power. If the latter is used, then the denominators in (5.1.1) and (5.1.2) become the same for stationary noise processes. If a time averaged noise power is used, the denominators are the same only for ergodic noise processes. For most practical applications the ergodic condition is generally assumed, and the denominators in (5.1.1) and (5.1.2) are often used interchangeably. When the signal $s(t)$ is a deterministic waveform, the definition in (5.1.1) is generally used, while (5.1.2) is preferred for stochastic signals. Note that for random signals the numerator in SNR_t becomes a random variable, and SNR_t loses its significance.

The instantaneous SNR_t may be related to a notion of output waveform distortion, when the output waveform is composed of a deterministic signal $s(t)$ with additive zero mean noise. In this latter case, the combined waveform has, at any t, a mean value $s(t)$ with variance given by the mean square noise value. Thus the parameter SNR_t is also the ratio of the square of the signal mean to the variance of the combined process at t. Since the square root of the variance is an indication of the "spread" of the noise about the mean, the term $1/(\text{SNR}_t)^{1/2}$ is basically a "percentage" spread of the signal at time t about the desired value, because of the additive noise. This then can serve as a measure of the distortion of the desired signal waveform $s(t)$ due to the noise. That is, large output SNR_t implies low signal distortion at that t. Conversely, SNR_t can be interpreted as a reciprocal distortion factor for the same reason.

When one is confronted with shot noise processes, as is the case following photodetection of optical fields, the relation of SNR parameters and distortion effects is no longer obvious. In particular, the process is no longer the sum of a desired signal plus an additive noise. Instead, the desired signal is immersed within the shot noise process. Even if no additive noise were included, distortion of the desired signal would still occur. That is, even though the desired signal corresponds to the mean of the shot noise, the process still has a variance that must be considered. Furthermore, when additive noise is present the variance involves not only the noise power but includes terms generated from signal and noise cross products, due to the operation of the photodetector. We can, however, continue to compute a SNR_t for shot noise processes by redefining

$$\text{SNR}_t = \frac{[\text{shot noise mean signal value at time } t]^2}{\text{shot noise variance at time } t} \tag{5.1.3}$$

The key point is that (5.1.3) is no longer a ratio of signal terms to noise terms (since the signal affects the variance) but still retains its meaning as a reciprocal distortion parameter.

When dealing with shot noise processes with random intensities, it is shown in Section 4.5 that the time averaged spectral density decomposes into a signal portion plus a noise portion. Thus a power SNR can be computed using (5.1.2) by determining the signal power and total noise power separately. We therefore conclude that when dealing with shot noise processes, an output SNR can be defined using (5.1.2) or (5.1.3) and each is a well-defined parameter that can be used to assess waveform distortion. It should be emphasized that these SNR can be computed using only second moments of the process, and complete shot noise statistics are not needed.

5.2 THE NONCOHERENT COMMUNICATION SYSTEM MODEL

The typical model of a noncoherent (direct detecting) communication system is shown in Figure 5.1. The desired information is intensity modulated onto an optical source and transmitted to the receiver. The optical receiver intercepts the field with the receiver area. After preprocessing in the receiver front end (frequency and spatial filtering, as described in Chapter 1), the collected field is focused through the receiver lens system onto the photodetecting surface in the focal plane. Background radiation is collected also by the antenna and processed along with the transmitted field. The photodetector, being basically a power detecting device, responds to the instantaneous field intensity collected over the receiver area. Its output appears as a shot noise process whose count intensity is proportional to the collected power. This output process represents the demodulated optical signal. For the receiver to recover the desired signal, it is necessary that the transmitted information be associated with the intensity variation of the transmitted field. The information waveform may be directly modulated onto the intensity of the transmitted field, or it may be subcarrier modulated prior

144

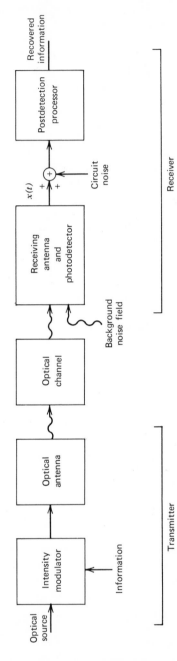

Figure 5.1. Intensity modulated, direct detection optical system model.

to optical intensity modulation. In the latter case, further receiver processing is necessary, following optical detection, to recover the information waveform.

To analyze the receiver operation it is necessary to describe the received field intensity. As discussed in Section 1.6 this description can be equivalently done either at the photodetector surface in the focal plane or at the receiver antenna surface in the aperture plane. It is common practice in the literature to refer analysis to the receiver surface. Here the received radiation field can be written in terms of wave functions as

$$f(t, \mathbf{r}) = \text{Real}\{a(t, \mathbf{r})e^{j\omega_1 t}\}, \qquad \mathbf{r} \in \mathscr{A}_r \qquad (5.2.1)$$

where $a(t, \mathbf{r})$ is the complex field envelope, $\mathscr{A}_r$ is the receiver surface, and ω_1 is the transmitter optical frequency in radians per second. For additive background noise, we can further expand

$$a(t, \mathbf{r}) = s(t, \mathbf{r}) + b(t, \mathbf{r}) \qquad (5.2.2)$$

where $s(t, \mathbf{r})$, $b(t, \mathbf{r})$ are the complex envelopes due to the transmitter signal and background noise, respectively. The dependence on the spatial coordinate $\mathbf{r}$ indicates the variation of the optical field over the detector surface area. Propagation losses, antenna effects, and coupling attenuation determine the received power level of $s(t, \mathbf{r})$, as described in Section 1.3. The background noise process is assumed to be a zero mean, blackbody radiation signal, and as such is described by its modal properties. If, in addition, the fields are assumed coherence-separable, the field envelopes in (5.2.2) can be mathematically expanded into orthonormal spatial functions, $\{w_j(\mathbf{r})\}$, defined over the receiver area $\mathscr{A}_r$. Thus, the received field can always be written as

$$f(t, \mathbf{r}) = \sum_{j=0}^{\infty} f_j(t) w_j(\mathbf{r}) \qquad (5.2.3)$$

where $f_j(t)$ describes the temporal variation in the spatial mode defined by the function $w_j(\mathbf{r})$.

For an ideal point source transmitter and a nonturbulent propagation path, the transmitter field appears as a plane wave normal to the receiver, spatially coherent over the receiver surface area. The expansion in (5.2.3) reduces to a single term and the receiver operates as a point detector. Thus only the temporal effects of the received radiation need be considered in this instance, and the $\mathbf{r}$ dependence can be dropped in our field description. The noise interference involves only the background effects of a single spatial mode. However, when transmission turbulence is encountered, the coherence function over the receiver surface is degraded, and more spatial modes are produced, forcing one to consider the source as an extended source rather than as a point source. This means that more terms in (5.2.3) must be used for field description. Both of these models are investigated. Subsequently, the detected field produces the shot noise processes described in Chapter 4 at the photodetector output. This shot noise process is governed by the received intensity arising from both signal and background noise collected in the optical front end.

The shot noise count process is controlled by the count intensity process $n(t)$, which is related to the received field intensity by (2.1.19),

$$n(t) = \alpha \int_{\mathscr{A}_r} |f(t, \mathbf{r})|^2 \, d\mathbf{r}$$

$$= \alpha A_r \sum_{j=0}^{\infty} |f_j(t)|^2 \qquad (5.2.4)$$

where A_r is the area of the receiver area $\mathscr{A}_r$. Thus all spatial modes contribute to the instantaneous detector intensity. In addition, circuit (thermal) noise may be added to the detector output prior to receiver postdetection processing. Signal-to-noise ratios associated with the receiver output aid in assessing system performance.

5.3 NONCOHERENT POINT DETECTION OF INTENSITY MODULATED FIELDS

Consider a direct detection receiver observing a single mode of the optical field so that a point detector receiver model can be used. The received field in (5.2.1) can be written as the scalar field

$$f(t) = \text{Real}\{[s(t) + b(t)]e^{j\omega_1 t}\} \qquad (5.3.1)$$

where $s(t)$ and $b(t)$ are the signal and noise processes. This description of the received field assumes the field is coherent over the receiver area, $\mathscr{A}_r$, and is equivalent to considering only the zero-order mode in (5.2.3) with $f_0(t) = \sqrt{A_r} f(t)$ and letting $w_0(\mathbf{r}) = 1/\sqrt{A_r}$, $\mathbf{r} \in \mathscr{A}_r$. The signal complex envelope $s(t)$ is assumed to be that of an intensity modulated optical field and therefore has the form

$$|s(t)|^2 = I_s[1 + m(t)] \qquad (5.3.2)$$

where I_s is the average field intensity and $m(t)$ is the intensity modulating waveform. We assume that the baseband $m(t)$ has a power spectrum $S_m(\omega)$, which occupies a one-sided bandwidth of B_m Hz, and a total power

$$P_m = \frac{1}{2\pi} \int_{-2\pi B_m}^{2\pi B_m} S_m(\omega) \, d\omega \qquad (5.3.3)$$

We further assume $m(t)$ is normalized so that $|m(t)| \leq 1$ to prevent over-modulation, which immediately implies $P_m \leq 1$. [If $m(t)$ is considered as a random process, then P_m must be small enough to ensure that $|m(t)| \leq 1$ with a high probability. For example, if $m(t)$ is a Gaussian process, P_m must be less than 0.14 to guarantee $|m(t)| \leq 1$ with a probability of 0.99.]

When the optical field in (5.3.1) impinges on the receiver, the shot noise process at the detector output is

$$x(t) = \sum_{j=1}^{k(0,\,t)} h(t - t_j) \tag{5.3.4}$$

where the component functions $h(t)$ define the receiver bandwidth. The function $k(0, t)$ is the stochastic count process over $(0, t)$. This count process has the intensity in (5.2.4) which, for the point detector model, simplifies to

$$n(t) = \alpha A_r |s(t) + b(t)|^2 \tag{5.3.5}$$

The detected shot noise process has the time averaged power spectral density derived in (4.5.10):

$$S_x(\omega) = |H(\omega)|^2 [\overline{E(n)} + S_n(\omega)] \tag{5.3.6}$$

where $H(\omega)$ is the transform of $h(t)$, $S_n(\omega)$ is the intensity spectrum of $n(t)$, and

$$\overline{E(n)} = \lim_{T \to \infty} \frac{1}{2T} \int_{-T}^{T} E[n(t)]\, dt \tag{5.3.7}$$

Since the signal and noise processes are statistically uncorrelated in (5.3.1), we can expand and average to obtain

$$E[n(t)] = \alpha A_r E[|s(t)|^2] + \alpha A_r E[|b(t)|^2] \tag{5.3.8}$$

Each term on the right corresponds to power collected over the receiver area at time t. For stationary background noise, the last term does not depend on t, and is given by the modal noise power of the background source. For the blackbody model in the optical frequency range, this noise has a flat power spectrum of level N_{ob} [given in (1.7.8)]. The noise power collected over the optical receiver bandwidth B_0 is then

$$\sigma_b^2 = \lambda^2 N(f) B_0 = N_{ob} B_0 \tag{5.3.9}$$

For the intensity modulated signal in (5.3.2), with $m(t)$ having a zero average value, $E|s(t)|^2 = I_s$. Therefore, (5.3.7) becomes

$$\overline{E[n]} = \alpha P_s + \alpha \sigma_b^2 \tag{5.3.10}$$

where $P_s = I_s A_r$ is the average signal power collected at the receiver. The power spectrum of $n(t)$ can be determined as the transform of the correlation function of $n(t)$. From (5.3.5) this becomes

$$R_n(\tau) = E[n(t)n(t + \tau)]$$
$$= (\alpha A_r)^2 E[(s(t) + b(t))(s^*(t) + b^*(t))$$
$$\cdot (s(t + \tau) + b(t + \tau))(s^*(t + \tau) + b^*(t + \tau))] \qquad (5.3.11)$$

Expanding the brackets and averaging yields

$$R_n(\tau) = \alpha^2[R_s(\tau) + R_b(\tau) + R_{sb}(\tau) + R_0] \qquad (5.3.12)$$

where

$$R_s(\tau) = A_r^2 E[|s(t)|^2 |s(t + \tau)|^2]$$
$$R_b(\tau) = A_r^2 E[|b(t)|^2 |b(t + \tau)|^2]$$
$$R_{sb}(\tau) = 2A_r^2 \, \text{Real}\{R_s(\tau)R_b^*(\tau)\}$$
$$R_0 = 2P_s \sigma_b^2$$

The first term is the correlation of the intensity of the signal envelope, the second is the correlation of intensity of the noise envelope, the third term is the cross correlation caused by the cross product between signal and noise, and the last term R_0 is a constant power level. The power spectral density of $n(t)$ is then the transform of (5.3.12). Hence,

$$S_n(\omega) = \alpha^2[S_s(\omega) + S_b(\omega) + S_{sb}(\omega) + R_0 2\pi \, \delta(\omega)] \qquad (5.3.13)$$

where each $S_i(\omega)$ is the Fourier transform of the corresponding correlation denoted by the subscript. For the intensity modulation of (5.3.2),

$$S_s(\omega) = P_s^2[2\pi \, \delta(\omega) + S_m(\omega)] \qquad (5.3.14)$$

and (5.3.6) is then

$$S_x(\omega) = |H(\omega)|^2 \{\alpha(P_s + \sigma_b^2) + \alpha^2 S_b(\omega) + \alpha^2 S_{sb}(\omega)$$
$$+ \alpha^2(R_0 + \alpha^2 P_s^2)2\pi \, \delta(\omega) + \alpha^2 P_s^2 S_m(\omega)\} \qquad (5.3.15)$$

Thus the detector output contains spectral components due to the shot noise (first term), the noise background (second term), the signal-noise cross products (third term), the zero frequency power (fourth term), and the desired signal modulation (last term). Note that the resulting spectrum contains specifically the spectral density of the desired intensity modulation $S_m(\omega)$ as a separate additive component. If amplitude modulation of the optical source had been used, the desired signal spectrum would be contained in the $S_{sb}(\omega)$ term in (5.3.13), which appears only as a convolution with the noise spectrum, and does not appear explicitly. This is why direct detection of an amplitude modulated (instead of intensity modulated) optical carrier cannot yield an undistorted signal component.

In considering the noise terms, we recall that in practical systems the bandwidth B_0 of the optical filter (and therefore the bandwidth of the background noise) is generally far greater than the bandwidth of the signal intensity $s(t)$. This means the correlation time of $b(t)$ is much shorter than that of $s(t)$, so that we can accurately approximate

$$R_{sb}(\tau) \approx R_s(0)R_b(\tau) \tag{5.3.16}$$

This implies that $S_{sb}(\omega)$, the Fourier transform of $R_{sb}(\tau)$, is approximately flat over the optical bandwidth, with spectral level $(\alpha P_s)^2 N_{0b}$. Similarly, the background intensity, $|b(t)|^2$, has a spectrum given by the convolution of the envelope process $|b(t)|$ with itself. This produces a resulting spectrum that is spread over twice the optical bandwidth with peak value $(\alpha N_{0b})^2 B_0$. Over typical signal bandwidths, much less than the optical bandwidths, the envelope squared noise process therefore appears also to have flat noise spectrum of level equal to its peak value. Thus the shot noise spectrum at the photodetector output is given, to a good approximation, as

$$S_x(\omega) = |H(\omega)|^2 \{ \alpha P_s(1 + \alpha N_{0b}) + \alpha N_{0b} B_0 (1 + \alpha N_{0b})$$
$$+ \alpha^2 P_s^2 S_m(\omega) + \alpha^2(R_0 + P_s^2) 2\pi\, \delta(\omega) \} \tag{5.3.17}$$

This power spectrum is shown in Figure 5.2, indicating the various detector output components. In addition, typical blackbody radiation in the optical region has a αN_{0b} value much less than unity. Hence, the shot noise spectral level can be further simplified to $\alpha(P_s + N_{0b} B_0)$. Remember that multiplying power values by the parameter α converts to effective count rate in the photodetector. Thus the parameters $\alpha N_{0b} B_0$ and αP_s, appearing in these

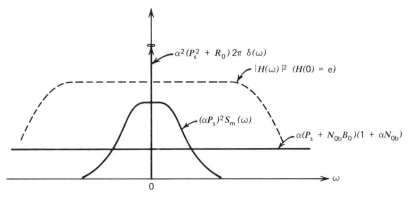

Figure 5.2.   Power spectrum of intensity modulated shot noise process.

spectral densities, also indicate the rate at which noise and signal electrons are being produced.

The photodetector current has added to it the receiver circuit noise $c(t)$, as shown in Figure 5.1. This circuit current noise represents thermal noise having spectral level $N_{0c} = 2\kappa \mathcal{T}^\circ / R_L$, where $\mathcal{T}^\circ$ is the receiver temperature in degrees Kelvin and R_L is the effective output impedance of the photodetector. The spectral density of the combined shot noise process and thermal noise at the receiver processor input is then

$$S_y(\omega) = S_x(\omega) + N_{0c} \tag{5.3.18}$$

with $S_x(\omega)$ given in (5.3.17). The output noise now consists of the shot noise level and the additive circuit noise level. The desired signal spectrum $S_m(\omega)$ appears explicitly as an additive term. The zero frequency term, indicated by the delta function in $S_x(\omega)$, can theoretically be filtered out from the desired signal.

If the postdetection processing consists of simple low pass filtering with transfer function $F(\omega)$, then the receiver output spectrum is given by $|F(\omega)|^2 S_y(\omega)$. The power in the output signal and noise components can be determined by integrating over the proper densities. If the output filter is rectangular, unit height over the bandwidth of the baseband modulation B_m, and is much less than the detector bandwidth specified by $H(\omega)$, the output signal power is then

$$P_{s0} = \frac{\alpha^2 P_s^2 e^2}{2\pi} \int_{-2\pi B_m}^{2\pi B_m} S_m(\omega)\, d\omega$$

$$= e^2 \alpha^2 P_s^2 P_m \tag{5.3.19}$$

The output noise power is

$$P_{n0} = [\alpha(P_s + \sigma_b^2)e^2 + N_{0c}]2B_m \tag{5.3.20}$$

where σ_b^2 is given in (5.3.9). The output power SNR is then

$$\mathrm{SNR}_p = \frac{P_{s0}}{P_{n0}} = \frac{\alpha^2 P_s^2 P_m}{[\alpha(P_s + \sigma_b^2) + (N_{0c}/e^2)]2B_m}$$

$$= \frac{[\alpha P_s P_m / 2B_m]}{1 + (\sigma_b^2/P_s) + (N_{0c}/\alpha P_s e^2)} \tag{5.3.21}$$

This equation represents the achievable SNR_p with direct intensity modulation onto an optical carrier, and use of direct detection at the receiver. If $e^2 \alpha P_s \gg N_{0c}$ (i.e., if the average count of the signal current greatly exceeds the thermal noise spectral level), the third term in the denominator can be dropped, and we say the receiver is *shot noise limited*. For this case,

$$\text{SNR}_p \approx \frac{\alpha P_s P_m/2B_m}{[1 + (\sigma_b^2/P_s)]}, \qquad \text{shot noise limited} \qquad (5.3.22)$$

If, in addition, the average received signal power P_s greatly exceeds the background noise power σ_b^2, and if the modulating baseband has its maximum available power so that $P_m = 1$, then the SNR_p is given by $\alpha P_s/2B_m$. This latter term is always an upper bound to the achievable SNR_p with noncoherent point detection. Substituting for α, we can rewrite this bound as

$$\frac{\alpha P_s}{2B_m} = \frac{(\eta/hf_1)P_s}{2B_m} = \frac{\eta P_s}{2hf_1 B_m} \qquad (5.3.23)$$

where f_1 is the optical frequency ($f_1 = \omega_1/2\pi$). This upper bound now has the appearance of a signal-to-noise ratio in which the signal power is given by ηP_s, while the noise appears as the accumulated power in a bandwidth B_m due to a two-sided noise level of hf_1. This latter term is often called the *quantum spectral level* and we say the SNR_p is *quantum limited*. Thus the quantum limited bound in direct detection systems, operating at optical frequencies f_1 with receiver power P_s and information bandwidth B_m, is given by

$$\text{SNR}_p = \frac{\eta P_s}{2hf_1 B_m}, \qquad \text{quantum limited} \qquad (5.3.24)$$

Note that SNR_p does not increase without bound as the background and circuit noise are weakened but rather attains the quantum limited SNR_p. This is because of the properties of the shot noise detection process, and it is this fact that distinguishes the optical communication system from its microwave counterpart. From the point of view of SNR_p, the optical system appears to have an additive quantum noise added to the standard background noise during detection. This quantum noise has a spectral level that is proportional to the optical frequency f_1, and places an ultimate limit on detector performance. Often an effective quantum temperature, $\mathscr{T}_q^\circ = hf_1/\kappa$, is associated with this quantum noise, and the latter is treated as an equivalent thermal noise in SNR analysis. When the signal power P_s is much less than the background noise power we have a *backgrounded limited condition*. In this case (5.3.22) becomes

$$\text{SNR}_p \cong \frac{\alpha P_s P_m/2B_m}{\sigma_b^2/P_s}$$

$$= \frac{\alpha P_s^2 P_m}{2\sigma_b^2 B_m}, \qquad \text{background noise limited} \qquad (5.3.25)$$

instead of (5.3.24).

Let us also determine the instantaneous SNR_t of the receiver output. If we consider the postdetection filter as a short-term integrator of length T sec (basically a low pass filter of bandwidth $1/T$), then the integrated shot noise component is

$$x(t) = \frac{e}{T} k(t - T, t) \tag{5.3.26}$$

where $k(t - T, t)$ is again the detector electron count over the interval $(t - T, t)$. The integrated circuit white noise has zero mean and variance N_{0c}/T. The mean value of the shot noise at any t is then

$$E[x(t)] = \frac{e}{T} [\alpha P_s T + \alpha \sigma_b^2 T]$$

$$= e(\alpha P_s + \alpha \sigma_b^2) \tag{5.3.27}$$

The variance of the receiver output process is given by the variance of the integrated circuit noise plus the variance of the shot noise; the latter is given by the variance of the count. Using (4.3.14), we therefore have

$$Var[x(t)] = \left(\frac{e}{T}\right)^2 Var[k(t - T, t)]$$

$$= \left(\frac{e}{T}\right)^2 [\alpha \sigma_b^2 T (1 + \alpha \sigma_b^2 T) + \alpha P_s T(1 + \alpha \sigma_b^2 T)] \tag{5.3.28}$$

$$+ \frac{N_{0c}}{T}$$

The instantaneous SNR_t of the receiver output, after integrating for T sec, is then

$$SNR_t = \frac{(\alpha P_s)^2}{(1/T)[(\alpha P_s + \alpha \sigma_b^2)(1 + \alpha \sigma_b^2 T)] + (N_{0c}/e^2 T)} \tag{5.3.29}$$

It is interesting to compare the expression for SNR_t above with that of SNR_p in (5.3.21). If we invoke the condition

$$\alpha \sigma_b^2 T \ll 1 \tag{5.3.30}$$

then (5.3.29) becomes

$$SNR_t = \frac{\alpha P_s T}{[1 + (\sigma_b^2/P_s) + (N_{0c}/\alpha P_s e^2)]} \tag{5.3.31}$$

This is identical to SNR_p with $P_s P_m$ and $1/2B_m$ replaced by P_s and T, respectively. Thus the weak background noise condition implied in the derivation of SNR_p takes on the form of (5.3.30) when viewed in terms of SNR_t. That is, the weak background noise assumption is valid for SNR_t if the detected background energy during the integration time T is much less than unity. This noise energy is specifically $\alpha \sigma_b^2 T = \alpha N_{0b} B_0 T = (\alpha N_{0b}) B_0 T$, where again αN_{0b} is the background noise average count. As pointed out earlier $\alpha N_{0b} \ll 1$, so the condition in (5.3.30) is typically true for moderate values of $B_0 T$. The parameter P_m does not appear in SNR_t, since the latter uses the square of the signal mean value in its numerator, while SNP_p uses the signal mean square value.

From (5.3.31) one can again define for SNR_t a shot noise limited condition $e^2 \alpha P_s \gg N_{0c}$ and a quantum limited condition $P_s \gg \sigma_b^2$. The corresponding SNR_t are then

$$SNR_t = \frac{\alpha P_s T}{1 + \sigma_b^2 / P_s} \tag{5.3.32}$$

and

$$SNR_t = \alpha P_s T = \frac{\eta P_s}{hf(1/T)} \tag{5.3.33}$$

The above are the counterparts to (5.3.22) and (5.3.24) and indicate system bounds when SNR_t is used to assess performance.

5.4 PHOTOMULTIPLICATION TO IMPROVE INCOHERENT DETECTION

We have seen in the preceding section that operation under a shot noise limited condition implies an improvement in SNR since the additive circuit noise can be neglected. In order to achieve this condition, however, the average signal power at the receiver must suitably exceed the circuit noise level. In typical operation transmitter power is often limited, and the shot noise limited condition may be difficult to achieve. Consequently, we may seek other means of obtaining shot noise limited performance. One procedure is to utilize several stages of high gain photomultiplication in the photodetection process.

The photomultiplier model is discussed in Section 4.10. Recall that the effect of photomultiplication is to alter the spectral density of the detected shot noise process from that in (5.3.6) to

$$S_x(\omega) = Q(\omega)|H(\omega)|^2[\overline{G^2}E[\overline{K}] + \overline{G}^2 S_n(\omega)] \tag{5.4.1}$$

where $\overline{G}$ is the mean photomultiplier gain, $\overline{G^2}$ is its mean square gain, and $Q(\omega)$ is an effective baseband filtering introduced by the transit time dispersion within the multiplier device. For an ideal device, $Q(\omega) = 1$ and the gain G is constant, so that $\bar{G} = G$ and $\overline{G^2} = G^2$. For a satisfactorily designed device, we only need $Q(\omega) \approx 1$ over the bandwidth of the transmitted information [i.e., over the bandwidth of $S_m(\omega)$]. In this case, the effect of photomultiplication is to increase the shot noise by the mean squared gain and increase the desired output spectrum by the squared mean of the gain.

With the substitution above the resulting power SNR is then

$$SNR_p = \frac{(\bar{G})^2 \alpha^2 P_s^2 P_m}{[\alpha \overline{G^2}(P_s + \sigma_b^2) + (N_{0c}/e^2)]2B_m}$$

$$= \frac{\alpha P_s P_m}{\{[\overline{G^2}/(\bar{G})^2][1 + (\sigma_b^2/P_s)] + [N_{0c}/e^2\alpha(\bar{G})^2 P_s]\}2B_m} \quad (5.4.2)$$

Now we see that the term due to circuit noise can be neglected, and the receiver operates in a shot noise limited condition, if the average gain of the photomultiplier is sufficiently large. This requires that

$$\alpha(\bar{G})^2 P_s e^2 \gg N_{0c} \quad (5.4.3)$$

or equivalently,

$$\bar{G} \gg \left[\frac{N_{0c}}{\alpha P_s e^2}\right]^{1/2} \quad (5.4.4)$$

The term in the bracket is the ratio of circuit noise spectral level to shot noise level of the signal radiation process. Thus shot noise limited behavior can be attained by using large multiplier gains.

When (5.4.3) is true, the shot noise limited SNR_p becomes

$$SNR_p = \frac{(\bar{G})^2}{\overline{G^2}}\left[\frac{\alpha P_s P_m}{(1 + \sigma_b^2/P_s)2B_m}\right], \qquad \text{shot noise limited} \quad (5.4.5)$$

This is simply the SNR_p in (5.3.22), derived without photomultiplication, multiplied by the factor in the first bracket. It is convenient to define the reciprocal of this factor as the parameter

$$F \equiv \frac{\overline{G^2}}{(\bar{G})^2} = \frac{(\bar{G})^2 + \text{Var}(G)}{(\bar{G})^2}$$

$$= 1 + \frac{\text{Var}(G)}{(\bar{G})^2} \quad (5.4.6)$$

where Var G is the variance of G. The parameter F, substituted into (5.4.5), plays the role of a photomultiplier *noise figure* in that it indicates the factor by which the shot noise limited SNR_p is reduced when a multiplier is used.

It is clear that F is always greater than unity and increases with the variance, or spread, of the device. Note that an ideal multiplier (constant gain) has $F = 1$, while multipliers for which the variance is a percentage of the square of the mean gain [Var $G = \delta^2(\bar{G})^2$] have $F = 1 + \delta^2$. Thus photomultiplier noise figure is directly related to its gain spread.

From a systems point of view, we see from this discussion that the use of a high gain multiplier generally implies shot noise limited behavior at the multiplier output, while introducing an effective noise figure of F, defined in (5.4.6). This can be accounted for in the system model by eliminating the circuit noise in Figure 5.1, and introducing a power gain device with gain $\bar{G}$ and noise figure F, as shown in Figure 5.3. Recall from the analysis of noisy

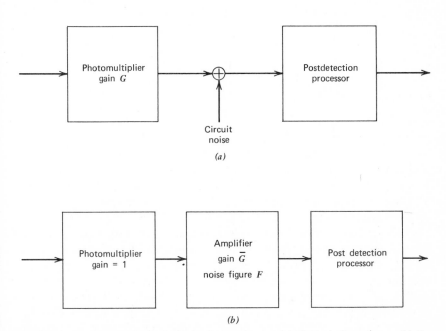

Figure 5.3. Photomultiplier model. (a) Typical receiver model. (b) Equivalent photomultiplier model.

electronic devices that a value of noise figure greater than unity is caused by added internal noise of the device. Relating this to (5.4.6) shows that the "spread" of a multiplier caused by randomness in its gain, appears as an internal noise source in overall system analysis. This noise is often referred to as photomultiplier *excess noise*. This analogy is explored further in Chapter 6.

5.5   OPTIMAL POSTDETECTION PROCESSING

The SNRs of the preceding section were derived assuming that the output filters in Figure 5.1 have filter functions that basically encompass the desired signal spectrum. It is well known, however, that these filters can be designed based upon more rigorous mathematical approaches. Two of the more common methods are the use of minimum mean squared error filtering and the maximization of SNR. We consider mean squared error filtering first.

Consider the receiver in Figure 5.4. The photodetector produces the shot noise process $x(t)$, containing the intensity modulation $m(t)$. We desire a linear filter that will operate upon $x(t)$ to recover, in the best possible way,

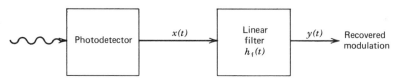

Figure 5.4.   Linearly filtered detector output process.

the modulation $m(t)$. When $m(t)$ is a stationary random process, we may interpret "best" in the sense of mean squared error reduction. That is, we seek the filter that minimizes the mean squared error between $m(t)$ and the filter output, at any t. In line with this problem we may recall that the effect of linear filtering a shot noise process is to produce another shot noise process with the same count intensity, but with filtered component functions. Knowing the component functions $h(t)$ at the filter input (these are due to the detector characteristics), the problem of finding the optimum filter is equivalent to finding the optimum component functions of a shot noise process so that the mean square error, with respect to its own intensity process, is minimized. If we can identify these component functions, then the optimal filter is simply that which produces this optimizing function when the detector component function is considered the input. That is, if $h_0(t)$ is the optimizing function, and if $h_f(t)$ is the desired filter impulse response, then the two are related by

$$h_0(t) = \int_{-\infty}^{\infty} h_f(t - \tau)h(\tau)\, d\tau \qquad (5.5.1)$$

We therefore consider the following design problem. Let the detector shot noise process be given as in (5.3.4),

$$x(t) = \sum_{i=0}^{k} h(t - t_i) \qquad (5.5.2)$$

where now k represents the count over $(-\infty, t)$. For simplicity, we assume this count has intensity $n(t) = \alpha|s(t)|^2 A_r = \alpha P_s[1 + m(t)]$, neglecting background and additive circuit noise. We further assume that the intensity modulation, $m(t)$, is a zero mean, stationary modulating process. The mean square error between the desired modulation and the actual detector output is then defined as

$$J \equiv E[\alpha P_s m(t) - y(t)]^2 \qquad (5.5.3)$$

where the average is to be taken over the random process $m(t)$ and the shot noise statistics of $x(t)$. We seek the component function $h_0(t)$ that minimizes J. Proceeding formally, we expand J to

$$J = \left[\alpha^2 P_s{}^2 m^2(t) - 2\alpha P_s m(t) \sum_{j=0}^{k} h_0(t - t_j) + \sum_{j=0}^{k} \sum_{i=0}^{k} h_0(t - t_i) h_0(t - t_j) \right]$$

$$(5.5.4)$$

The first average is simply the power in the signal $\alpha P_s m(t)$. The expectation of the second term can be computed by averaging, first with a condition on $m(t)$, then averaging over $m(t)$. This yields

$$E[m(t)y(t)] = E_m \left\{ m(t) E \sum_{j=0}^{k} h_0(t - t_j) \right\}$$

$$= \alpha E_m \left\{ m(t) \int_{-\infty}^{\infty} h_0(t - t_j)(P_s + P_s m(t_j))\, dt_j \right\}$$

$$= \alpha \int_{-\infty}^{\infty} h_0(t - t_j) P_s R_m(t - t_j)\, dt_j \qquad (5.5.5)$$

where $R_m(\tau)$ is the correlation function of the intensity process $m(t)$. The third term in (5.5.4) can be separated into terms for which $i = j$ and $i \neq j$, similar to (4.3.11). The subsequent average is then

$$E[y^2(t)] = \alpha P_s \int_{-\infty}^{\infty} h_0{}^2(t - x)\, dx$$

$$+ \alpha^2 \int_{-\infty}^{\infty} \int h_0(t - t_i) h_0(t - \tau_j)[P_s{}^2 + P_s{}^2 R_m(t_i - t_j)]\, dt_i\, dt_j \qquad (5.5.6)$$

Substituting (5.5.5) and (5.5.6) into (5.5.4) gives an expanded version of J. However, this result can be further simplified by noting that the first term on the right in (5.5.6) can be written as

$$\alpha P_s \int_{-\infty}^{\infty} h^2(t - x)\, dx = \alpha P_s \int_{-\infty}^{\infty} \int h(t - x) h(t - z)\, \delta(x - z)\, dx\, dz \qquad (5.5.7)$$

This then can be combined with the remaining term to yield a final form for J:

$$J = E[\alpha^2 P_s^2 m^2(t)] - 2\alpha^2 P_s^2 \int_{-\infty}^{\infty} h_0(t - x)R_m(t - x)\,dx$$

$$+ \alpha \iint_{-\infty}^{\infty} h_0(t - x)h_0(t - z)[\alpha P_s^2 + \alpha P_s^2 R_m(x - z) + P_s\,\delta(x - z)]\,dx\,dz \tag{5.5.8}$$

We now desire the $h_0(t)$ that minimizes J for any t. This can be obtained by straightforward application of the calculus of variations [1]. We replace $h_0(t)$ by $h_0(t) + \varepsilon\Delta(t)$ in J, then note that the solution $h_0(t)$ is that given by the functional $h_0(t)$ for which

$$\left.\frac{dJ}{d\varepsilon}\right|_{\varepsilon=0} = 0 \tag{5.5.9}$$

for any arbitrary perturbing function $\Delta(t)$. This leads to the fact that the optimal solution must satisfy the equation (Problem 5.5)

$$\alpha^2 P_s^2 R_m(\tau) = \int_{-\infty}^{\infty} h_0(\tau - x)[(\alpha P_s)^2 + (\alpha P_s)^2 R_m(x) + \alpha P_s\,\delta(x)]\,dx \tag{5.5.10}$$

The $h_0(t)$ satisfying (5.5.10) is the optimal component function. The equation, a linear integral equation, is a modified form of the Wiener–Hopt equation [2, 3]. We can easily solve for $H_0(\omega)$, the Fourier transform of $h_0(t)$, by transforming both sides of (5.5.10). Defining $S_m(\omega)$ as the transform of $R_m(\tau)$, we obtain

$$H_0(\omega) = \frac{\alpha^2 P_s^2 S_m(\omega)}{\alpha^2 P_s^2 S_m(\omega) + \alpha P_s + \alpha^2 P_s^2\,\delta(\omega)} \tag{5.5.11}$$

as the component function solution. The postdetection filter impulse response $h_f(t)$ in (5.5.1) must be that which produces $h_0(t)$ from $h(t)$. Its transform $H_f(\omega)$ therefore satisfies $H_f(\omega) = H_0(\omega)/H(\omega)$, where $H(\omega)$ is the transform of the photodetector component function $h(t)$. Thus,

$$H_f(\omega) = \frac{1}{H(\omega)}\left[\frac{\alpha^2 P_s^2 S_m(\omega)}{\alpha^2 P_s^2 S_m(\omega) + \alpha P_s + \alpha^2 P_s^2\,\delta(\omega)}\right] \tag{5.5.12}$$

is the desired filter transfer function. If we take into account the delta function at $\omega = 0$, we can rewrite $H_f(\omega)$ as

$$H_f(\omega) = \begin{cases} 0, & \omega = 0 \\ \dfrac{1}{H(\omega)}\left[\dfrac{\alpha^2 P_s^2 S_m(\omega)}{\alpha^2 P_s^2 S_m(\omega) + \alpha P_s}\right] \triangleq \tilde{H}_f(\omega), & \omega > 0 \end{cases} \tag{5.5.13}$$

The requirement for zero transfer at $\omega = 0$ means the optimal filter must "notch out" zero frequency, while providing $\tilde{H}_f(\omega)$ at other frequencies. This zero can be produced by following the filter function $H_f(\omega)$ with a subtraction, or by providing zero "dc" gain. Note that the optimal filter depends only on the power levels and the spectrum of $m(t)$, and not on the particular statistics of the modulation $m(t)$. With background noise included, a slightly more complicated denominator occurs (Problem 5.6). The filter corresponding to (5.5.13) may however, not be causally realizable in the sense that its inverse transform $h_f(t)$ may not be zero for negative t. This could have been avoided by incorporating this requirement into the original calculus of variations solution. This is standard procedure in Wiener filter theory and leads to a modified version of $H_f(\omega)$ that guarantees causal realizability. The modification, however, requires factoring $\tilde{H}_f(\omega)$ into zeros and poles. The reader is referred to more rigorous treatments on Wiener filter theory (References 2–5) for further explanation of this procedure.

The factor $1/H(\omega)$ in $H_f(\omega)$ can be interpreted as a "whitening filter" for the photodetection shot noise—converting its component functions to delta functions. The remaining factor then properly shapes these delta functions. As is true for most whitening filters, $1/H(\omega)$ alone may not be theoretically realizable [requiring infinite gain as $H(\omega) \to 0$] but embedding in $\tilde{H}_f(\omega)$ often avoids this problem. Generally, photodetectors have wider bandwidths than the intensity variation, and $H(\omega) \approx 1$ over the spectral extent of $S_m(\omega)$. In this case, the filter transfer function essentially matches the spectrum shape of the optimal component function. That is, if $H(\omega) \cong 1$, $H_f(\omega) \cong \alpha^2 P_s^2 S_m(\omega)/[\alpha^2 P_s^2 S_m(\omega) + \alpha P_s]$. Recall αP_s is also the spectral level of the shot noise spectrum associated with the spectral density of the photodetector output, while $S_m(\omega)$ is the desired information spectrum. Thus the optimal filter is of the form of a signal spectrum divided by a signal plus a shot noise spectrum. This is identical to the result that would be obtained for the linear mean square filter for the case of additive noise. In other words, the optimal linear filter treats the photodetector output as if it were the sum of the desired signal and additive white noise of spectral level αP_s. This, of course, relates to our earlier discussions on shot noise spectrums in Chapter 4 and, in a sense, justifies system design based upon the treatment of shot noise as an additive receiver noise.

We may instead seek an output filter that maximizes SNR, rather than minimizing a mean squared error. Consider the postdetection correlator in Figure 5.5 that must operate on the photodetected output for T sec and maximize the instantaneous SNR_t at time $t = T$, when a field of known intensity is being received. The problem can be formulated as follows. We consider the interval $(0, T)$ to be divided into $D_t = 2B_0 T$ time modes, each of width $\Delta t = 1/2B_0$ sec. We denote the count during the ith such modal

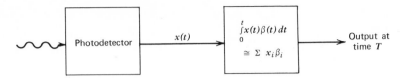

Figure 5.5. Postdetection processing using a shot noise correlator (linearly weighted counts).

interval to be k_i, and denote the corresponding signal field intensity sample as $|s_i|^2$, which is the integrated signal intensity over the ith interval [see (3.7.10)]. A linear filter operating on the photodetector output can be considered as a sequence of linear weightings $\{\beta_i\}$ upon the detected output sequence $\{k_i\}$. The problem is to determine the weighting coefficients β_i so that the modified count

$$y \triangleq \sum_{i=1}^{D_t} \beta_i k_i \tag{5.5.14}$$

has maximum SNR_T, where now

$$\mathrm{SNR}_T = \frac{(E[y])^2}{\mathrm{Var}(y)} \tag{5.5.15}$$

Following our procedure leading to (5.3.27) and (5.3.28), we have the signal mean

$$E[y] = \alpha A_r \sum_{i=0}^{D_t} \beta_i |s_i|^2 \tag{5.5.16}$$

The variance from this mean is

$$\mathrm{Var}(y) = \sum_{i=0}^{D_t} \beta_i^2 [\alpha |s_i|^2 A_r + \alpha N_{ob}(1 + \alpha N_{ob}) + \alpha^2 N_{ob}|s_i|^2 A_r] \tag{5.5.17}$$

where αN_{ob} is again the noise count per mode. Thus,

$$\mathrm{SNR}_T = \frac{[\alpha A_r \sum_{i=0}^{D_t} \beta_i |s_i|^2]^2}{\sum_{i=0}^{D_t} \beta_i^2 c_i} \tag{5.5.18}$$

where c_i is the bracket in (5.5.17). We desire the $\{\beta_i\}$ set maximizing (5.5.18) for the signal intensity set $\{|s_i|^2\}$. If we rewrite the numerator summation as

$$\sum_i \beta_i |s_i|^2 = \sum_i (\beta_i \sqrt{c_i})\left(\frac{|s_i|^2}{\sqrt{c_i}}\right) \tag{5.5.19}$$

then apply the Schwarz inequality and substitute back into (5.5.18), we immediately establish that

$$\text{SNR}_T \le \sum_{i=0}^{D_t} \frac{[\alpha |s_i|^2 A_r]^2}{c_i} \tag{5.5.20}$$

Furthermore, the upper bound occurs when the equality in the Schwarz inequality holds, that is, when $\beta_i = |s_i|^2/c_i$, or when

$$\beta_i = \frac{|s_i|^2}{\alpha |s_i|^2 A_r [1 + \alpha N_{0b}] + \alpha N_{0b}[1 + \alpha N_{0b}]} \tag{5.5.21}$$

This equation represents the desired weighting for the ith time mode required to maximize SNR_T while processing for T sec. This sequence of $\{\beta_i\}$ therefore determines the weighting function of the desired processor. Note that the weighting depends on both the signal and noise intensities during each mode. If the background noise is weak, β_i is essentially constant and the optical processor collects counts equally in time; that is, it simply integrates the detector output for T sec. If the background noise is strong (relative to the received signal power), β_i is proportional to the signal intensity, and counts are collected according to the variation of this intensity over $(0, T)$. In essence, the weighting "follows" the signal intensity variation, and the weighting is said to be *matched* to the signal intensity.

5.6 INTENSITY MODULATED SUBCARRIER SYSTEMS

The previously derived SNR equations pertain to an optical system in which the desired information waveform is intensity modulated directly onto the optical carrier. An alternative procedure is to make use of an auxiliary subcarrier to carry the information signal. The overall system would appear as in Figure 5.6. At the transmitter the information waveform is modulated onto an RF or IF subcarrier using standard subcarrier modulation formats such as AM, FM, and so on. The subcarrier is then intensity modulated onto the main optical carrier. At the receiver the modulated subcarrier signal is

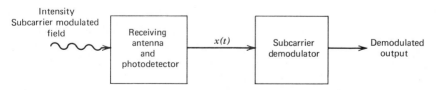

Figure 5.6. Receiving system for intensity subcarrier modulation.

first recovered by photodetection of the optical beam. The recovered sub-carrier is then fed into a subcarrier demodulation channel to demodulate the information signal. The primary advantage of such an operation is the possible signal processing improvement obtained during the subcarrier demodulation that otherwise may not be possible with intensity modulation alone. In the following we examine some common types of subcarrier modulating formats and their applicability to optical systems.

AM/IM Systems. Consider a system employing amplitude modulation (AM) of the data onto the subcarrier, prior to intensity modulation (IM) of the subcarrier on the optical carrier. Such systems are denoted AM/IM systems. The baseband signal $m(t)$ that we previously defined now has the form

$$m(t) = C(1 + d(t)) \sin \omega_{sc} t \qquad (5.6.1)$$

where ω_{sc} is the subcarrier frequency and $d(t)$ is the data signal waveform, having a bandwidth B_d Hz and normalized such that $|d(t)| \leq 1$, with power $P_d \leq 1$. The subcarrier signal above occupies a bandwidth $B_m = 2B_d$ Hz. To satisfy the intensity overmodulation requirement $[|m(t)| \leq 1]$, we see that in (5.6.1) we require $C < \frac{1}{2}$ so that $P_m = C^2(1 + P_d)/2 \leq (1 + P_d)/8$. The shot noise limited SNR_p that occurs in the subcarrier bandwidth at the input to the AM demodulator is obtained directly from (5.3.22) as

$$(SNR_p)_{sc} = \frac{\alpha P_s (1 + P_d)/8}{(1 + \sigma_b^2/P_s)2B_m} \qquad (5.6.2)$$

where we have simply used the fact that $m(t)$ now corresponds to the subcarrier signal. Equation 5.6.2 implies a power of $\alpha P_s/8$ in the AM subcarrier and a power of $\alpha P_s P_d/8$ in the AM modulation. An ideal AM demodulator would then yield an output data SNR_p of

$$(SNR_p)_d = \frac{\alpha P_s C^2 P_d}{(1 + \sigma_b^2/P_s)2B_b} = \frac{1}{4}\left[\frac{\alpha P_s P_d}{(1 + \sigma_b^2/P_s)2B_d}\right] \qquad (5.6.3)$$

It is interesting to compare (5.6.3) to result one would obtain if the data $d(t)$ were intensity modulated directed onto the optical carrier. In this case the SNR_p would be given by (5.3.22) with $B_m = B_d$ and $P_m = P_d$, which is identical to the bracketed term in (5.6.3). Thus the SNR_p is degraded by a factor of 4 (6 dB) if an auxiliary AM subcarrier is used. This poorer perform-ance can be explained by the fact that relatively little processing improvement is obtained in the AM subcarrier channel. This can be seen by writing (5.6.3) in terms of (5.6.2) so that we have

$$(SNR_p)_d = 2(SNR_p)_{sc} \qquad (5.6.4)$$

if we use $P_d = 1$. Thus an improvement of only 3 dB (factor of 2) is obtained during AM processing, even in an ideal AM demodulator. For this reason AM/IM systems are generally not considered efficient for optical systems, and more interest is devoted to FM/IM systems.

FM/IM Systems. In this communication format, the subcarrier is frequency modulated (FM) with the data instead of amplitude modulated. The baseband signal that intensity modulates the optical carrier is now

$$m(t) = C \sin\left[\omega_{sc}t + \Delta_f \int d(t) \, dt\right] \quad (5.6.5)$$

where Δ_f is now the subcarrier frequency deviation due to the data signal $d(t)$. We now restrict $C = 1$ to prevent intensity overmodulation, and the subcarrier now occupies a bandwidth

$$B_m = 2(\beta + 1)B_d \quad (5.6.6)$$

where $\beta = \Delta_f/2\pi B_d$ is the FM modulation index. The resulting subcarrier SNR_p is then

$$(SNR_p)_{sc} = \frac{\alpha P_s(C^2/2)}{(1 + \sigma_b^2/P_s)2B_m} = \frac{\alpha P_s/2}{(1 + \sigma_b^2/P_s)2B_m} \quad (5.6.7)$$

The FM subcarrier demodulating channel operating above threshold yields a data output SNR_p of

$$(SNR_p)_d = 3\beta^2 \frac{\alpha P_s C^2}{(1 + \sigma_b^2/P_s)2B_d}$$

$$= \left[6\beta^2\left(\frac{B_m}{B_d}\right)\right](SNR_p)_{sc} \quad (5.6.8)$$

The bracketed term indicates the improvement possible during the FM subcarrier demodulation, and depends explicitly on the transmitted subcarrier deviation Δ_f. Thus by using wide band (large) FM subcarriers an FM/IM system can produce larger output SNR_p than one using direct intensity modulation.

5.7 MULTIMODE DIRECT DETECTION

All of the preceding results are based on the assumption of a point detector, which basically assumes that the optical field is completely coherent over the photodetector aperture and the receiver observes a single field mode. When the transmitter signal field has been transmitted through a turbulent medium prior to optical reception, the spatial coherence of the function is distorted so that the field is no longer coherent over the receiver surface. The effect is to generate more spatial modes in the received field, and the source appears to be larger. The receiver must then open its field of view to see more of the source, and no longer operates as a point detector. Mathematically, the received field expansion in (5.2.3) contains more than a single term, each such term defining a spatial mode. The total count intensity over all modes collected by the receiver is given in (5.2.4). Each model envelope, $f_j(t)$, is assumed to have within it a signal component $s_j(t)$, with average power $P_{sj} = E[|s_j(t)|^2]A_r$, and a background noise of power σ_b^2. Each mode contributes to the total shot noise power. The total average signal power over D_s spatial modes is then

$$P_s = \sum_{j=1}^{D_s} P_{sj} \tag{5.7.1}$$

We may first examine if an optimal procedure exists for collecting spatial mode power over the receiver area. Let us assume that each signal mode at the receiver can be weighted by a spatial weighting factor q_j in collecting total power from that mode. In essence, a linear weighting is presumed for the spatial modes, just as a linear weighting was considered for the time modes. The weighted average signal power collected over all modes is

$$\hat{P}_s = \alpha \sum_{j=1}^{D_s} q_j P_{sj} \tag{5.7.2}$$

The total weighted shot noise power in a bandwidth B_m hz is then

$$\hat{P}_n = 2\alpha B_m \sum_{j=1}^{D_s} q_j^2 (P_{sj} + \sigma_b^2) \tag{5.7.3}$$

The resulting SNR_p for the weighted modes is then

$$SNR_p = \frac{\hat{P}_s}{\hat{P}_n} = \frac{[\alpha \sum_{j=1}^{D_s} q_j P_{sj}]^2}{2\alpha B_m \sum_{j=1}^{D_s} q_j^2 (P_{sj} + \sigma_b^2)} \tag{5.7.4}$$

This is the resulting SNR_p that will be developed from the mode collection. In determining the $\{q_j\}$ that maximize this SNR_p, we immediately see that the problem is identical in form to our earlier analysis in Section 5.5, in which we sought optimal time weightings $\{\beta_i\}$. In fact (5.7.4) is identical in form to (5.5.18). The solution therefore follows as before, with proper conversion of parameters, yielding

$$q_j = \left(\frac{\alpha}{2B_m}\right)\left[\frac{P_{sj}}{P_{sj} + \sigma_b^2}\right] \tag{5.7.5}$$

as the optimal spatial weighting. This indicates the manner in which modal power should be collected if we wish to maximize SNR_p in multimode reception. In particular, it states that no mode should be collected if it contains no signal power, while very little weight should be associated with modes having strong noise, or weak signal power. Modes having equal power should be collected with equal weightings. Note that the weightings $\{q_j\}$ can be considered a form of spatial filtering applied to the source modes, just as the $\{\beta_i\}$ represent time filtering applied to the temporal modes. In practical systems, spatial weighting can be accomplished by spatial filtering (field stops) placed in the receiver focal plane. However, it is difficult to match exactly the optimal spatial weighting required in (5.7.5), since power levels are generally not known exactly. It does tell us, however, that field stops should be placed on all modes not containing desired signal power.

Let us apply these results to a practical example. Let the received field from an extended source be described by M_s spatial modes at the receiver, where M_s is given by the ratio of the source irradiance solid angle to the receiver diffraction limited field of view. We assume the signal mode power is identical for all modes, so that $P_{sj} = P_{s0}$ for $j = 1, 2, \ldots, M_s$. The receiver field of view is assumed to encompass M_d spatial modes. The optimal weighting in (5.7.5) requires that the M_s signal modes be weighted equally, and no modes be included beyond M_s. Thus SNR_p in (5.7.4) becomes

$$SNR_p = \begin{cases} \dfrac{(\alpha M_d P_{s0})^2}{M_d(P_{s0} + \sigma_b^2)2\alpha B_m}, & M_d < M_s \quad (5.7.6) \\[3ex] \dfrac{(\alpha M_s P_{s0})^2}{[\alpha M_s P_{s0} + \alpha M_d \sigma_b^2]2B_m}, & M_d \geq M_s \quad (5.7.7) \end{cases}$$

Rewriting, this is equivalent to

$$SNR_p = \begin{cases} M_d\left[\dfrac{\alpha P_{s0}^{\,2}}{(P_{s0} + \sigma_b^2)2B_m}\right], & M_d < M_s \quad (5.7.8) \\[3ex] \left[\dfrac{\alpha P_s^2}{[P_s + (M_d)\sigma_b^2]2B_m}\right], \\[3ex] = \dfrac{\alpha P_s^2}{\left[P_s + \left(\dfrac{\Omega_{fv}}{\Omega_{dL}}\right)\lambda^2 N(f)B_0\right]2B_m}, & M_d \geq M_s \quad (5.7.9) \end{cases}$$

where (5.79) is obtained by substituting from (1.7.6). The bracket in (5.7.8) is precisely the SNR_p for an optical detector observing only a single spatial mode of the distorted field. Hence, (5.7.8) indicates a linear improvement in SNR_p over single mode detection as the number of observed spatial modes (i.e., the receiver field of view) increases. Basically, we are saying that if the turbulent channel has spread the signal energy over more than one spatial mode, then opening the field of view to observe more of these modes yields a linear increase in SNR_p over single mode detection. As shown in Problem 5.10, typical detectors may observe as many as 10^6 to 10^7 spatial modes, so this improvement may be as significant as 60 to 70 dB. However, we see from (5.7.9) that this improvement occurs only so long as $M_d \leq M_s$. Beyond this point, increasing field of view no longer observes more signal energy, but continues to add more background noise energy, with an overall effect of reducing SNR_p. Thus, SNR_p has a maximum at $M_d = M_s$, with value

$$(SNR)_p)_{max} = \frac{M_s \alpha P_{s0}^2}{[P_{s0} + \sigma_b^2]2B_m}$$

$$= \frac{\alpha P_s/2B_m}{1 + (\sigma_b^2/P_s)M_s} \tag{5.7.10}$$

where $P_s = M_s P_{s0}$ is the total signal power available in all modes, as given in (5.7.1). Equation 5.7.10 represents the maximum SNR_p possible, while observing M_s spatial modes of total power P_s. Note that $(SNR_p)_{max}$ achieves its largest value when $M_s = 1$, that is, when the total available source power P_s is transmitted in a single mode, and a single mode detector is used. In this case (5.7.10) reduces to (5.3.22) previously calculated for a shot noise limited single mode detector. This means the multimode SNR_p can never exceed the system performance achievable with a single mode link having the same source power. We also note that the quantum limited bound in (5.3.24) is still achievable with multimode operation if $\sigma_b^2 M_s/P_s \ll 1$. That is, we do not need diffraction limited (single mode) receivers to achieve quantum limited performance, as long as the signal power dominates the total noise.

In deriving (5.7.6) to (5.7.10) we have considered that the sum of the average signal power over all modes is useful power contributing directly to the desired signal term. This is true, however, for a given transmitted intensity $m(t)$ only if the instantaneous power variation in all modes is in time synchronism. That is, the original signal modulation spread over all modes by the channel must still be temporally coherent. Adding power from all modes at each t then reinforces the total signal intensity at the detector output. (In the literature this is referred to as *coherent combining*.) In practice, the channel can often introduce random delays to the modulation in each mode. (This is the mode delay dispersion referred to in Section 1.5.) This delay, if excessive, may destroy the coherent combining. For example, consider again (5.7.1) for the case where the transmitted power variation is $P(t)$, which arrives as many independent spatial modes at the receiver. The power collected from the ith mode becomes $G_i P(t - \tau_i)$, where G_i is the channel gain of the ith mode and τ_i is its delay. The total power collected over all such modes is then

$$P_s(t) = \sum_i G_i P(t - \tau_i) \qquad (5.7.11)$$

instead of (5.7.1). We see that if the $\{\tau_i\}$ are sufficiently dispersed (separated), the mode power variation is collected with different delays, causing the individual terms in (5.7.11) to interfere with, rather than reinforce, each other. Thus the coherent combining advantage is lost. The required condition is that the dispersion among the $\{\tau_i\}$ be much smaller than the time variations in the intensity modulation. The overall effect is that the dispersion places a limit on the maximum modulating frequency that can be used (Problem 5.9). Delay dispersions in a weak scattering space channel are reported to be quite small (on the order of fractions of a nanosecond) and the upper frequency value is about several hundred gigahertz. Fiberoptic channels, however, may have dispersions that limit frequencies to the megahertz range. Some consideration has been given to developing methods to equalize delay dispersion over transmission paths. Theoretically, one can attempt to separate the modes by detecting separately the various directions of arrival using an array of detectors. Then, if sufficient information was available to determine, or estimate, a priori the expected delay of each mode, each separate delay can be corrected prior to combining to form (5.7.11). This requires parallel processing at the receiver, and the receiver complexity is greatly increased in order to achieve the multimode advantage. In the weak scattering space channel where delay dispersions are not prohibitive, this added complexity may not be cost effective. However, delay equalization may have its most practical application to the fiberoptic channel. [14, 15] Here less expensive fibers tend to produce many signal modes, with significant delay dispersion. One therefore has a design alternative of either improving the fiber (reducing its modes) or attempting to direct detect and equalize the delays. However, it is much more difficult to separate modes spatially in an optical guide, and equalization is generally applied directly to the detected signal in (5.7.11). Various types of equalizers have been suggested for this purpose [6–15].

5.8 SCINTILLATION AND APERTURE AVERAGING

In addition to generating an apparent extended source by increasing the number of spatial modes, and injecting possible delay effects, a transmission channel also may introduce random scintillation to the intensity of each mode. This means that the instantaneous intensity will have an added degree of randomness due to the scintillation. Let us examine again (5.7.11) with the delays neglected ($\tau_i = 0$) and consider the channel gains $\{G_i\}$ to be identically distributed random mode gains having mean $\bar{G}$ and variance σ_G^2. At each t, the ith spatial mode will therefore have a mean power $\bar{G}P_s$ and a variance $\sigma_G^2 P_s^2$ about the mean. A single mode detector observing only the ith mode will therefore see a normalized scintillation variance of

$$\frac{\text{power variance}}{(\text{mean power})^2} = \frac{\sigma_G^2 P_s^2}{(\bar{G}P_s)^2} = \frac{\sigma_G^2}{\bar{G}^2} \tag{5.8.1}$$

Now suppose we collect power from M_r independent receiver modes, the ith having power $G_i P_s$. The resulting scintillation variance in (5.8.1) is now

$$\frac{M_r \sigma_G^2 P_s^2}{(M_r \bar{G}P_s)^2} = \frac{1}{M_r}\left(\frac{\sigma_G^2}{\bar{G}^2}\right) \tag{5.8.2}$$

We have therefore reduced the scintillation variance by the number of observed modes. In essence, we have averaged the random mode power over a set of independent modes, and averaged out the scintillation variance. Thus, increasing the number of modes (opening up the receiver field of view or increasing the receiver area, or both) tends to reduce spatial scintillation. This operation is referred to as "aperture averaging." We point out that the procedure is valid only if the source always fills the receiver field of view, that is, each receiver mode contains signal energy. Thus, (5.8.2) achieves its minimum value when the number of receiver modes equals the number of effective source modes, that is, the receiver field of view has been opened up until it encompasses the source area.

REFERENCES

[1] Gelfand, I. and Fomin, S., *Calculus of Variations*, Prentice-Hall, Englewood Cliffs, New Jersey, 1963.

[2] Thomas, J., *Statistical Communication Theory*, John Wiley and Sons, New York, 1969, Chap. 5.

[3] Lee, Y. W., *Statistical Theory of Communication*, John Wiley and Sons, New York, 1960, Chap. 14.

[4] Van Trees, H., *Detection, Estimation, and Modulation*, Part I, John Wiley and Sons, New York, 1968, Chap. 6.

[5] Viterbi, A., *Principles of Coherent Communications*, McGraw-Hill Book Co., New York, 1966, Chap. 5, Appendix D.

[6] Lucky, R. W., "Automatic Equalization for Digital Communication," *Bell Systems Tech. J.*, 547–588 (April 1965).

[7] George, D. A., "Matched Filters for Interfering Signals," *IEEE Trans. Inf. Theory*, 153 (January 1965).

[8] Aaron, M. R. and Tufts, D. W., "Intersymbol Inteference and Error Probability," *IEEE Trans. Inf. Theory*, 26–35 (January 1966).

[9] Lucky, R. W., "Techniques for Adaptive Equalization of Digital Communication Systems," *Bell Systems Tech. J.*, 255–286 (February 1966).

[10] Lucky, R. W. and Rudin, H. R., "Generalized Automatic Equalization for Communication Channels," *Proc. IEEE*, 439 (March 1966).

[11] Niessen, C. W. and Drouilhet, P. R., "Adaptive Equalizer for Pulse Transmission," 1967 IEEE International Conference on Communications.

[12] Austin, M. E., "Decision-Feedback Equalization for Digital Communication Over Dispersive Channels," M.I.T. Research Laboratory of Electronics Tech. Report 461, August 11, 1967.

[13] DiToro, M. J., "A New Method of High-Speed Adaptive Serial Communication Through Any Time-Variable and Dispersive Transmission Medium," Conference Record, IEEE Communication Convention, Boulder, Colorado, June 1965, p. 763.

[14] Personick, S. D., "Baseband Linearity and Equalization in Fiber Optic Digital Systems," Bell Systems Tech. J., **52** (no. 7), pp. 1175–1194 (September 1973).

[15] Henderson, D. M., "Dispersion and Equalization in Fiber Optic Systems," Bell Systems Tech. J., **52** (no. 10), pp. 1867–1876 (December 1973).

PROBLEMS

1. An optical system operates at 0.6 microns wavelength with a photodetector of efficiency 50%, a load resistance of 100 ohms, and is maintained at 300°K.
(a) Determine the approximate range of signal count rate needed to achieve shot noise limited conditions.
(b) Convert (a) to power in watts.
(c) If a blackbody background has an effective temperature of 1000°K, determine the received signal power needed to achieve quantum limited performance.
(d) Determine the resulting quantum limited SNR in a bandwidth of 1 MHz.

2. An optical carrier is intensity modulated with the signal $m(t)$ and transmitted through a fading channel. The effect of the fading is to multiply the transmitted intensity by the multiplicative term $q(t) = q_0(1 + r(t))$, where q_0 is a constant and $r(t)$ has spectrum $S_r(\omega)$. The signal $m(t)$ has spectrum $S_m(\omega)$ which overlaps that of $q(t)$.
(a) How will the fading manifest itself in terms of the detected shot noise spectrum?
(b) Suppose the signal $m(t)$ is first amplitude modulated onto a subcarrier, and the subcarrier intensity modulates the carrier. Determine the spectrum in this case.
(c) It has been suggested that if the fading can be "observed" at the detector output, it can be "divided out" (automatic gain control) of the signal term. Comment if this seems possible in (a) and (b) above.

3. In photomultipliers it is common practice to specify the multiplier gain variance as a fraction δ of its mean gain. Derive the effective noise figure of the

photomultiplier in terms of the parameter δ using the results in Problem 2.11b.

4. An avalanche photomultiplier generates a signal-to-noise ratio

$$\text{SNR}_\text{p} = \left(\frac{\alpha P}{2B}\right) \frac{G^2}{G^{2+q}(1 + N_\text{ob}/P) + N_\text{oc}/\alpha P}$$

where G is the detector gain, q is a positive number in $(0, 1)$, and the remaining parameters are defined in (5.3.21).

(a) Determine the gain value G that maximizes SNR_p for a given value of q.

(b) Write the resulting $(\text{SNR}_\text{p})_\text{max}$ for the cases where $q = 1$, $P \gg N_\text{ob}$, $q = 1$, $P \ll N_\text{ob}$, and $q < 1$.

5. Beginning with (5.5.8), generate (5.5.10), using (5.5.9). That is, carry out the required minimization by the calculus of variations.

6. Repeat the derivation of the optimal minimal mean square (Wiener filter) in Section 5.5 with additive background noise. That is, begin with (5.5.2) and assume the count has intensity $n(t) = \alpha P_\text{s}[1 + m(t)] + n_\text{b}$, where n_b accounts for the addition of background with intensity level n_b over the detector area.

7. An optical system uses an intensity modulation $m(t)$ with power spectrum $S_\text{m}(\omega)$, and employs a Wiener filter following direct detection. Neglect background noise.

(a) Assume

$$S_\text{m}(\omega) = S_0, \qquad |\omega| \leq 2\pi B_\text{m}$$

Sketch the Wiener filter amplitude function. Consider both the case where $S_0 B_\text{m} \gg 1$ and $S_0 B_\text{m} \ll 1$.

(b) Assume instead

$$S_\text{m}(\omega) = \frac{C}{1 + (\omega/2\pi B_\text{m})^2}$$

Sketch again the Wiener filter amplitude function, showing the peak value of the filter response and the half-power frequency.

8. A baseband signal with unit power and bandwidth 1 MHz is to be RF subcarrier modulated, with the latter modulated onto an optical carrier at 10^{14} Hz. The system is operated quantum limited, and a subcarrier SNR threshold of 20 dB is required for subcarrier demodulation.

(a) How much received optical power is needed to operate the subcarrier system?

(b) What is the baseband SNR after subcarrier demodulation if AM/IM is used?

(c) What is the baseband SNR after subcarrier demodulation if FM/IM is used with a subcarrier deviation of 10 MHz?

(d) What are the subcarrier bandwidths needed in parts (b) and (c)?

(e) If the ratio of background to received signal power is 0.5 (no longer quantum limited), how much more optical power is needed in part (a)?

9. Consider the reception of only two field modes in (5.7.11). Let $G_1 = G_2 = \frac{1}{2}$, $\tau_2 - \tau_1 = \tau_d$, and let $P(t) = P_s(1 + \sin 2\pi f_s t)$.

(a) Determine the collected power in a direct detection receiver observing both modes.

(b) For a delay difference of $\tau_d = 10^{-9}$ sec, find the maximum frequency f_s that can be sent such that the sinusoidal power variation will have an amplitude whose degradation is less than 3 dB.

10. A source is effectively spread by turbulence to a solid angle of 2.5×10^{-4} sr when viewed from the receiver. Let the receiver lens have area 0.01 meters, operating at optical frequency $f = 6 \times 10^{14}$ Hz.

(a) Determine the number of spatial modes into which the extended source is effectively divided.

(b) Assuming a constant irradiance source, compute the possible decibel improvement in SNR_p over single mode reception, obtained by opening the receiver field of view to that of the source.

11. (a) Determine the scintillation variance in (5.8.1) for the case of log-normal fading (Equation 1.5.2) in each mode, when receiving M_r modes.

(b) Find how large M_r should be relative to the log variance of the channel in order for the scintillation noise to be less than 1%. Evaluate this for the case when the variance is 0.4.

12. Several definitions are used in the discussion of optical photodetectors. *Responsivity* is the ratio of detected output voltage to input voltage. *Noise Equivalent Power* (NEP) is the amount of incident power per unit bandwidth needed to produce an output power equal to the detector output noise power. *Detectivity* (D) is defined as 1/NEP. The parameter D^* (called D-*star*) is the detectivity of a one square centimeter detector.

(a) Derive the relation between NEP and responsivity.

(b) Show that $D^* = (A_r B_0)^{1/2}/\text{NEP}$, where B_0 is the optical bandwidth and A_r is the receiver area.

13. The bandwidth B_0 of an optical filter is generally related to the allowable field of view. For an interference filter this can be expressed as

$$B_0 = f \sin^2(\Omega_{fv}^{1/2})$$

where f is the optical frequency. Use this result in (5.7.9) to maximize SNR_p with respect to Ω_{fv}, when the source field of view is Ω_s.

14. Compute the bound in equation (5.5.20) when the signal is a rectangular function in time occupying S intervals. How does this compare with the quantum limited value when $S = 1$.

15. Using (4.10.7) determine the effect on the detected spectrum due to a random delay dispersion in an optical channel that is uniformly distributed over $(0, T)$. What would be the output SNR_p if we use a post detection filter inversely related to the effective dispersive filtering out to a frequency of q/T?

6 Coherent (Heterodyne) Detection

In the preceding chapter we studied direct, or incoherent, detection of the received optical field. In such a system we found that the receiver must combat the background radiation and the internal receiver thermal noise while attempting to recover the intensity modulation of the transmitted beam. The receiver noise is overcome by achieving shot noise limited conditions, and the resulting performance then depends only on an ability to transmit signals whose power level dominates that of the received background. Shot noise limited operation is achieved with exceptionally strong received signal fields, or by the use of high quality photomultipliers. Unfortunately, the former is difficult to achieve in long-range links, and high performance photomultipliers cannot be constructed at all frequencies in the optical range. This means that when channel properties force operation at optical frequencies for which no suitable photomultiplier exists, other means must be used to improve detectability. An alternative method is the use of heterodyning.

In heterodyne detection the receiver operates by optically adding a locally generated field to the received field, prior to photodetection. The prime objective is to use the added local field to improve the detection of the weaker received field in the presence of the interval receiver thermal noise. The combined field is then photodetected as if it were a single received optical field. Since the addition of two electromagnetic fields requires spatial alignment of the fields, the use of heterodyne detection is often called (spatial) coherent detection.

6.1 THE HETERODYNE RECEIVER

A typical heterodyning receiver is shown in Figure 6.1. The received optical field is projected onto the photodetector surface by the receiver lens and front end system. A local optical field, generated by a receiver source, is diffracted by a receiver lens and aligned by means of a mirror with the

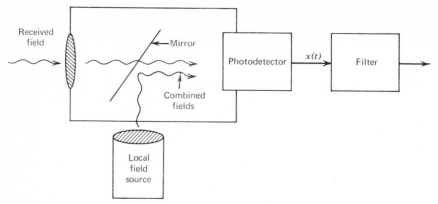

Figure 6.1. The heterodyne receiver.

received field in the photodetector. The detector responds to the combined field of the received and local sources by producing a detector shot noise process in the usual way. The mixing of the two fields can be described in terms of diffraction patterns in the focal plane, or by referring each diffracted field back to the receiver surface as plane waves. The equivalence of the two procedures is guaranteed by the transformation relationship between orthonormal functions in the two planes, as discussed in Chapter 1. The most common method is to refer analysis to the receiver (aperture) plane, and treat both the received and local fields at that point. We emphasize that in practical systems the field mixing is almost always done in the focal plane at the detector surface, so receiver plane analysis is purely for mathematical convenience. We find, however, that certain types of heterodyne distortion are best described in terms of diffraction patterns at the detector.

We first analyze a perfectly aligned heterodyne receiver involving only received plane waves arriving normal to the receiver area. The received plane wave is projected to the detector as a single mode diffraction pattern. The local field is also generated as a plane wave and passed through an identical lens, producing an identical diffraction pattern. The local field is perfectly aligned by the mirror so that its diffraction pattern is exactly superimposed upon that of the received field. When referred back to the receiver plane both fields appear as normal plane waves in the same spatial mode. The resulting heterodyne operation can therefore be discussed entirely in the time domain, and we need only describe the fields over this one mode. Let the received field in this mode be written as

$$f_R(t) = \text{Real}\{a_R(t)e^{j\omega_1 t}\}; \qquad \mathbf{r} \in \mathscr{A}_r \tag{6.1.1}$$

where $a_R(t)$ is again the complex received field envelope, ω_1 is the received optical frequency, and $\mathscr{A}_r$ is the receiving lens area. The total received field is composed of the sum of the transmitted signal field and the background noise field within the same mode. Hence, the received field envelope can be expanded as

$$a_R(t) = s(t) + b(t) \qquad (6.1.2)$$

where $s(t)$ is the signal envelope and $b(t)$ the complex noise envelope at frequency ω_1, similar to (5.3.1). The form of the signal envelope, and its associated frequency spectrum, depends on the manner in which the transmitter information has been modulated onto the optical carrier. This is examined in Section 6.2.

We consider the local source in Figure 6.1 to generate an optical plane wave in the same spatial mode when referred back to the receiver plane. We write this as

$$f_L(t) = \text{Real}\{a_L e^{j\omega_2 t}\}; \qquad \mathbf{r} \in \mathscr{A}_r \qquad (6.1.3)$$

representing a monochromatic normal plane wave at frequency ω_2. The combined receiver field is then

$$f(t) = \text{Real}\{a_R(t) e^{j\omega_1 t} + a_L e^{j\omega_2 t}\} \qquad (6.1.4)$$

which is spatially coherent over the receiver area $\mathscr{A}_r$.

Let us now designate each complex envelope in (6.1.4) in terms of an amplitude and a phase function so that we can write

$$f(t) = \text{Real}\{|a_R(t)| \exp[j(\omega_1 t + \theta_R(t))] + |a_L| \exp[j(\omega_2 t + \theta_L)]\} \qquad (6.1.5)$$

where the subscripts R and L refer to the received and local fields. This combined field therefore has the intensity per unit receiver area given by

$$I(t) = |f(t)|^2$$
$$= |a_R(t)|^2 + |a_L|^2 + 2|a_L||a_R(t)| \cos[(\omega_1 - \omega_2)t + \theta_R(t) + \theta_L] \qquad (6.1.6)$$

over the receiver area $\mathscr{A}_r$. This intensity function is then integrated over the receiver surface to yield the count intensity $n(t)$. This has the form

$$n(t) = n_R(t) + n_L(t) + n_{RL}(t) \qquad (6.1.7)$$

where

$$n_R(t) = \alpha \int_{\mathscr{A}_r} |a_R(t)|^2 \, d\mathbf{r} = \alpha A_r |a_R(t)|^2 \qquad (6.1.8a)$$

$$n_L(t) = \alpha \int_{\mathscr{A}_r} |a_L|^2 \, d\mathbf{r} = \alpha A_r |a_L|^2 \qquad (6.1.8b)$$

$$n_{RL}(t) = \alpha \int_{\mathscr{A}_r} 2|a_L||a_R(t)| \cos(\omega_{12} t + \theta_R(t) - \theta_L) \, d\mathbf{r}$$

$$= 2\alpha A_r |a_L||a_R(t)| \cos(\omega_{12} t + \theta_R(t) - \theta_L) \qquad (6.1.8c)$$

where A_r is again the area of $\mathscr{A}_r$ and $\omega_{12} = \omega_1 - \omega_2$. Note that the first term is the intensity of the received field, while the remaining terms are due to the presence of the local oscillator. The last term $n_{RL}(t)$ appears as the "beat" effect between the two fields. This beat term has an amplitude directly proportional to the received amplitude variation, and a phase that is linearly related to the received phase variation. Hence, amplitude and phase information in the received signal are preserved in the intensity of the heterodyned field. Recall that noncoherent detection never allowed an undistorted version of the received envelope to appear in the intensity function.

The important application of (6.1.6) occurs when the local source signal is made much stronger than the received field power, a condition relatively easy to attain in practice. In this case we see that if

$$|a_L| \gg |a_R(t)| \tag{6.1.9}$$

for almost all t, then (6.1.6) becomes

$$I(t) \cong |a_L|^2 + 2|a_L||a_R(t)| \cos(\omega_{12} t + \theta_R(t) - \theta_L) \tag{6.1.10}$$

Note that now the transmitted information present in $|a_R(t)|$ and/or $\theta_R(t)$ appears only in the cross product term, and the intermodulation terms generated in noncoherent detection (Equation 5.3.12) are neglected because of (6.1.9). Thus the structure of the heterodyned intensity takes on a completely different form from that following direct optical detection. Note also that the local field phase θ_L adds directly to the phase of the transmitted field $\theta_R(t)$, and any time variations in θ_L are indistinguishable from phase variations of the received field itself. This means that strict control of undesired local field phase variations is needed when heterodyning phase modulated fields.

After photodetection of the combined field, the resulting shot noise will have the count intensity function $n(t)$ in (6.1.7). The corresponding power spectrum of the shot noise, given by (5.3.6), is

$$S_x(\omega) = |H(\omega)|^2 [\overline{E[n]} + S_n(\omega)] \tag{6.1.11}$$

where now $\overline{E[n]} \cong \alpha A_r |a_L|^2$. It is convenient to define the local source power as

$$P_L = |a_L|^2 A_r \tag{6.1.12}$$

The spectrum $S_n(\omega)$ in (6.1.11) is then

$$S_n(\omega) = \alpha^2 [(P_L)^2 2\pi \, \delta(\omega) + 4A_r P_L S_m(\omega)] \tag{6.1.13}$$

where $S_m(\omega)$ is the intensity spectrum of

$$m(t) = |a_R(t)| \cos[\omega_{12} t + \theta_R(t) - \theta_L] \tag{6.1.14}$$

If $S_R(\omega)$ is the intensity spectrum of the complex envelope $a_R(t)$, and if θ_L is considered to be uniformly distributed over $(0, 2\pi)$ relative to $\theta_R(t)$, then

$$S_m(\omega) = \tfrac{1}{4}S_R(\omega - \omega_{12}) + \tfrac{1}{4}S_R(-\omega - \omega_{12}) \tag{6.1.15}$$

That is, $S_m(\omega)$ is the spectrum of $a_R(t)$ shifted to $\pm \omega_{12}$. Furthermore, from (6.1.2) we have

$$S_R(\omega) = S_s(\omega) + S_b(\omega) \tag{6.1.16}$$

where now $S_s(\omega)$ and $S_b(\omega)$ are the intensity spectra of the received signal envelope $s(t)$ and background envelope $b(t)$, respectively. Thus (6.1.11) becomes

$$S_x(\omega) = |H(\omega)|^2 \{\alpha P_L + (\alpha P_L)^2 2\pi\,\delta(\omega) + (4\alpha^2 A_r P_L)[\tfrac{1}{4}S_s(\omega - \omega_{12})$$
$$+ \tfrac{1}{4}S_s(-\omega - \omega_{12}) + \tfrac{1}{4}S_b(\omega - \omega_{12}) + \tfrac{1}{4}S_b(-\omega - \omega_{12})]\} \tag{6.1.17}$$

This total spectrum is sketched in Figure 6.2. Note that the spectral distribution in the vicinity of ω_{12} is composed of three components: (1) shot noise of level αP_L, (2) background noise of spectral shape $(\alpha^2 A_r P_L)S_b(\omega - \omega_{12})$, and (3) signal of spectrum $(\alpha^2 A_r P_L)S_s(\omega - \omega_{12})$. The signal spectrum is simply the spectrum of the transmitted optical field,

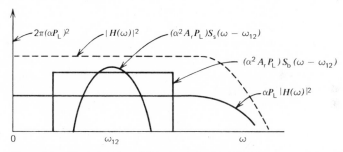

Figure 6.2. The heterodyned detector output spectrum.

except it is now located at the beat frequency ω_{12}. Note that the background term is proportional to the receiver area A_r, having been heterodyned in a manner similar to the signal. For spectrally white background noise intensity [i.e., $S_b(\omega) = S_b$ for all ω] it is convenient to substitute $N_{0b} = S_b A_r$ as the background noise power level per mode, as in Section 5.3, and rewrite the background noise spectrum in (6.1.17) as $2\alpha^2 P_L N_{0b}$. When written this way, the heterodyned background noise appears to be proportional to the power spectrum level. Notice that this is a flat spectrum, and appears at all frequencies.

In optical homodyning the local source is at the frequency of the transmitted optical beam, and $\omega_{12} = 0$. In this case, the detected spectrum is given by (6.1.17) with ω_{12} set equal to zero. Hence,

$$S_x(\omega)\bigg|_{\omega_{12}=0} = |H(\omega)|^2 \left\{ \alpha P_L + (\alpha P_L)^2 2\pi\,\delta(\omega) + \left(\frac{4\alpha^2 A_r P_L}{2}\right)(S_s(\omega) + S_b(\omega)) \right\}$$

(6.1.18)

The homodyned spectrum is now centered at $\omega = 0$ rather than at ω_{12}. In particular, we note that homodyning produces an intensity whose spectrum contains the exact signal spectrum $S_s(\omega)$. However, this corresponds to an undistorted version of the data waveform only if the data is amplitude modulated onto the optical carrier. If angle modulation had been used, homodyning no longer produces an undistorted data spectrum but rather a low frequency version of the optical carrier spectrum (Recall that angle modulation spreads the carrier spectrum over a bandwidth much larger than that of the data.) It should also be noted that the spectral component at zero frequency [delta function in (6.1.18)], though far removed from the signal spectrum after heterodyning, lies within the signal spectrum after homodyning. Since this corresponds to a known constant value, however, it can theoretically be removed. We should also point out that any low frequency variations in $|a_L|$ will fall directly in the signal band after homodyning. We see, therefore, that requirements must often be placed on local field amplitude stability when homodyning.

The detected shot noise spectrum $S_x(\omega)$ has added to it the receiver thermal noise current, with spectral level N_{0c}, as in (5.3.18). When the detector output is filtered at the beat frequency ω_{12} with filter transfer function $F(\omega)$, as shown in Figure 6.3, the filtered spectrum appears as

$$S_f(\omega) = |F(\omega)|^2 [S_x(\omega) + N_{0c}]$$

(6.1.19)

This filter is generally tuned to the bandpass characteristic of the desired signal spectrum. If we assume a wide optical bandwidth $[H(\omega) = e]$ and a

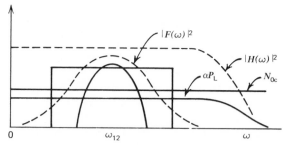

Figure 6.3. The filtered detector spectrum. $F(\omega)$ is the postdetection filter.

flat filter function $[F(\omega) = 1]$ over the bandwidth of the signal, the filtered signal power after heterodyning is

$$P_s = e^2(4\alpha^2 A_r P_L) \cdot \frac{1}{2\pi} \int_{-\infty}^{\infty} |F(\omega)|^2 [\tfrac{1}{4}S_s(\omega - \omega_{12}) + \tfrac{1}{4}S_s(-\omega - \omega_{12})] \, d\omega$$

$$= e^2(4\alpha^2 A_r P_L) \frac{I_s}{2} \tag{6.1.20}$$

where I_s is the signal intensity in the received optical field. The bandpass noise in the heterodyned signal bandwidth can be determined by collecting the noise terms. If the background is assumed white, the total noise power in the signal bandwidth B_s is then

$$P_n = [e^2\alpha P_L + 2e^2\alpha^2 P_L N_{0b} + N_{0c}]2B_s \tag{6.1.21}$$

The terms in brackets represent the two-sided noise levels due to shot noise, background, and circuit noise, respectively. The heterodyned SNR_p is then

$$SNR_p = \frac{2(\alpha^2 A_r P_L I_s)}{[\alpha P_L + 2\alpha^2 P_L N_{0b} + (N_{0c}/e^2)]2B_s}$$

$$= \frac{2\alpha A_r I_s/2B_s}{1 + 2\alpha N_{0b} + (N_{0c}/\alpha P_L e^2)} \tag{6.1.22}$$

This represents the detected SNR_p developed at the output of a heterodyning receiver operating within a single spatial mode, over a bandwidth B_s about frequency $\omega_{12} = \omega_1 - \omega_2$. Note that SNR_p depends directly on the signal power $P_s = I_s A_r$ collected over the receiver area. Clearly, this area should be as large as possible, provided it always corresponds to a single spatial mode. This premise is inherent in the single mode model used in deriving SNR_p. We find in Section 6.5, however, that for typical operating conditions this coherent area A_r cannot be made arbitrarily large.

The heterodyned SNR_p is also interesting when compared with the corresponding result for direct detection (Equation 5.3.21). We note that in both cases P_s represents the received averaged signal field power. The direct detected power, however, is reduced by the factor P_m ($P_m \le 1$), because the baseband signal must be intensity modulated onto the optical carrier rather than directly modulated onto it. The bandwidths B_s and B_m each represent bandwidths occupied after modulation with the information waveform, except B_s refers to the optical intensity modulated bandwidth, while B_m is the modulated bandwidth prior to intensity modulation. Hence, (6.1.22) and (5.3.21) can be compared directly in terms of power and bandwidth. Second, we note that the circuit noise can be eliminated, and shot noise limited behavior achieved, by using a strong local source such that $e^2 \alpha P_L \gg N_{0c}$. In this sense the local source power P_L is playing the role of the photomultiplied signal power $\bar{G}P_s$ in (5.4.2). Thus effective signal amplification, as far as achieving shot noise limited operation, is provided by a strong local source. Third, we notice, however, that this local source cannot eliminate the effect of the background. In fact, we notice that even with a strong source (6.1.22) becomes

$$\text{SNR}_p = \frac{2\alpha P_s/2B_s}{1 + 2\alpha N_{0b}} \tag{6.1.23}$$

and we never actually reach the quantum limited condition of (5.3.24). In practice $\alpha N_{0b} \ll 1$ and it is usually argued that (6.1.23) is approximately a quantum limited result, and we often write

$$\text{SNR}_p \approx \frac{2\alpha P_s}{2B_s}$$

$$= \frac{2\eta P_s}{2hf B_s} \tag{6.1.24}$$

as the quantum limited bound. On the other hand, we should be aware that for high temperature background sources, such that $\alpha N_{0b} \gg 1$, (6.1.23) is instead

$$\text{SNR}_p = \frac{\alpha P_s/B_s}{1 + 2\alpha N_{0b}}$$

$$= \frac{P_s}{2N_{0b}B_s} \tag{6.1.25}$$

The receiver is therefore thermally limited, and not quantum limited, even though a strong local source was used. Thus "hot" background sources prevent attainment of the quantum limited bound of heterodyne detection predicted by (6.1.24). We may also point out that at lower frequencies the thermally limited bound in (6.1.25) is also attained, which explains why quantum limited operation cannot be achieved by heterodyning at microwave frequencies, except perhaps at extremely low temperatures.

The detected shot noise, following heterodyning with a strong local oscillator, has been shown to have the power spectrum in (6.1.17) and the field intensity function of (6.1.10). The corresponding count intensity is then

$$n(t) = \alpha A_r |a_L|^2 \left\{ 1 + \frac{|a_R(t)|}{|a_L|} \cos[\omega_{12}t + \theta_R(t) - \theta_L] \right\} \tag{6.1.26}$$

When the condition of (6.1.9) is imposed, it is clear that with strong local sources ($|a_L| > |a_R(t)|$), we have $n(t) \gg 1$ for almost all t. [We say "almost all t" since $|a_R(t)|$ may be a random process, and with a small probability $n(t)$ in (6.1.26) may take on arbitrarily small values at some t.] From our earlier discussion in Section 4.7, we know that the resulting high intensity shot noise has, at almost all t, a first-order probability density closely approximated as a Gaussian density, with mean and variance proportional to the instantaneous intensity. Thus the heterodyned shot noise produced by a strong local source can be statistically modeled as a Gaussian random process. In this sense, the heterodyned shot noise appears as if it were the sum of a signal $n(t)$ and a nonstationary Gaussian noise process whose variance is also related to $n(t)$. In using this model, however, it must be remembered that the shot noise is only conditionally Gaussian, so that the "signal" and additive "noise" in this type of representation are not truly independent processes. However, since the dominant contribution to the noise term comes from the local oscillator, the assumption of independence is generally accepted.

6.2 DATA SNR$_p$ WITH SINGLE MODE HETERODYNE DETECTION

The SNR$_p$ derived in (6.1.22), (6.1.23), and (6.1.25), for heterodyning in a single spatial mode, refers to signals at the output of the photodetector filter $F(\omega)$ (see Figure 6.1). Of ultimate interest, however, is the SNR$_p$ of the recovered data waveform. To determine this, we must take into account the postdetection processing and its effect on the previously derived heterodyned SNR$_p$. This processing depends on the manner in which the data waveform $d(t)$ is demodulated from the optical carrier envelope.

AM with Heterodyne Detection.   If amplitude modulation (AM) is used to modulate a data waveform $d(t)$ onto the optical carrier, we can write $s(t)$ in (6.1.2) as

$$s(t) = \sqrt{I_s}\,[1 + d(t)] \tag{6.2.1}$$

where we constrain $|d(t)| \leq 1$ to prevent overmodulation and let I_s be the average intensity of the optical signal field. Note that the optical envelope is real and its intensity spectrum is

$$S_s(\omega) = I_s[2\pi\,\delta(\omega) + S_d(\omega)] \tag{6.2.2}$$

where $S_d(\omega)$ is the spectrum of $d(t)$. [This is valid whether $d(t)$ is deterministic or random, so long as $S_d(\omega)$ is properly defined.] The total power in the received optical signal field is then

$$P_s = \frac{I_s A_r}{2}(1 + P_d) \tag{6.2.3}$$

where P_d is the power in $d(t)$. The bandwidth occupied by the envelope $s(t)$, and therefore the bandwidth occupied by the heterodyned signal term at frequency ω_{12}, is then

$$B_s = 2B_d \tag{6.2.4}$$

where B_d is the bandwidth of $d(t)$. After photodetecting and filtering, the heterodyned signal, in the absence of noise, appears as an AM signal at carrier frequency ω_{12} and with bandwidth $2B_d$. Ideal AM demodulation now affords an improvement factor of 2 from the AM SNR_p to the data SNR_p [see (5.6.4)]. From (6.1.24), the AM quantum limited SNR_p [for the shot noise limited case, the denominator must be modified according to (6.1.23)] is then

$$
\begin{aligned}
[SNR_p]_{\omega_{12}} &= \frac{2\alpha P_s P_d}{2B_s} \\
&= \frac{2\alpha P_s P_d}{4B_d}
\end{aligned}
\tag{6.2.5}
$$

where $P_s = I_s A_r/2$, and the ω_{12} subscript indicates SNR_p following heterodyning. For the case $P_d = 1$, the data SNR_p is then

$$
\begin{aligned}
[SNR_p]_d &= 2[SNR_p]_{\omega_{12}} \\
&= 2\left[\frac{\eta P_s}{hf(2B_d)}\right]
\end{aligned}
\tag{6.2.6}
$$

which represents the achievable (upper bound) SNR_p for the data waveform following heterodyning and ideal AM demodulation. Note that when (6.2.6) is compared to the equivalent result for AM–IM operation with direct detection (Equation 5.6.3) we see that the AM $(SNR_p)_d$ is improved by a factor of 8 with perfect heterodyning. This can be directly related to the fact that the data are amplitude modulated directly onto the optical carrier rather than onto its intensity.

An alternative method is to AM directly onto the optical carrier and use optical homodyning at the receiver, obviating the need for AM demodulation following photodetection. In this case, the filtered output of the photodetector yields directly the spectrum of (6.2.2), in the absence of background noise. The recovered quantum limited SNR_p is then

$$
\begin{aligned}
[SNR_p]_d &= \frac{\alpha(2|a_L|)^2(A_r I_s/2)P_d}{|a_L|^2 2B_d} \\
&= 2\left[\frac{\eta P_s P_d}{hf(2B_d)}\right]
\end{aligned}
\tag{6.2.7}
$$

which is identical to (6.2.6) for $P_d = 1$. Thus we see that there is no difference in data SNR_p between homodyning and heterodyning with AM demodulation. Some confusion often occurs on the point, primarily because there is a tendency to compare data SNR_p in (6.2.7) with heterodyned SNR_p in (6.2.5). For the case $P_d = 1$, $P_s = I_s A_r/2$, and we see that the former is larger than the latter by a factor of 2. The difference, of course, is that in the homodyne case this 2 factor is already included, whereas in heterodyning this improvement is yet to be recovered (by the ideal AM detector).

FM with Heterodyne Detection.   When the data $d(t)$ is frequency modulated (FM) onto the optical carrier, the complex envelope now has the form

$$s(t) = \sqrt{I_s} \exp\left[-j \, \Delta\omega \int d(t) \, dt \right] \tag{6.2.8}$$

Here $\Delta\omega$ is the transmitted frequency deviation and $P_s = I_s A_r/2$ is the power in the received signal mode. The bandwidth occupied by the envelope $s(t)$ is approximately

$$B_s = 2(\beta + 1)B_d \quad \text{Hz} \tag{6.2.9}$$

where $\beta = \Delta\omega/2\pi B_d$ is the modulation index and B_d is the data bandwidth. After heterodyning and filtering, the quantum limited SNR_p is then

$$[SNR_p]_{\omega_{12}} = 2\left[\frac{\alpha P_s}{2B_s} \right] \tag{6.2.10}$$

Subsequent FM demodulation of the frequency modulated beat term at frequency ω_{12} generates the well-known FM improvement, and yields an output data SNR_p of

$$[SNR_p]_d = 3\beta^2 \left(\frac{B_s}{B_d} \right) [SNR_p]_{\omega_{12}}$$

$$= 6\beta^2 \left(\frac{\eta P_s}{hf(2B_d)} \right) \tag{6.2.11}$$

provided (6.2.10) is above the FM threshold. The output SNR_p can therefore be increased by using large values of β, that is, wide band FM operation. This result is larger by a factor of 2 over that possible with direct detection [see (5.6.8)].

6.3   THE ALIGNMENT AND FIELD MATCHING PROBLEM

In the preceding sections we considered an idealized heterodyning operation in which both the received field and the local field are perfectly aligned, and each generate the identical diffraction pattern in the focal plane. In practice, it is extremely difficult to obtain this ideal heterodyning condition, and effects of misalignment and local field distortion must be considered. Minimizing these anomalies therefore becomes an important part of system design.

Let us reconsider the basic heterodyning system in Figure 6.1. We again consider each field to generate identical diffraction patterns, but we now assume that the local field is not perfectly aligned with the received field. As a result the diffraction pattern on the detector surfaces do not overlap, as shown in Figure 6.4. Recall that in Section 1.6, it is shown that a plane wave

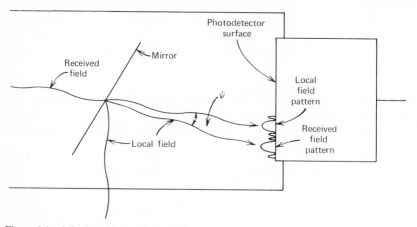

Figure 6.4.   Misaligned heterodyne receiver.

of wavelength λ, arriving normal to a square aperture of width d, is projected to a modal function (diffraction pattern) given by

$$\phi_0(x, y) = \frac{A_r}{\lambda f_c} \left[\frac{\sin(\pi \, dx/\lambda f_c)}{(\pi \, dx/\lambda f_c)} \cdot \frac{\sin(\pi \, dy/\lambda f_c)}{(\pi \, dy/\lambda f_c)} \right] \tag{6.3.1}$$

where x and y are coordinates in the detector plane measured from the center and f_c is the focal length. If the same plane wave arrived at an offset angle from the normal, the diffraction pattern would be approximately the same, except it would be displaced.

Now consider the case of the heterodyne operation of the received field at wavelength λ_1 and the local field at wavelength λ_2, each misaligned from the normal by angles ψ_{xR}, ψ_{yR}, ψ_{xL}, and ψ_{yL}, respectively, as shown in Figure 6.5. The resulting terms in (6.1.8c) now become

$$n_{RL}(t) = [2\alpha a_L m(t)] \, \text{Real} \left\{ \int_{\mathscr{A}_d} \phi_0(x - x_R, y - y_R)\phi_0^*(x - x_L, y - y_L) \, dx \, dy \right\} \tag{6.3.2}$$

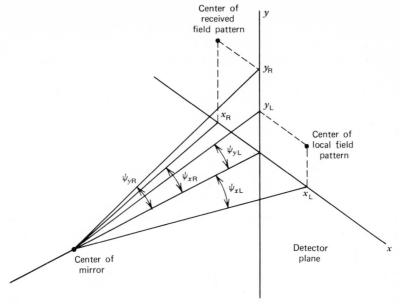

Figure 6.5. Heterodyned offset angles and diffraction pattern locations.

where $x_R = \pi d \sin(\psi_{xR})/\lambda_1$, $y_R = \pi d \sin(\psi_{yR})/\lambda_1$, $x_L = \pi d \sin(\psi_{xL})/\lambda_2$, $y_L = \pi d \sin(\psi_{yL})/\lambda_2$, and $m(t)$ is the modulated intensity term in (6.1.14). For the rectangular aperture this now integrates to

$$n_{RL}(t) = [2\alpha a_L m(t)]A_r\left[\frac{\sin \psi_x}{\psi_x} \cdot \frac{\sin \psi_y}{\psi_y}\right] \qquad (6.3.3)$$

where

$$\psi_x \triangleq x_R - x_L \qquad (6.3.4a)$$

$$\psi_y \cong y_R - y_L \qquad (6.3.4b)$$

Thus the previously derived cross product term with perfect alignment is reduced by the misalignment angles of the two fields, according to (6.3.3). For acceptable heterodyning it is generally required that ψ_x and ψ_y be less than unity. This means, for example, that if the received field is perfectly aligned ($\psi_{xR} = \psi_{yR} = 0$), the local field must be aligned such that $\sin(\psi_{xL})$ and $\sin(\psi_{yL})$ are less than $\lambda_2/\pi d$. As the aperture size d is increased, the alignment requirement becomes tighter. A similar argument can be developed for the circular apertures as well (see Problem 6.4).

In the idealized model we also assumed that, in addition to perfect alignment, the diffraction pattern of the two fields is identical. When separate lens are used to focus each field onto the photodetector, it is difficult to produce identical patterns. Suppose, for example, we assume that the received field is a single mode diffraction pattern $\sqrt{A_r}\,\phi_{OR}(\mathbf{r})$, while the local field has a pattern $\sqrt{A_r}\,\phi_{OL}(\mathbf{r})$. The heterodyned cross term now becomes

$$n_{RL}(t) = A_r 2\alpha a_L\, m(t)\, \text{Real}\left\{\int_{\mathscr{A}_d} \phi_{OR}(x, y)\phi_{OL}^*(x, y)\, dx\, dy\right\} \qquad (6.3.5)$$

The power in this time function generates the signal power of the heterodyned detector output. Thus,

$$P_{s0} = (2eA_r\alpha a_L)^2\left(\frac{I_s}{2}\right)\left|\int_{\mathscr{A}_d} \phi_{OR}(x, y)\phi_{OL}^*(x, y)\, dx\, dy\right|^2 \qquad (6.3.6)$$

where $I_s/2$ is again the intensity of the received signal field envelope $m(t)$. The additive shot noise in a bandwidth $2B_s$ due to a strong local source is then

$$P_n = \alpha e^2(1 + 2\alpha N_{Ob})2B_s \int_{\mathscr{A}_d} |a_L\sqrt{A_r}\,\phi_{OL}(\mathbf{r})|^2\, d\mathbf{r}$$

$$= \alpha e^2(1 + 2\alpha N_{Ob})2B_s A_r a_L{}^2 \int_{\mathscr{A}_d} |\phi_{OL}(\mathbf{r})|^2\, d\mathbf{r} \qquad (6.3.7)$$

instead of (6.1.21). The corresponding SNR_p then becomes

$$\text{SNR}_p = \left[\frac{2\alpha P_s/2B_s}{1 + 2\alpha N_{Ob}}\right]\frac{\{\text{Real}\int_{\mathscr{A}_d} \phi_{OR}(\mathbf{r})\phi_{OL}^*(\mathbf{r})\, d\mathbf{r}\}^2}{\int_{\mathscr{A}_d} |\phi_{OL}(\mathbf{r})|^2\, d\mathbf{r}} \qquad (6.3.8)$$

This result can be compared directly to (6.1.23). We see that the previously derived SNR_p has been multiplied by the ratio of integrals above. By the use of the Schwarz inequality on the numerator we immediately establish that the ratio is bounded by one, and therefore appears as a SNR_p suppression factor of the heterodyne operation. Furthermore, we also see that the ratio has its maximum value only if the diffraction patterns are exactly proportional, that is,

$$\phi_{OL}(\mathbf{r}) = C\phi_{OR}(\mathbf{r}) \qquad (6.3.9)$$

Thus the local field must be spatially matched to the received field at the detector surface. This means, of course, they are also perfectly aligned.

The same conclusion is reached if we refer both detector fields back to the aperture plane. Recall that the diffraction pattern is related to the Fourier transform of the aperture field. We can therefore associate an aperture function $w_0(\mathbf{r})$ at the receiver surface $\mathscr{A}_r$ with each focal plane function

$\phi_0(\mathbf{r})$ at the detector surface $\mathcal{A}_d$. However, by Parceval's theorem we know that

$$\int_{\mathcal{A}_r} w_{0R}(\mathbf{r}) w_{0L}^*(\mathbf{r}) \, d\mathbf{r} = \int_{\mathcal{A}_d} \phi_{0R}(\mathbf{r}) \phi_{0L}^*(\mathbf{r}) \, d\mathbf{r} \qquad (6.3.10)$$

Thus the same suppression ratio in (6.3.8) would be produced if we considered the summing of the fields to have occurred at the aperture using the aperture functions rather than at the detector surface.

It is not always possible to match exactly the spatial patterns of two separate fields, and as a result a degradation in idealized SNR_p almost always occurs. Consider the case where the received field has the diffraction pattern (6.3.1), but the additive source field is represented as a pure plane wave. (This would occur if a smaller lens is used to focus the local field or, in the limit, no lens is used at all and the local field is projected directly on the detector.) The local field function then becomes $\phi_{0L}(\mathbf{r}) = \exp[-j\mathbf{z} \cdot \mathbf{r}]$, representing a plane wave arriving at the detector from an off-axis ray direction $\mathbf{z}$. The suppression term in (6.3.8) is then

$$\frac{\{\text{Real} \int_{\mathcal{A}_d} \phi_{0R}(\mathbf{r}) \exp[j\mathbf{z} \cdot \mathbf{r}] \, d\mathbf{r}\}^2}{\int_{\mathcal{A}_d} |\exp[-j\mathbf{z} \cdot \mathbf{r}]|^2 \, d\mathbf{r}} \qquad (6.3.11)$$

For a square aperture of dimension d and a square detector of dimension b, (6.3.11) becomes

$$\left(\frac{\lambda_1 f_c}{2\pi^2 \, db}\right)^2 \left\{ S_i\left[\frac{\pi b}{\lambda_1}\left(\frac{d}{2f_c} + \frac{\lambda_1}{\lambda_2} \sin \psi_{xL}\right)\right] + S_i\left[\frac{\pi b}{\lambda_1}\left(\frac{d}{2f_c} - \frac{\lambda_1}{\lambda_2} \sin \psi_{xL}\right)\right] \right\}^2$$

$$\cdot \left\{ S_i\left[\frac{\pi b}{\lambda_1}\left(\frac{d}{2f_c} + \frac{\lambda_1}{\lambda_2} \sin \psi_{yL}\right)\right] + S_i\left[\frac{\pi b}{\lambda_1}\left(\frac{d}{2f_c} - \frac{\lambda_1}{\lambda_2} \sin \psi_{yL}\right)\right] \right\}^2 \qquad (6.3.12)$$

where (ψ_{xL}, ψ_{yL}) are again the off-axis angles of the local field and

$$S_i(x) = \int_0^x \frac{\sin t}{t} \, dt \qquad (6.3.13)$$

is the sine integral. If the misalignment angles (ψ_{xL}, ψ_{yL}) are small (such that $\sin \psi \ll d/2f_c$), then (6.3.12) reduces to

$$\left(\frac{2\lambda_1 f_c}{\pi^2 db}\right)^2 \left\{ S_i\left(\frac{\pi db}{2\lambda_1 f_c}\right) \right\}^4 = \left(\frac{1}{\pi}\right)^2 \left\{ \frac{S_i^2(b/d_0)}{(b/d_0)} \right\}^2 \qquad (6.3.14)$$

where we have defined $d_0 \triangleq 2\lambda_1 f_c/\pi d$, for convenience. The SNR_p suppression is therefore a function of the normalized detector size, b/d_0, when heterodyning with a local plane wave. The maximum value of the braces in (6.3.14) occurs when $b/d_0 \approx 2.2$, for which the associated suppression factor is $(1.29)^2/\pi^2$ $= 0.168$. Thus the heterodyned SNR_p is reduced by approximately 7.7 dB when plane wave heterodyning is used and the detector area is properly selected. Note that the required detector size is $b = 2.2\, d_0 \approx 1.4(\lambda_1 f_c/d)$, where we recognize $(2\lambda_1 f_c/d)$ as the distance between zeros in the received field diffraction pattern. The suppression loss is directly attributable to the fact that the local field is spread out over a wider area of the detector surface than that which is actually necessary for the received field. This excess field

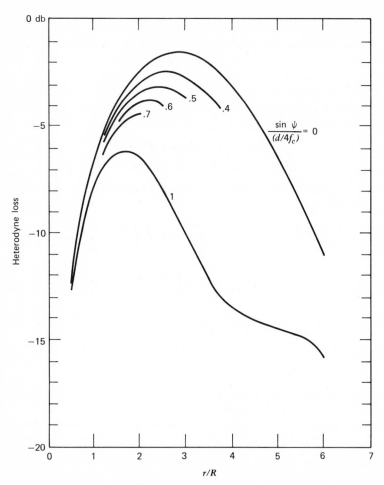

Figure 6.6. Heterodyne loss due to misalignment of plane wave local field, circular aperture and detector. (Equation 6.3.15). r = detector radius, d = aperture diameter, $R_0 = 2\lambda f_c/\pi d$, ψ = offset angle.

power is therefore wasted power in terms of the heterodyne operation; that is, it should have been concentrated over the received field pattern.

For the circular aperture and detector, the misalignment suppression loss in (6.3.11) is instead (Problem 6.5)

$$\left[\frac{2}{(r/R)} \int_0^{r/R} J_1(x) J_0\left(\frac{4f_c x \sin \psi}{d}\right) dx \right]^2 \tag{6.3.15}$$

where r is the detector radius, d the aperture diameter, $R = 2\lambda f_c/\pi d$, and ψ is the local oscillator misalignment angle. A plot of (6.3.15) is shown in Fig. 6.6. Note again that alignment losses can be reduced by proper selections of detector size, the latter dependent on the degree of misalignment. It is interesting to note that the losses with misalignment angles out to about $\psi \leq \sin^{-1}(d/4f_c)$ are only several decibels greater than that for a perfectly aligned plane wave ($\psi = 0$), with detectors having $r \approx 3R$. This means that except for the loss factor of using a mismatched plane wave, (6.3.15) shows an insensitivity to alignment errors for this range of ψ. This implies the feasibility of using a single local plane wave for heterodyning over an entire array.

6.4 MULTIMODE HETERODYNING

In an earlier discussion we examined a receiver employing idealized heterodyning over a single spatial mode. When more than one mode is involved, or when conflicting modes between received and local fields occur, the results must be properly modified. In this section we attempt to develop a more general mathematical model for handling these multimode cases.

Consider a general coherence-separable optical field, composed of the sum of a transmitted optical signal field and a background white Gaussian noise field. From our earlier discussion, we know we can expand such fields over the receiver surface according to (5.2.3), as

$$f_R(t, \mathbf{r}) = \sum_{j=0}^{\infty} f_{Rj}(t) w_j(\mathbf{r}) \tag{6.4.1}$$

where $f_{Rj}(t)$ is the received field time variation in the jth spatial mode. The functions $\{w_j(\mathbf{r})\}$ are the orthonormal spatial functions associated with the field area $\mathscr{A}_r$. The $\{f_{Rj}(t)\}$ are uncorrelated random processes, containing both signal and noise terms, that describe the optical field in each mode. For the special case of an optical field occupying a single spatial mode (6.4.1) takes the form of (6.1.1). When the single mode assumption is not valid the field must be modeled in the more general expression of (6.4.1), with the obvious addition of more spatial modes.

Consider now a deterministic local field expanded in the same functions as (6.4.1). (We are assuming perfect alignment and matched diffraction patterns.) Thus we may write

$$f_L(t, \mathbf{r}) = \sum_{j=0}^{\infty} f_{Lj}(t) w_j(\mathbf{r}) \tag{6.4.2}$$

to represent the local field modal expansion. We assume each field mode has power P_L. Again, a local field occupying a single spatial mode represents a specialization of (6.4.2) to (6.1.3). When more local modes exist, (6.4.2) must be used to describe its field. In the heterodyning operation the received and local fields are again summed at the receiver surface. The resulting photodetected count intensity is again given by

$$n(t) = \alpha \int_{\mathscr{A}_r} |f_R(t, \mathbf{r}) + f_L(t, \mathbf{r})|^2 \, d\mathbf{r} \tag{6.4.3}$$

Substituting from (6.4.1) and (6.4.2), expanding, and invoking the orthonormality of the eigenfunctions, then yields

$$n(t) = n_R(t) + n_L(t) + n_{RL}(t) \tag{6.4.4}$$

where now

$$n_R(t) = \alpha A_r \sum_{j=0}^{\infty} |f_{Rj}(t)|^2 \tag{6.4.5a}$$

$$n_L(t) = \alpha A_r \sum_{j=0}^{\infty} |f_{Lj}(t)|^2 \tag{6.4.5b}$$

$$n_{RL}(t) = 2\alpha A_r \, \text{Real}\left\{ \sum_{j=0}^{\infty} f_{Rj}(t) f_{Lj}^*(t) \right\} \tag{6.4.5c}$$

This represents the generalized intensity produced from multimode heterodyning. When expanded in this way, we see that the cross product term is the sum of products of time functions, each time function associated with the same spatial mode, where the sum is over all spatial modes common to each field. The term $n_L(t)$, which represents the primary shot noise term for strong local fields, is composed of the sum of the intensity components of each mode. Equation 6.4.4 reduces in certain special cases of interest. If the local field exists only in the (j_0)th mode, then (6.4.5c) simplifies to

$$n_{RL}(t) = 2\alpha A_r \, \text{Real}\{ f_{Rj_0}(t) f_{Lj_0}^*(t) \} \tag{6.4.6}$$

Note that the cross product involves only the received time function in the (j_0)th mode. Hence, heterodyning with a single mode local source can be interpreted as mathematically selecting only the received field in the corresponding eigenmode. The selection is achieved by summing and integrating out the orthonormal cross product terms. Physically, the summing is performed by aligning the fields, and the spatial integration is performed by the photodetection operation. When the received field itself contains an identical single mode, as assumed in (6.4.3), then (6.4.4) yields our earlier result in (6.1.7). When the received field contains more than one mode (due, for example, to imperfect transmission or turbulent channels) the heterodyned intensity is related only to that component of the local source mode.

Now consider a strong local field occupying M spatial modes, heterodyned with a single mode signal field in additive multimode background noise. We assume each local field mode has equal amplitude a_L, and the signal is confined to the (j_0)th mode. As before, the signal component in the (j_0)th mode generates a power term of

$$P_s = (4e^2\alpha^2 A_r P_L)\left(\frac{I_s}{2}\right) \tag{6.4.7}$$

For each of the local field modes we generate a noise term

$$P_n = e^2[\alpha P_L + 2(\alpha^2 P_L)N_{0b}]2B_s \tag{6.4.8}$$

over a bandwidth B_s. The total noise is that contributed from all modes on the receiver surface plus that added by the circuit noise to the detector output. The heterodyned SNR becomes

$$\begin{aligned}
\mathrm{SNR}_p &= \frac{2(\alpha^2 P_L)A_r I_s}{MP_n + (N_{0c}2B_s/e^2)} \\
&= \frac{2\alpha P_s/2MB_s}{1 + 2\alpha N_{0b} + (N_{0c}/M\alpha P_L e^2)}
\end{aligned} \tag{6.4.9}$$

Clearly, we have reduced the resulting SNR_p from the idealized case in (6.1.22) by the factor M, due to the number of unnecessary local field modes. These excess modes are heterodyning background noise, while producing additive shot noise. We are, however, using the excess mode power to aid in overcoming the circuit noise, that is, to produce shot noise limited operation. Note that the division by M can be considered a form of signal power suppression, similar to the suppression effect in (6.3.6). In the latter case, the cause is field mismatch, whereas in the present case the fields are matched but unnecessary modes are used. In essence, the local field source is overspread for the received field.

Now extend to the case where both the local and received fields occupy more than one mode. Let the local field have M_L modes, and the received field M_R modes. We describe the transmitted signal field as $s(t)\exp(j\omega_1 t)$, where $s(t)$ is the complex envelope with intensity I_s referred to the receiver. We assume, for simplicity, the received field is distributed equally over all M_R modes, and we describe the received signal field in the ith mode as $(1/\sqrt{M_R})s(t - \tau_i)\exp[j\omega_i(t - \tau_i)]$ where τ_i is the mode delay and ω_i is the mode frequency. The difference in delays $\{\tau_i\}$ over the modes is due to the channel delay dispersion, and the difference in the frequencies is due to possible Doppler spreading over the modes. Each mode has intensity I_s/M_R, so that the total available signal intensity I_s is spread equally over all modes. We consider the local field to be $a_L \exp[j\omega_2 t]$ in each mode. Hence, the heterodyned signal in (6.4.6) is

$$n_{RL}(t) = 2\alpha a_L A_r \, \text{Real} \sum_{i=1}^{M_L} (1/\sqrt{M_R})s(t - t_i) \exp[j\omega_{i2} t - \omega_i \tau_i]$$

$$= 2\alpha a_L A_r \sum_{i=1}^{M_L} (1/\sqrt{M_R})s(t - t_i) \cos(\omega_{i2} t - \omega_i \tau_i) \qquad (6.4.10)$$

where $\omega_{i2} = \omega_i - \omega_2$. Equation 6.4.10 corresponds to the heterodyned collection of delayed field modes, and therefore represents the heterodyne counterpart to our multimode direct detection result in (5.7.11). If the dispersion in frequency (referred to the difference frequency ω_{i2}) and delay is excessive, (6.4.10) corresponds to interfering waveforms and the structure of the waveform $s(t)$ is destroyed. On the other hand, if we assume the frequencies and delays are not widely dispersed, the terms in (6.4.10) combine to reinforce the signal. For example, if we assume that the frequency dispersion is negligible ($\omega_i = \omega_1$) and the delays are all identical, ($\tau_i = \tau$), then the heterodyned SNR can be computed as in (6.4.9). For the case $M_R > M_L$

$$P_s = (2e\alpha a_L A_r)^2 E\left[\sum_{i=1}^{M_L} \frac{s(t - \tau)}{\sqrt{M_R}} \cos[\omega_{12}(t - \tau)]\right]^2$$

$$= 2(e^2\alpha^2 A_r P_L)M_L{}^2\left(\frac{I_s}{M_R}\right) \qquad (6.4.11)$$

while for $M_R \leq M_L$

$$P_s = 2(\alpha^2 e^2 P_L A_r)M_R I_s \qquad (6.4.12)$$

In either case, the noise power is given by

$$P_n = e^2[M_L \alpha P_L + 2M_L(\alpha^2 P_L)N_{0b} + (N_{0c}/e^2)]2B_s \qquad (6.4.13)$$

and therefore increases with M_L. In (6.4.11) signal power is lost by not using all available signal modes. In (6.4.13), the performance can clearly be improved by reducing M_L to M_R, which does not affect P_s but reduces P_n. When $M_L = M_R$, the resulting SNR is

$$\text{SNR}_p = \frac{2\alpha A_r I_s / 2 B_s}{1 + 2\alpha N_{0b} + (N_{0c}/M_L \alpha P_L e^2)} \tag{6.4.14}$$

If single mode heterodyning had been used, the SNR_p would be given by (6.4.9) with the mode intensity I_s replaced by I_s/M_R. Thus (6.4.14) is M_R times larger than if single mode heterodyning were used. The system is therefore benefiting from the coherent combining of the signal modes when the frequency and delay dispersion are negligible. We emphasize that the delay dispersions in (6.4.10) must be small relative to a period of the heterodyne frequency $\omega_{12}/2\pi$. [Recall that in direct detection the delay spread is relative to the time variations of the intensity modulation $s(t)$.] For example, if we assume the $\{\tau_i\}$ are independent delays, and each produce uniformly distributed phases over a period $\omega_{12}/2\pi$, then (6.4.11) becomes

$$P_s = 2(e\alpha a_L A_r)^2 M_L \left(\frac{I_s}{M_R} \right) \tag{6.4.15}$$

and (6.4.14) is reduced by the factor $1/M_L$. In this case, we are no longer coherently combining the heterodyned modes and the multimode advantage is lost. Thus an advantage can be attained in heterodyning with many field modes only if the mode delay dispersion is small, or can be reduced. This suggests again the possible use of delay equalization over the heterodyned modes as a tradeoff for field power and bandwidth in multimode detection. Just as in direct detection, this can theoretically be achieved in a space channel by spatially separating the heterodyned modes (heterodyne and detect each mode separately with a detector array) and attempting to phase correct (perhaps from phase measurements obtained by previous channel probing, or by estimating each mode phase using the methods in Chapter 9) before combining. The nonstationarity of the space channel may seriously hinder the practical implementation of this type of equalizer. The local field still has the requirement of matching and aligning to each field mode, but in addition it must now have enough strength in each mode to overcome the circuit noise of each individual detector. If the field is not properly matched, a power suppression factor must be included for each detector, which may negate the coherent combining advantage if too severe. Again, delay equalization may have its most practical application to the fiberoptic channel, where fibers tend to produce many signal modes. Heterodyning over a single spatial mode produces power loss because of the unusable modes. Therefore, one must improve the fiber (reduce its number of modes) or attempt to regain lost power by multimode heterodyning with delay equalization. Since it is difficult to separate modes spatially in an optical guide, equalization must be achieved directly with the heterodyned signal in (6.4.10). This operation is hindered by the fact that when the modes are in "equilibrium," each mode itself has a delay spread that must undergo a comparable equalization.

6.5 EFFECT OF ATMOSPHERIC TURBULENCE ON HETERODYNING

We pointed out that if an optical field is transmitted over a turbulent path, the effect is to "break up" the optical beam spatially. In discussing non-coherent detection in Chapter 5 we considered this effect as an apparent extension of the optical source. In dealing with coherent detection it is more convenient to consider the turbulent effect as one of converting a coherent field to a random field. This randomness is over the spatial variable $\mathbf{r}$, as well as t, and corresponds to a point to point loss in coherence over the receiver area. In addition, the turbulence causes the field to evolve as a random process in time at each point. We are here interested in assessing the spatial effects of turbulence on heterodyning. Let us first examine the detected field at a fixed time t. Subsequent time averaging will allow comparison to our earlier SNR_p results.

Consider the received signal field to be represented as $s(t, \mathbf{r}) \exp(j\omega_1 t)$ at the receiver surface, where $s(t, \mathbf{r})$ is again the complex signal envelope. A local heterodyning field source, producing a single mode diffraction pattern at the photodetector surface, will project back to the aperture as an equivalent plane wave function. We again describe this local field function as $\sqrt{A_r} a_L \exp(j\omega_2 t)w_0(\mathbf{r})$, where $|w_0(\mathbf{r})|^2$ integrates to unity over $\mathscr{A}_r$. Optical mixing followed by photodetection produces the intensity cross product term

$$n_{\text{RL}}(t) = \sqrt{A_r}(2\alpha a_L) \, \text{Real}\left\{ e^{j\omega_{12}t} \int_{\mathscr{A}_r} s(t, \mathbf{r})w_0^*(\mathbf{r}) \, d\mathbf{r} \right\} \tag{6.5.1}$$

If the received signal field and the local field are aligned plane waves in the same mode, that is, $s(t, \mathbf{r}) = \sqrt{A_r} \, s(t)w_0(\mathbf{r})$ then (6.5.1) reduces to (6.1.8c), with I_s the average intensity of $s(t)$. The resulting SNR_p is then identical to (6.1.23) when operating under shot noise limited conditions.

Now let us reconsider this result when the received field has undergone turbulent transmission, so that the received plane wave condition is no longer true, and $s(t, \mathbf{r})$ must be considered a random field envelope. Since the integral in (6.5.1) now involves a spatially random field at each t, $n_{\text{RL}}(t)$

is now a random process in t. Its mean squared value (total spatial power at time t) is then

$$E[n_{\mathrm{RL}}{}^2(t)] = A_{\mathrm{r}}(2\alpha a_{\mathrm{L}})^2 E \left| \int_{\mathscr{A}_r} s(t, \mathbf{r}) w_0^*(\mathbf{r})\, d\mathbf{r} \right|^2$$

$$= A_{\mathrm{r}}(2\alpha a_{\mathrm{L}})^2 \int_{\mathscr{A}_r} \int_{\mathscr{A}_r} E[s(t, \mathbf{r}_1)s^*(t, \mathbf{r}_2)] w_0(\mathbf{r}_1) w_0^*(\mathbf{r}_2)\, d\mathbf{r}_1\, d\mathbf{r}_2 \quad (6.5.2)$$

Recall that for coherence-separable fields, the field coherence can be written as

$$E[s(t, \mathbf{r}_1)s^*(t, \mathbf{r}_2)] = R_t(t, t)\tilde{R}_s(\mathbf{r}_1, \mathbf{r}_2) \quad (6.5.3)$$

where $\tilde{R}_s(\mathbf{r}_1, \mathbf{r}_2)$ is the normalized mutual coherence function and $R_t(t, t)$ is the field intensity at time t. Equation 6.5.2 becomes

$$E[n_{\mathrm{RL}}{}^2(t)] = A_{\mathrm{r}}(2\alpha a_{\mathrm{L}})^2 R_t(t, t) \int_{\mathscr{A}_r} \int_{\mathscr{A}_r} \tilde{R}_s(\mathbf{r}_1, \mathbf{r}_2) w_0(\mathbf{r}_1) w_0^*(\mathbf{r}_2)\, d\mathbf{r}_1\, d\mathbf{r}_2 \quad (6.5.4)$$

The time averaged heterodyned signal power [time average of $E(n_{\mathrm{RL}}(t)^2)$] is then the numerator of the SNR_p parameter, replacing (6.1.20). Hence,

$$P_s = 2(e\alpha a_{\mathrm{L}})^2 I_s A_{\mathrm{r}} \int_{\mathscr{A}_r} \int_{\mathscr{A}_r} \tilde{R}_s(\mathbf{r}_1, \mathbf{r}_2) w_0(\mathbf{r}_1) w_0^*(\mathbf{r}_2)\, d\mathbf{r}_1\, d\mathbf{r}_2 \quad (6.5.5)$$

where I_s is again the time averaged signal field intensity. The shot noise power collected over the receiver area A_{r} can be computed as before. For a strong local source the shot noise contributes a power of αP_{L}. The effect of additive background noise appears only in the portion of the noise heterodyned by the local source. Hence, the total noise power collected at the detector (shot noise plus heterodyned background) is identical to that of single mode detection. Therefore, the turbulent SNR_p is then

$$\mathrm{SNR}_p = \frac{2\alpha^2 P_{\mathrm{L}} I_s}{[\alpha P_{\mathrm{L}} + 2\alpha^2 P_{\mathrm{L}} N_{0\mathrm{b}}]2B_s} \left[\iint_{\mathscr{A}_r} \tilde{R}_s(\mathbf{r}_1, \mathbf{r}_2) w_0(\mathbf{r}_1) w_0^*(\mathbf{r}_2)\, d\mathbf{r}_1\, d\mathbf{r}_2 \right] \quad (6.5.6)$$

If we compare (6.5.6) to our earlier result in (6.1.23), we see that the bracketed term above plays the role of an effective receiver area. Thus we write

$$\mathrm{SNR}_p = \frac{2\alpha I_s A_{\mathrm{re}}}{[1 + 2\alpha N_{0\mathrm{b}}]2B_s} \quad (6.5.7)$$

where we define

$$A_{\mathrm{re}} \triangleq \int_{\mathscr{A}_r} \int_{\mathscr{A}_r} \tilde{R}_s(\mathbf{r}_1, \mathbf{r}_2) w_0(\mathbf{r}_1) w_0^*(\mathbf{r}_2)\, d\mathbf{r}_1\, d\mathbf{r}_2 \quad (6.5.8)$$

We can immediately establish that this effective area is never greater than the true area A_r. This follows since $\tilde{R}_s(\mathbf{r}_1, \mathbf{r}_2)$ is a normalized coherence function and therefore bounded by one for all $\mathbf{r}_1, \mathbf{r}_2$. This means

$$A_{re} \leq \left[\int_{\mathscr{A}_r} |w_0(\mathbf{r})| \, d\mathbf{r} \right]^2 \leq A_r \int_{\mathscr{A}_r} |w_0(\mathbf{r})|^2 \, d\mathbf{r} = A_r \qquad (6.5.9)$$

We therefore conclude that as far as waveform distortion is concerned, the effect of transmission turbulence on heterodyning can be accounted for by replacing true receiver area by an effective receiver area in our previous equations. The size of this effective area depends on the operating conditions of the system model.

Consider a circular receiver of diameter d and an impinging homogenious signal field, so that

$$A_r = \frac{\pi d^2}{4} \qquad (6.5.10a)$$

$$\tilde{R}_s(\mathbf{r}_1, \mathbf{r}_2) = \tilde{R}_s(|\mathbf{r}_1 - \mathbf{r}_2|) \qquad (6.5.10b)$$

The local source is assumed to have a single mode diffraction pattern and therefore a receiver function

$$w_0(\mathbf{r}) = \begin{cases} \dfrac{1}{\sqrt{A_r}}, & |\mathbf{r}| \leq \dfrac{d}{2} \\[2ex] 0, & |\mathbf{r}| > \dfrac{d}{2} \end{cases} \qquad (6.5.11)$$

where the spatial axis origin is taken as the detector center. Making the substitutions

$$\boldsymbol{\rho}_1 = \mathbf{r}_1 - \mathbf{r}_2 \qquad (6.5.12)$$

$$\boldsymbol{\rho}_2 = \tfrac{1}{2}(\mathbf{r}_1 + \mathbf{r}_2) \qquad (6.5.13)$$

the effective heterodyning area in (6.5.8) becomes

$$A_{re} = \int_{\mathscr{A}_r} \int_{\mathscr{A}_r} w_0(|\boldsymbol{\rho}_2 + \tfrac{1}{2}\boldsymbol{\rho}_1|) w_0(|\boldsymbol{\rho}_2 - \tfrac{1}{2}\boldsymbol{\rho}_1|) \tilde{R}_s(|\boldsymbol{\rho}_1|) \, d\boldsymbol{\rho}_1 \, d\boldsymbol{\rho}_2$$

$$= \int_{\mathscr{A}_r} \tilde{R}_s(|\boldsymbol{\rho}_1|) \, d\boldsymbol{\rho}_1 \int_{\mathscr{A}_r} w_0(|\boldsymbol{\rho}_2 + \tfrac{1}{2}\boldsymbol{\rho}_1|) w_0(|\boldsymbol{\rho}_2 - \tfrac{1}{2}\boldsymbol{\rho}_1|) \, d\boldsymbol{\rho}_2 \qquad (6.5.14)$$

The inner integration† in $\boldsymbol{\rho}_2$ can be related to the overlapping area of two equal diameter circles located at $\pm\tfrac{1}{2}\boldsymbol{\rho}_1$. This integral therefore depends only

† The inner integral is called the *operating transfer function* (OTF) of the receiver. See Equation D.16 of Appendix D.

on $|\boldsymbol{\rho}_1|$, the distance between circle centers. The inner integration was determined by Fried [1] (see Problem 6.9) to be

$$
R_0(|\boldsymbol{\rho}_1|) = \begin{cases} \dfrac{2}{\pi d^2}\left[d^2\cos^{-1}\left(\dfrac{|\boldsymbol{\rho}_1|}{d}\right) - |\boldsymbol{\rho}_1|(d^2 - |\boldsymbol{\rho}_1|^2)^{1/2}\right], & |\boldsymbol{\rho}_1| \le d \\ 0, & |\boldsymbol{\rho}_1| > d \end{cases}
$$

$$(6.5.15)$$

The effective area is therefore

$$
A_{\mathrm{re}} = \int_0^d \tilde{R}_s(|\boldsymbol{\rho}_1|)R_0(|\boldsymbol{\rho}_1|)\,d\boldsymbol{\rho}_1 \tag{6.5.16}
$$

The presence of only the magnitude of $\boldsymbol{\rho}_1$ in the integrand immediately suggests the use of polar coordinates, and (6.5.16) can be converted to the scalar integral

$$
A_{\mathrm{re}} = 2\pi\int_0^d \tilde{R}_s(\rho)R_0(\rho)\rho\,d\rho \tag{6.5.17}
$$

where $\rho = |\boldsymbol{\rho}_1|$. This establishes the form of the heterodyning area due to turbulence for circular receivers and homogeneous signal fields. The result of this example, though restricted to the stated assumptions, does allow us to examine qualitatively the relationship between the spatial coherence function of the signal and the detector size. If $\tilde{R}_s(\rho)$ has a flat coherence over r_0 [i.e., $\tilde{R}_s(\rho) = 1$, $\rho \le r_0$], then (6.5.17) becomes

$$
A_{\mathrm{re}} = 2\pi\int_0^d R_0(\rho)\rho\,d\rho, \qquad r_0 \ge d \tag{6.5.18a}
$$

and

$$
A_{\mathrm{re}} = 2\pi\int_0^{r_0} R_0(\rho)\rho\,d\rho, \qquad r_0 < d \tag{6.5.18b}
$$

These areas can be evaluated by substituting from (6.5.15) and applying standard definite integrals. In particular, we have

$$
2\pi\int_0^d R_0(\rho)\rho\,d\rho = (4d^2)\int_0^1 [u\cos^{-1}u - u^2(1-u^2)^{1/2}]\,du
$$

$$
= \frac{\pi d^2}{4} \tag{6.5.19}
$$

and therefore

$$
A_{\mathrm{re}} = \frac{\pi d^2}{4} = A_{\mathrm{r}} \tag{6.5.20}
$$

Thus the effective area is the true area as long as the coherence distance exceeds the receiver diameter. The integration in (6.5.18b) is more difficult but can be bounded by making use of the monotonic decreasing characteristic of $R_0(\rho)$. This yields

$$A_{re} \leq 2\pi R_0(0) \int_0^{r_0} \rho \, d\rho = \pi r_0^2, \qquad r_0 < d \qquad (6.5.21)$$

Now we see that the effective area is no larger than the coherence area when the latter width is less than the receiver diameter. The area A_{re} is sketched in Figure 6.7 as a function of receiver diameter d. For $d < r_0$ the effective area (and therefore the heterodyned SNR_p) is directly related to d^2 and increases as the receiver size increases, as it should for single mode heterodyning. When d exceeds r_0 the effective area no longer depends on d, and never exceeds the value at $d = r_0$. Hence, no further improvement is possible by increasing d beyond the coherence distance. Physically, the turbulence breaks up the optical beam to such an extent that increasing receiver size no longer collects useful (coherent) signal energy. The principal conclusion here is that when performing single mode heterodyning there is no advantage in using receiver diameters larger than the coherence distance of the received optical field. Furthermore, the maximum attainable signal power is limited to that collectable over such a coherence area.

The implications of the preceding discussion are useful in system design, although the coherence function used is somewhat idealized. When more realistic coherence functions are used the conclusions are essentially the same. For example, consider Tatarski's model [2] for a turbulent field coherence function:

$$\tilde{R}_s(\rho) = \exp[-3.44(\rho/r_0)^{5/3}] \qquad (6.5.22)$$

where r_0 is an experimentally measured coherence distance, depending on atmospheric conditions. The effective area now takes the integral form

$$A_{re} = 2\pi \int_0^d \exp\left[-3.44\left(\frac{\rho}{r_0}\right)^{5/3}\right] R_0(\rho)\rho \, d\rho \qquad (6.5.23)$$

Fried [1] has numerically computed (6.5.23) and conversion of his results yield the curve shown in Figure 6.7. The curve has essentially the same shape as the idealized bounds plotted previously, and similar conclusions can be made concerning the receiver area and the measured parameter r_0.† Typical

---

† These conclusions follow from the fact that the coherence functions considered decay to zero for large ρ. In certain scattering channel models, the coherence functions may actually approach a low level, constant value as $\rho \to \infty$ (see the reported results in Section D.2 of Appendix D). This implies the existence of a weak coherent component within the distorted field. In this case, increasing the receiver diameter beyond the "basic" coherence distance r_0 may actually lead to a further increase in collected energy.

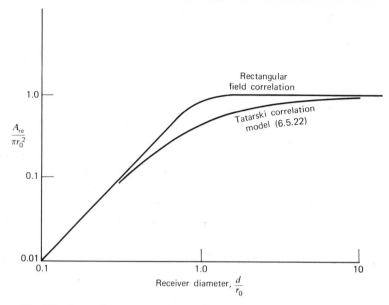

Figure 6.7. Effective receiver area A_{re}: d = receiver diameter, r_0 = coherence distance of the received field.

values for the r_0 parameter have been discussed at length in the literature [3–9]. The important point is that r_0 is generally quite small (on the order of centimeters or less for typical scattering media) which places a severe restriction on the ability to heterodyne through a scattering channel.

It would be convenient to interpret turbulent heterodyning completely in terms of field modes. This can be accomplished by recalling that the mutual coherence function $\tilde{R}_s(\mathbf{r}_1, \mathbf{r}_2)$ in (6.5.3) can always be expanded as

$$\tilde{R}_s(\mathbf{r}_1, \mathbf{r}_2) = \sum_{i=0}^{\infty} \gamma_i W_i(\mathbf{r}_1) W_i^*(\mathbf{r}_2) \tag{6.5.24}$$

where the $\{W_i(\mathbf{r})\}$ are its eigenfunctions and $\{\gamma_i\}$ its corresponding eigenvalues. The latter is normalized so that

$$\sum_{i=0}^{\infty} \gamma_i = A_r \tag{6.5.25}$$

If we substitute the expansion into (6.5.5), we obtain

$$P_s = (2e\alpha a_L)^2 I_s A_r \sum_{i=0}^{\infty} \gamma_i \left| \int_{\mathscr{A}_r} W_i(\mathbf{r}) w_0^*(\mathbf{r}) \, d\mathbf{r} \right|^2 \tag{6.5.26}$$

as the available heterodyned signal power. This means the effective receiver area is

$$A_{re} = \sum_{i=0}^{\infty} \gamma_i \left| \int_{\mathscr{A}_r} W_i(\mathbf{r}) w_0^*(\mathbf{r}) \, d\mathbf{r} \right|^2 \qquad (6.5.27)$$

and depends only on the relation of the local field aperture function and the turbulence-induced eigenfunctions. We see immediately that if $w_0(\mathbf{r}) = W_0(\mathbf{r})$ (i.e., the local field mode exactly matches the zero-order mode of the received field), then $A_{re} = \gamma_0$, and the detector has extracted all the available power from this mode. If the zero-order modes are not matched, we extract less than this available power (as is evident by application of the Schwarz inequality to the integral). We also see that if more spatial modes were matched by the local aperture field, considerable improvement over our previous results is possible. That is, if the local field had the aperture expansion as in (6.4.2)

$$f_L(t, \mathbf{r}) = \sqrt{A_r} \, a_L e^{j\omega_2 t} \sum_{i=0}^{\infty} W_i(\mathbf{r}) \qquad (6.5.28)$$

then (6.5.27) becomes

$$A_{re} = \sum_{i=0}^{\infty} \gamma_i = A_r \qquad (6.5.29)$$

and the full aperture would be used for heterodyning rather than only the r_0 area. In theory the turbulent system performance approaches that of idealized single mode heterodyning. This, of course, requires the local field to match exactly the turbulent field modes as they arrive at the receiver. A priori knowledge of their structure would be required. Use of "pilot tone fields" transmitted along with the modulated fields, followed by spatially coherent amplification of the pilot field as a local field, has been suggested for this purpose [10]. It has also been shown that there exists a reciprocity relationship between an optical heterodyne receiving system and a transmitter using the same aperture through the same medium [11]. Consequently, a received field can be used to properly phase a transmitted field which can transit the medium and present a completely integrated plane wave at the original point. This can be accomplished whenever the medium is "frozen" long enough in time to make the two-way transit. The extension of this concept to general scattering media is straightforward.

REFERENCES

[1] Fried, D., "Optical Heterodyne Detection of an Atmospherically Distorted Wavefront," *Proc. IEEE*, **55**, No. 1, 57–67 (January 1967).

[2] Tatarski, V., *Wave Propagation in Turbulent Medium*, McGraw-Hill Book Co., New York, 1961.

[3] Fried, D., "Statistics of Geometric Representation of Wavefront Distortion," *J. Opt. Soc.*, **55**, No. 11, 1427 (November 1965).

[4] ———, "Aperture Averaging of Scintillation," *J. Opt. Soc.*, **57**, No. 2 (February 1967).

[5] ———, "Limiting Resolution Through the Atmosphere," *J. Opt. Soc.*, **56**, No. 10, 1380 (October 1966).

[6] ———, "Optical Resolution Through a Random Medium for Long and Short Exposures," *J. Opt. Soc.*, **56**, No. 10, 1372 (October 1966).

[7] Fried, D. and Cloud, J., "Propagation of an Infinite Plane Wave in Random Medium," *J. Opt. Soc.*, **56**, No. 12 (December 1966).

[8] Brookner, E., "Atmospheric Propagation and Channel Models for Laser Wavelengths," *IEEE Trans. Commun. Tech.*, **COM-18**, 396 (August 1970).

[9] Pratt, W., *Laser Communications*, John Wiley and Sons, New York, 1969, Chap. 7.

[10] Proceedings of the MIT-NASA Workshop, Williamston, Massachusetts, *Optical Space Communication* (R. Kennedy and S. Karp, eds.), August 1968.

[11] Shapiro, J. H., "Reciprocity of the Turbulent Atmosphere," *J. Opt. Soc. Am.*, **61**, No. 4, pp. 49, April 1971.

PROBLEMS

1. A heterodyne system uses an RF bandpass filter following photodetection. The filter is tuned to a frequency of 1 GHz with a bandwidth of 10 MHz. The local oscillator operates at frequency 10^{14} Hz and the transmitter carrier is at an optical frequency such that oscillator mixing produces the proper RF for the filter.

(a) How much optical Doppler (frequency) shift of the received carrier is acceptable in the system? Neglect carrier modulation.

(b) Convert (a) to normal velocity between transmitter and receiver.

(c) How much local oscillator frequency instability can occur?

2. A 1 mW optical transmitter uses a 6 in. transmitting lens at 10^{14} Hz. The beam is transmitted over a 500 mile free space path to a heterodyning receiver. The receiver has a collecting area of 4 in.

(a) How much local signal power is needed to achieve a strong oscillator condition, assuming a 20 dB local to received field power ratio is needed?

(b) How much is needed to overcome circuit noise with the same ratio as (a), when the detector load resistance is 100 ohms and operates at room temperature?

(c) If the detector had gain G, determine how the answers in (a) and (b) are affected.

3. In a heterodyne receiver the local field is perfectly aligned in the center of the detector. The received optical field operates at 10^{14} Hz and the receiver area is 6 in. square. What is the minimal offset angle that can be allowed an arriving plane wave in order to guarantee that the alignment power loss is less than 3 dB?

4. Show that if a circular receiver lens of radius R is used when heterodyning with two plane waves, the equivalent suppression factor to (6.3.2) for misaligned angles is proportional to

$$\int_{\mathscr{A}_d} \left(\frac{J_1|\mathbf{u}|}{|\mathbf{u}|}\right)\left(\frac{J_1(|\mathbf{u} - \mathbf{u}_0|)}{|\mathbf{u} - \mathbf{u}_0|}\right) d\mathbf{u}, \qquad |\mathbf{u}_0|^2 = (x_R - x_L)^2 + (y_R - y_L)^2$$

5. Consider a receiver with a circular aperture of radius d with the field focused on a circular detector having radius r. Assume a plane wave local oscillator field mixes with the field on the detector.

(a) Show that the heterodyne SNR_p is reduced by the factor

$$\left[\frac{2}{r/R}\int_0^{r/R} J_1(x)J_0\left(\frac{4f_c x \sin \psi}{d}\right) dx\right]^2$$

where ψ is the angle between the local oscillator direction and the main axis, and $R = 2\lambda f_c/\pi d$.

(b) Set $2f \sin \psi/d = 1$ and evaluate the integral by using the first two terms of an expansion of J_0 as a power series. Compare with Figure 6.6.

(c) Derive the equivalent integral for a circular aperture and a square detector.

6. A receiver performs multimode heterodyning with two separate modes (different directions of arrival) by using a local source plane wave over a square detector surface. Assume the plane wave is aligned normal to the detector and the modes occur off axis angles ψ_1 and ψ_2, respectively.

(a) Determine the heterodyned multimode SNR_p, usng coherent addition of the modal fields, as a function of the angles ψ_1 and ψ_2. Assume equal power per mode and neglect circuit noise.

(b) For what choice of angles ψ_1 and ψ_2 is the suppression loss minimal?

7. Given the heterodyned signal term in (6.4.10), assume the delays are small relative to time variations in $s(t)$, so that $s(t - \tau) \approx s(t)$. Define $\lambda_i = \omega_i \tau_i$.

(a) Write a general expression for the multimode signal power for an arbitrary joint distribution $p(\lambda_1, \lambda_2, \ldots, \lambda_{ML})$.

(b) Show that if the $\{\lambda_i\}$ are independent and each uniformly distributed over $(0, 2\pi)$, then (6.4.11) follows.

8. In a pilot tone system an unmodulated optical carrier is transmitted in addition to, but not interfering with, the modulated carrier. If channel distortion affects both carriers equally, the unmodulated carrier can be used as a space coherent reference for heterodyning with the distorted modulated carrier. Consider the distorted modulated field to be $a(t)f(\mathbf{r}) \exp[j\omega_1 t]$ and the received pilot carrier to be $f(\mathbf{r}) \exp[j\omega_1 t]$. The two fields are both received and summed at the detector.

(a) Determine the detected intensity of the two fields. Neglect background and circuit noise.

(b) Why is a *spatial amplifier* (a device that amplifies the field spatially) necessary in pilot tone heterodyning?

(c) If such an amplifier is available, what is the resulting SNR_p at the detector output? How does this compare to single mode heterodyning?

(d) If we assume a "frozen" channel, what can be done to "preemphasize" the two-way system transmitter field in both aperture and focal plane?

9. Consider the two overlapping circles, each of radius R, as shown in Figure 6.8.

(a) Compute the overlapping shaded area. [*Hint*: Determine the shaded area in P and multiply by 4. The area in b is the area of a sector minus the area of a triangle.]

(b) Show the answer can be put into the form

$$2\left[R^2 \cos^{-1}\left(\frac{a}{R}\right) - a(R^2 - a^2)^{1/2}\right]$$

(c) Relate $R_0(\mathbf{r})$ in (6.5.14) and derive (6.5.15).

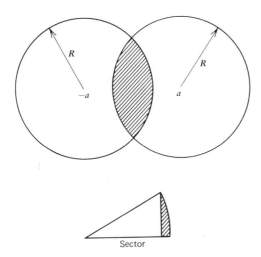

Sector

Figure 6.8. Problem 6.9.

10. A homodyne system has a received carrier with phase $\theta_1(t)$. The detector output is low pass filter with $F(\omega)$, and the filter output is fed back to control the phase of the local oscillator $\theta_2(t)$. (The oscillator is phase modulated with the filter output.) Derive an equivalent feedback system that shows the

manner in which the input phase variations $\theta_1(t)$ are tracked. Assume single mode homodyning and account for background noise and circuit noise at the detector output.

11. The heterodyne alignment condition is often stated as

$$(\mathbf{z}_L - \mathbf{z}_R) \cdot \mathbf{r} \leq 1$$

where $\mathbf{z}_L$ is the local field ray vector, $\mathbf{z}_R$ the received field ray vector, and $\mathbf{r}$ a point on the detector surface. Each $\mathbf{z}$ has magnitude $2\pi/\lambda$. Determine the received field bandwidth restriction imposed by the heterodyne alignment condition, when both fields are aligned.

7 Digital Communications— Binary Systems

In the analog transmission of information the primary objective is to transmit a given waveform from the transmitter to the destination with as little distortion as possible. In our earlier discussion the eventual distortion was measured in terms of a signal-to-noise ratio, which was basically an indication of the "variance spread" of the output message. Another way of sending information is by digital transmission, in which the desired message is coded into symbols and only the symbols are transmitted. The receiver no longer has the objective of preserving actual symbol waveshape, but rather has the role of simply detecting, that is, decoding, which symbol has been transmitted. This digital communication system is the object of study in this and the next chapter.

7.1 THE OPTICAL BINARY DIGITAL SYSTEM

In a digital communication system the desired information at the transmitter is generated as, or converted to, binary symbols prior to transmission. This is accomplished by standard analog-to-digital conversion circuits, which are generally composed of voltage samplers followed by a quantization of the sample values into binary symbols. These binary symbols are called bits and shall herein be designated as "zeros" and "ones." The resulting sequence of bits is then transmitted by encoding (i.e., mapping) the bits into analog waveforms, which are then transmitted to the receiver. This encoding can be accomplished either on a bit by bit basis, or by using blocks of bits. The former is called a *binary* digital system, the latter a *block coded* system. In this chapter we concentrate on binary systems only, and postpone discussion of block coded digital systems to Chapter 8.

A binary system model is shown in Figure 7.1. Each bit is encoded one at a time in sequence into one of two possible waveforms, one representing a binary one and the other a binary zero. Each waveform is of finite length, say T sec, and therefore one bit is transmitted every T sec. The waveform

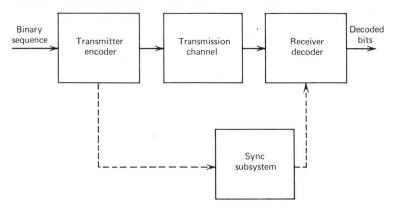

Figure 7.1. The digital binary system model.

sequence generated by encoding is transmitted by standard transmission methods to a receiver. At the receiver a decoder decides which of the two possible bits is being received during each T sec interval, and attempts to reconstruct the bit sequence from these decisions. Since the encoding operation is known at the receiver, a correct decision as to which waveform is being received will successfully decode the corresponding transmitted bit. The decoded bits represent the recovered symbol sequence, which can then be converted back to the desired information. Noise and distortion incurred during the transmission will cause some decoded bits to be in error, thereby degrading system performance. This operation implicitly requires the existence of exact bit timing between encoder and decoder. That is, the decoder must know exactly when each T sec bit interval begins and ends at the receiver. This time synchronization is accomplished by a timing subsystem, which becomes a vital part of the overall digital system design.

An optical version of the binary system just described operates by using the bit waveforms to generate an optical radiation field as a transmission signal during each T sec interval. The most common transmission method is to use the binary waveforms to intensity modulate a laser source, which is then transmitted as an optical field. At the receiver, a photodetector, or perhaps an array of photodetectors, detects the corresponding field intensity, which is then used for bit decoding. Such a system is called a direct detection, or spatial noncoherent, binary system, since it relies purely on intensity detection. The alternative system is one in which the bit waveforms are used to amplitude, phase, or frequency modulate the laser source, and spatially coherent heterodyning must be used to recover bit information.

A block diagram of a direct detection binary system using a single photodetector is shown in Figure 7.2. The bit encoded waveforms are transmitted

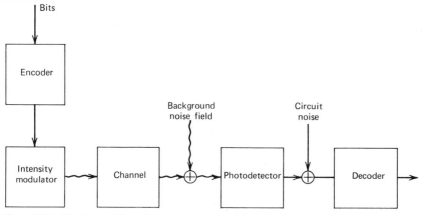

Figure 7.2. The binary direct detection system model.

as an intensity modulated field during each T sec interval. After optical channel propagation the radiation field is collected at the receiver along with additive background radiation. The photodetector at the receiver can be an extended detector, observing many spatial modes of the optical field, or it may be a diffraction limited (point) detector observing only a single spatial mode. In practical optical systems the detector is almost always an extended detector. The photodetector generates the detected output shot noise process. The decoder following the photodetector observes the output process over each T sec interval, and must decide which field intensity is being received. We first assume shot noise limited behavior, and postpone discussion of effects of decoder thermal noise to Section 7.8. Since a single detector is assumed, the field is integrated over the detector area, and the spatial effect of the field intensity $I(t, \mathbf{r})$ can be suppressed. Furthermore, during photodetection this field intensity is converted to a count intensity having the form

$$n(t) = \alpha \int_{\mathscr{A}} I(t, \mathbf{r}) \, d\mathbf{r} \tag{7.1.1}$$

where $\mathscr{A}$ is the detector area. Therefore, without loss of generality, we can assume the detected count intensity is in fact the transmitted intensity, and the general direct detection model of Figure 7.2 can be simplified to that in Figure 7.3a. In this model one of two count intensities, $n_1(t)$ or $n_2(t)$, generates a detected shot noise process

$$x(t) = \sum_{j=1}^{k(0, t)} h(t - t_j) \tag{7.1.2}$$

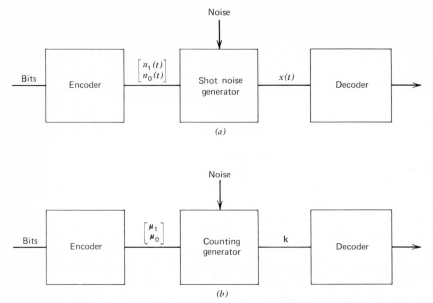

Figure 7.3. Binary direct detection system model: (*a*) intensity waveform model; (*b*) discrete counting model.

where $k(0, t)$ is the count process over area $\mathscr{A}$. This shot noise serves as the decoder input. Recall that an infinite bandwidth detector produces component functions $h(t)$ that are delta functions, while a finite bandwidth of B_h implies $\{h(t)\}$ that are approximately $\tau_h = 1/2B_h$ sec in width. If $h(t)$ is modeled as a rectangular function of width τ_h, then from (4.1.3)

$$x(t) = \left(\frac{e}{\tau_h}\right)k(t - \tau_h, t) \tag{7.1.3}$$

and the shot noise is replaced by the modified counting process as the decoder observable. We assume that the decoder is perfectly timed by an existing synchronization subsystem.

It is often analytically convenient to use a vector representation for the detector output. One procedure for vectorizing the decoder observable is to model it as a counting process, and represent it by its time samples taken every $\tau_s \triangleq 1/2B_n$ sec apart, where B_n is the approximate bandwidth of the intensity modulation. (In effect we are assuming the detector shot noise is filtered to the bandwidth B_n of the intensity. We then must sample at least this often in order to represent the field intensity accurately.) These time samples can then be used to derive a vector version of the photodetector output. Thus if we denote $k_j \triangleq k(t_j - \tau_s, t_j)$, $j = 1, 2, \ldots, 2B_n T$, where $t_j = j/2B_n$ is the jth time sample, the output can be represented by the count vector

$$\mathbf{k} = (k_1, k_2, \ldots, k_{2B_n T}) \tag{7.1.4}$$

Note that $2B_n T$ is the number of τ_s sec counting intervals occurring in the bit time T. Since the count vector components k_j are associated with disjoint time intervals, they represent mutually independent random count variables. Note also that if $n_i(t)$ is the count intensity of the transmitted ith bit, then over interval j its average count is given by

$$\mu_{ij} \triangleq \int_{t_j - \tau_s}^{t_j} n_i(t)\, dt \tag{7.1.5}$$

This can be used to define the signal count intensity vector $\boldsymbol{\mu}_i = (\mu_{i1}, \mu_{i2}, \ldots)$ for each bit $i = 0, 1$. This procedure leads to the overall vector model in Figure 7.3b as opposed to the waveform model in Figure 7.3a. In the vector model bits are encoded directly into signal energy vectors, which in turn generate count vectors for the decoder processing. We emphasize that these vector models serve purely as mathematical convenience, and do not necessarily mean that we time sample in the actual system implementation.

We would expect that from a practical point of view, constraints will be placed upon system parameters. Since energy in an interval of a transmitted field is proportional to the integral of the count intensity over that interval, a transmitter energy constraint is equivalent to an intensity integral constraint. Thus, if we constrain field energy over $(0, T)$, we equivalently require

$$\int_0^T n_i(t)\, dt = K_i \tag{7.1.6}$$

where K_i is the average signal count corresponding to the allowable receiver field energy for bit i. We emphasize that both $n_i(t)$ and K_i are referred to the receiver, and are related to transmitter parameters through the equations of Section 1.3. In the vector model, the equivalent to (7.1.6) is

$$\sum_{j=1}^{2B_n T} \mu_{ij} = K_i \tag{7.1.7}$$

which constrains the component summation. If constraints are placed on average power instead of average energy, the equivalent terms take the form

$$\frac{1}{T} \int_0^T n_i(t)\, dt \triangleq n_i, \qquad \frac{1}{2B_n T} \sum_{j=1}^{2B_n T} \mu_{ij} = n_i \tag{7.1.8}$$

where n_i is the allowable average count rate over $(0, T)$. As we shall see subsequently, the type of transmitter constraint often dictates the type of performance, and in certain cases, the ultimate system design.

7.2 MAXIMUM A POSTERIORI DECODING

In the digital system of Figure 7.2, and the equivalent models in Figure 7.3, the decoder has the task of observing the detector output and decoding the transmitted signal. The presence of added noise fields during photodetection, along with the added system noise prior to the decoding circuitry, causes errors in the ultimate bit decisioning. An accepted procedure for decoder design for generating these decisions makes the use of the concept of maximum a posteriori (MAP) detection. Under this criterion, the decoder upon observing the noisy detector output, selects the bit that has the highest probability of having been sent as the received bit. Mathematically, this requires the decoder, over each T sec interval, to compute the probability that the ith bit ($i = 0, 1$) was sent, and select the most probable bit. The decoder then proceeds to examine the detector output over the next T sec block interval and repeats the decisioning operation. For a given T sec interval, which we take as $(0, T)$ for convenience, the MAP decoder can therefore be mathematically modeled as follows. If we use the count vector model from (7.1.4) and denote the observed count vector over $(0, T)$ as $\mathbf{k}$, then the MAP decoder must compute $P(i|\mathbf{k})$, where

$$P(i|\mathbf{k}) \triangleq \text{the probability that the } i\text{th bit was transmitted,}$$
$$\text{given the observed count vector } \mathbf{k} \qquad (7.2.1)$$

The decoder decides a binary one was sent if $P(1|\mathbf{k}) > P(0|\mathbf{k})$, and decides a binary zero was sent if $P(0|\mathbf{k}) > P(1|\mathbf{k})$. If $P(1|\mathbf{k}) = P(0|\mathbf{k})$, then each bit is equally likely to have been sent, and a random choice can be made among the bits. For convenience, we denote this comparison test as simply

$$P(1|\mathbf{k}) \gtrless P(0|\mathbf{k}) \qquad (7.2.2)$$

Thus the decoder must compute these probabilities in a formal application of the bit test. However, by applying Bayes' rule, we note

$$P(i|\mathbf{k}) = \frac{P(\mathbf{k}|i)P(i)}{P(\mathbf{k})} \qquad (7.2.3)$$

where $P(\mathbf{k}|i) =$ probability of $\mathbf{k}$ occurring, given ith bit was sent
   $P(i) =$ a priori probability that the ith bit will be sent
   $P(\mathbf{k}) =$ probability that the observed $\mathbf{k}$ occurred

For equally likely bits, $P(i) = \frac{1}{2}$, and $P(1|\mathbf{k}) \gtrless P(0|\mathbf{k})$ if $P(\mathbf{k}|1) \gtrless P(\mathbf{k}|0)$, since $P(\mathbf{k})$ is the same for both bits. Thus the MAP decoder, after observing $\mathbf{k}$, equivalently can compute

$$\Lambda_i \triangleq P(\mathbf{k}|i), \qquad i = 0, 1 \tag{7.2.4}$$

and decide the bit by the test

$$\Lambda_1 \gtrless \Lambda_0 \tag{7.2.5}$$

The parameter Λ_i is called the *likelihood function* for bit i, and the test in (7.2.5) is sometimes referred to as the maximum likelihood (ML) test (since the bit corresponding to the maximum Λ_i is selected). Thus the MAP decoding test is the same as the ML test, and the terminology can be interchanged. We point out that this would not be true if the bits were not equally likely a priori. In general, the ML test in (7.2.5) is easier to implement than (7.2.2), since the term $P(\mathbf{k})$ in (7.2.3) need not be computed. It is important to recognize that the Λ_i are evaluated at a particular value of $\mathbf{k}$—the observed $\mathbf{k}$ during the bit interval—in carrying out the test in (7.2.5). Also, we point out that we could perform the equivalent comparison

$$\log \Lambda_1 \gtrless \log \Lambda_0 \tag{7.2.6}$$

if desired, since the log is monotonic increasing in its argument (i.e. $\log \Lambda_1 \gtrless \log \Lambda_0$ if $\Lambda_1 \gtrless \Lambda_0$). Since bit i is transmitted as a known intensity, we can reinterpret Λ_i as the probability of $\mathbf{k}$ given the ith intensity was received. If the received signal intensity happens to depend on a random parameter θ, having a priori density $p(\theta)$, then the previous discussion is still valid, provided that we remember $P(\mathbf{k}|i)$ is a conditional average over θ (see Problem 7.3). That is, we use $P(\mathbf{k}|i) = E_\theta[P(\mathbf{k}|i, \theta)]$, where $P(\mathbf{k}|i, \theta)$ is conditioned on the ith intensity being received with parameter value θ. Thus the likelihood function becomes

$$\Lambda_i = \int_{-\infty}^{\infty} P(\mathbf{k}|i, \theta)p(\theta)\, d\theta \tag{7.2.7}$$

This is called a *generalized likelihood functional* and can be used directly in the test of (7.2.5) or (7.2.6). In typical applications, θ may correspond to random phase, frequency, or amplitude parameters incurred during transmission of the ith intensity, and (7.2.7) can be used if the density $p(\theta)$ is known.

In the implementation of any digital decoder, we are always interested in its ultimate performance. An important and widely accepted measure of a binary decoder's operation is its probability of making a bit error during any bit interval. We formally define the average error probability PE of a binary system as

$$PE = \tfrac{1}{2}(PE|1) + \tfrac{1}{2}(PE|0) \tag{7.2.8}$$

where PE$|i$ is the probability that an error will be made when bit i is sent in a given interval. For the MAP decoder, the latter conditional probability is simply the probability that the test in (7.2.2), (7.2.5), or (7.2.6), whichever is used, will yield the incorrect bit decision. Since all three tests are equivalent, their error probabilities must be identical. It is evident that such probabilities depend on the statistics associated with the models in Figure 7.3. In subsequent discussion, we will find that in some cases, PE can be determined exactly, while in others we must resort to bounding arguments or simulation studies. We also point out that the MAP decoding procedure always yields the minimum PE over all possible decoding procedures (Problem 7.4).

7.3 MULTIMODE DIRECT DETECTION BINARY SYSTEMS—POISSON DETECTION

The mathematical formulation of the MAP decoder for an intensity modulated, direct detection optical system was presented in the preceding section. The specific implementation of this decoder depends on the counting statistics associated with the system model. When the background noise field is temporally and spatially white, the count statistics over any counting interval are governed by $(D - 1)$th order Laguerre probabilities, where D is the total number of time-space modes observed over the detector area and counting interval. We showed earlier that if $D \gg 1$, then these count probabilities are approximately Poisson in nature. We call this situation multimode detection, or simply refer to it as Poisson detection.

Multimode detection occurs if either many spatial modes exist (i.e., a large field of view) and/or many temporal modes exist. The latter is true if the optical detector bandwidth B_0 is such that $2B_0\tau_s \gg 1$, where τ_s is a counting interval. Since $\tau_s \approx 1/2B_n$, this requires $B_0/B_n \gg 1$; that is, the optical bandwidth is many times the intensity bandwidth. For the multimode case, the conditional probability of a count k_j occurring in interval j, when bit i is sent, is given by

$$P(k_j|i) = \text{Pos}(k_j, \mu_{ij} + \mu_b) \qquad (7.3.1)$$

where μ_{ij} is the signal count energy in (7.1.4), and μ_b is the average noise count per counting interval

$$\mu_b = (\alpha N_{0b})2B_0\tau_s D_s \qquad (7.3.2)$$

Here D_s is the number of spatial modes and αN_{0b} is the noise count per time-space mode. The count vector $\mathbf{k}$ therefore has probability

$$P(\mathbf{k}|i) = \prod_{j=1}^{2B_nT} \frac{(\mu_{ij} + \mu_b)^{k_j}}{k_j!} \exp[-(K_i + K_b)] \qquad (7.3.3)$$

where $2B_n T$ is the number of counting intervals in T sec, and K_i and K_b are the average signal and noise counts over $(0, T)$. That is,

$$K_i = \int_0^T n_i(t)\, dt, \qquad K_b = T\mu_b/\tau_s \qquad (7.3.4)$$

The log likelihood function is then

$$\log \Lambda_i = \sum_{j=1}^{2B_n T} k_j \log\left(1 + \frac{\mu_{ij}}{\mu_b}\right) - K_i$$

$$+ \left[\sum_{j=1}^{2B_n T} (k_j \log \mu_b - \log k_j!) - K_b \right] \qquad (7.3.5)$$

Note that the first two terms depend on i while the bracketed term does not. Thus the bracketed term will be the same for both bits, $i = 0, 1$, and therefore can be neglected in applying the test in (7.2.6). The MAP decoding procedure therefore requires only the computation of

$$\tilde{\Lambda}_i \triangleq \sum_{j=1}^{2B_n T} k_j \log\left(1 + \frac{\mu_{ij}}{\mu_b}\right) - K_i \qquad (7.3.6)$$

for $i = 0, 1$, and then comparing. The discrete version of the MAP decoder is immediately evident as a pair of weighted sums of the observed count components, followed by a comparison for the larger. That is, after observing **k** during each bit interval, the decoder simultaneously computes (7.3.6) for each bit i, and determines the maximum. The overall decoder structure is shown in Figure 7.4a for the counting model. Note that the summation in (7.3.6) is a discrete correlation of **k** with the appropriate log intensity vector. The subtraction of the K_i appears as a bias adjustment to account for energy differences in the bit transmissions.

An integral version of the MAP decoder can be obtained by replacing the discrete summation in (7.3.6) by an equivalent integral. This can be accomplished by considering the counting and intensity processes to be bandlimited to a bandwidth less than B_h, where $\tau_s = 1/2B_h$. We then can write $\mu_{ij} = n_i(t_j)\tau_s$, and the summation in (7.3.6) becomes

$$\sum_{j=1}^{T/\tau_s} k_j \log\left(1 + \frac{\mu_{ij}}{\mu_b}\right) = \sum_{j=1}^{T/\tau_s} k(t_j - \tau_s, t_j) \log\left(1 + \frac{n_i(t_j)}{n_b}\right) \qquad (7.3.7)$$

where $n_b = \mu_b/\tau_s$ is the average noise count rate. If we now expanded the bandlimited counting function $k(t - \tau_s, t)$ and the function

$$\log\left[1 + \frac{n_i(t)}{n_b}\right]$$

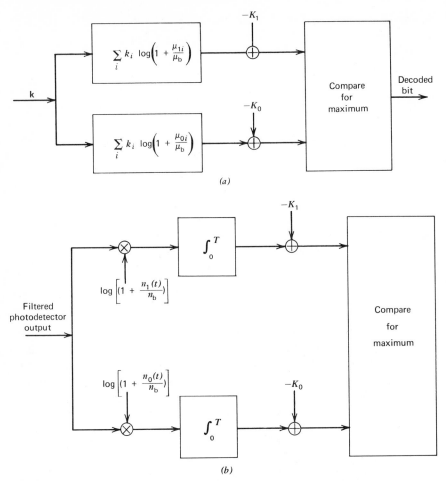

Figure 7.4. Map decoder models: (a) discrete model; (b) integral model.

into orthonormal sampling functions,† we see that the right side in (7.3.7) is equivalent to the integral

$$\int_0^T \frac{k(t - \tau_s, t)}{\tau_s} \log\left(1 + \frac{n_i(t)}{n_b}\right) dt \qquad (7.3.8)$$

† The sampling functions are the set $\phi_i(t) = \sin[2\pi Bt - i\pi]/[2\pi Bt - i\pi]$, which are orthogonal over $(-\infty, \infty)$ for all integers i. The functions have the property that when used to Fourier expand a function $s(t)$ that is bandlimited to $\pm B$ Hz, the Fourier coefficients are $s(i/2B)/\sqrt{2B}$. That is, the coefficients are proportional to time samples of $s(t)$ taken every $1/2B$ sec apart.

Using (7.1.3) and dropping the e factor, the integral version of $\tilde{\Lambda}_i$ is then

$$\tilde{\Lambda}_i = \int_0^T x(t) \log\left(1 + \frac{n_i(t)}{n_b}\right) dt - K_i \qquad (7.3.9)$$

This represents the time integral equivalent of (7.3.6), and involves a time correlation of the detector output $x(t)$ with the time function

$$\log\left(1 + \frac{n_i(t)}{n_b}\right).$$

The block diagram of the decoder structure is shown in Figure 7.4b. It requires the simultaneous generation of the log intensity time functions used for correlation. In general, the integral form of the decoder is more convenient for determining system implementation and design, whereas the discrete version is more suitable for analysis.

In the foregoing analysis, correlation is performed entirely in the time domain, since a single detecting surface was assumed. However, the results can be easily extended to a detector array, in which a raster of detectors is used in the detector plane. Recall that in Section 5.7 it was shown that signal-to-noise ratio can be improved by properly weighting the collected field over both time and space. Here we seek the array processing that would be used to perform MAP decisioning on the observed field. The latter field can be described by its counts over elemental volumes $\{v_i\}$ rather than time intervals $\{\tau_s\}$. The volumes are composed of the areas of the individual detectors making up the array, and the counting time intervals. If we assume each detector operates in a multimode condition, the corresponding MAP decoder processing is given by the extension of the Poisson vector model in (7.3.6) to

$$\tilde{\Lambda}_i = \sum_V k(v_i) \log\left(1 + \frac{\mu(v_i)}{\mu_b}\right) - K_i \qquad (7.3.10)$$

where V is the total volume over the array area and bit interval, K_i is the average signal count over V, $k(v_i)$ is the observed count over volume v_i, and $\mu(v_i)$ and μ_b are signal and noise counts over v_i. The signal count $\mu(v_i)$ is related to the signal count intensity by

$$\mu(v_i) = \iint_{v_i} n(t, \mathbf{r}) \, dt \, d\mathbf{r} \qquad (7.3.11)$$

Thus the MAP array would collect and weight the counts over the volume V according to (7.3.10) for each bit before performing the comparison test. It is interesting that MAP decoding adjusts the counts according to the log of the intensity. Contrast this result with (5.5.21) for maximizing signal-to-noise ratio, where the weighting was directly proportional to the intensity. This illustrates the fact that maximizing SNR is in general *not* synonomous with MAP decoding in Poisson detection. The two, however, become approximately the same when μ_b, the noise count is much greater than the signal count.

7.4 EXAMPLES OF BINARY DIGITAL SIGNALING

The preceding section presented general system models for MAP decoders in an intensity modulated binary optical system. In this section we apply these results to some common types of binary signaling formats used with intensity modulation and discuss specific systems. The examples selected are of practical importance in modern system design, primarily because of their ease of signal generation and simplicity in decoding procedures.

On-Off Keying (OOK). In this format a binary one is transmitted as a pulse of optical energy during $(0, T)$, while absence of signal energy represents a binary zero. Information is sent by pulsing a laser on and off during successive T sec bit times, according to the digital symbols. Such a modulator can be easily implemented by means of a simple transmitter switch, as in Figure 7.5a. Mathematically, this type of modulation corresponds to use of the binary intensity set:

$$
\begin{aligned}
n_1(t) &= n_s, & 0 \le t \le T \\
n_0(t) &= 0, & 0 \le t \le T
\end{aligned}
\tag{7.4.1}
$$

where n_s is the received count rate. The MAP decoder simplifies to a single correlator, since $\log(1 + n_0(t)/n_b) = 0$ for all $0 \le t \le T$. Furthermore, the correlator requires only the computation of

$$
\begin{aligned}
\tilde{\Lambda}_1 &= \sum_j k_j \log\left(1 + \frac{n_s}{n_b}\right) - n_s T \\
&= \log\left(1 + \frac{n_s}{n_b}\right) k(0, T) - n_s T
\end{aligned}
\tag{7.4.2}
$$

where $k(0, T)$ is the count over the bit interval. The corresponding MAP test in (7.2.6) simplifies to

$$
k(0, T) \gtrless \frac{n_s T}{\log\left(1 + \dfrac{n_s}{n_b}\right)}
\tag{7.4.3}
$$

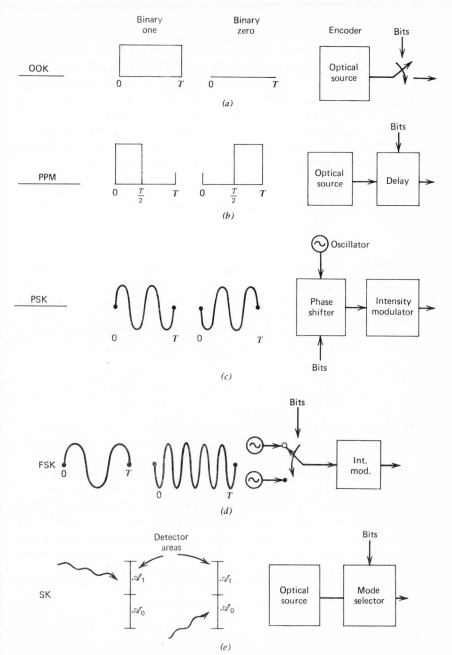

Figure 7.5. Examples of binary optical signaling. (*a*) On-off keying, (*b*) Pulse position modulation. (*c*) Phase shift keying. (*d*) Frequency shift keying. (*e*) Spatial keying.

The right-hand side is a constant, and MAP decisioning is achieved by a simple count comparison to a specified threshold. Since counting is related to energy detection, the test above is equivalent to a test of the collected field energy over $(0, T)$. The fact that the threshold depends on both received signal power and noise power, points out a basic disadvantage of the OOK; that is, the received signal and noise powers must be known exactly in order to properly set the threshold for MAP detection.

The error probability for OOK can be determined in a straightforward manner. Since the count $k(0, T)$ is a Poisson random variable whose mean is the total energy collected, the resulting PE in (7.2.8) becomes

$$\begin{aligned} \text{PE} = &\tfrac{1}{2}[\text{Prob } k(0, T) < k_T \text{ when one is sent}] \\ &+ \tfrac{1}{2}[\text{Prob } k(0, T) > k_T \text{ when zero is sent}] \\ &+ \tfrac{1}{4}[\text{Prob } k(0, T) = k_T \text{ when one is sent}] \\ &+ \tfrac{1}{4}[\text{Prob } k(0, T) = k_T \text{ when zero is sent}] \end{aligned} \qquad (7.4.4)$$

where k_T is the threshold used. The last two terms account for the equalities in the binary test, in which case a random choice is made, with probability $\frac{1}{2}$ of being incorrect. Note that if k_T is not an integer, the last two terms will be zero. When Poisson probabilities are inserted, PE becomes

$$\begin{aligned} \text{PE} &= \frac{1}{2} \sum_{k=0}^{k_T} \gamma_{kk_T} \text{ Pos}(k, K_s + K_b) + \frac{1}{2} \sum_{k=k_T}^{\infty} \gamma_{kk_T} \text{ Pos}(k, K_b) \\ &= \frac{1}{2} \sum_{k=0}^{k_T} \gamma_{kk_T} \frac{(K_s + K_b)^k}{k!} e^{-(K_s + K_b)} + \frac{1}{2} \sum_{k=K_T}^{\infty} \gamma_{kk_T} \left(\frac{K_b{}^k}{k!} \right) e^{-K_b} \end{aligned} \qquad (7.4.5)$$

where K_s and K_b are the average signal and noise counts over the total bit interval, and $\gamma_{kk_T} = \frac{1}{2}$ for $k = k_T$ and is 1 otherwise. If k_T is not an integer, then the summation index $k = k_T$ must be interpreted as the nearest integer to k_T. For an upper (lower) index we take the nearest lower (upper) integer. Equation 7.4.5 can be easily evaluated from readily available Poisson summation tables. [1, 2]. Some typical plots are shown in Figure 7.6 as a function of K_s for several values of K_b, using the threshold in (7.4.3). It can be shown (Problem 7.9) that this choice of threshold produces the minimal PE for a given value of K_s and K_b. The OOK error probability can also be written directly in terms of the incomplete gamma function

$$\Gamma(n, x) \triangleq \int_0^x e^{-t} t^{n-1} \, dt \qquad (7.4.6)$$

by using the identity:

$$\sum_{n=c}^{\infty} \frac{x^n}{n!} e^{-x} = \frac{\Gamma(c, x)}{\Gamma(c, \infty)} \qquad (7.4.7)$$

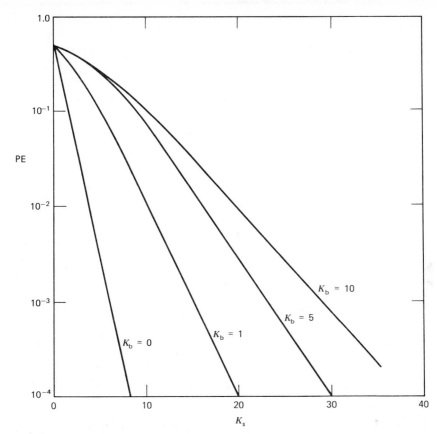

Figure 7.6. Bit error probability, OOK binary systems: K_s = average signal count, K_b = average noise count per bit interval.

When used in (7.4.5) the OOK error probability becomes

$$PE = \frac{1}{2}\left[1 + \frac{\Gamma(k_T, K_s) - \Gamma(k_T, K_s + K_b)}{\Gamma(k_T, \infty)}\right] \qquad (7.4.8)$$

When k_T is an integer, $\Gamma(k_T, \infty) = (k_T - 1)!$. Equation 7.4.8 is a convenient functional form for the summations in (7.4.5), although the latter is preferable for actual evaluation. The OOK error probability can also be written in terms of other familiar functions (Problem 7.8).

Pulse Position Modulation (PPM). In this case binary information is sent as an optical pulse in one of two adjacent pulse positions within a bit interval, as shown in Figure 7.5b. Thus binary symbols are sent by proper positioning of an optical pulse within each bit interval. This can be generated at the transmitter by properly delaying a laser pulse into either bit position. The PPM format is represented mathematically by the intensity set

$$n_1(t) = n_s, \qquad 0 \le t < \frac{T}{2}$$

$$= 0, \qquad \frac{T}{2} \le t \le T$$

$$n_0(t) = 0, \qquad 0 \le t < \frac{T}{2} \qquad (7.4.9)$$

$$= n_s, \qquad \frac{T}{2} \le t \le T$$

Note the signals each have average count energy $K_s = n_s T/2$. The MAP decoder performs the test in (7.2.6). Using (7.4.2), this generates the test

$$k\left(0, \frac{T}{2}\right) \gtrless k\left(\frac{T}{2}, T\right) \qquad (7.4.10)$$

which is simply a count comparison over each pulse position, deciding the bit by the largest count. Note that knowledge of the received power levels is not required to perform the test, as in OOK, indicating a basic advantage of PPM operation. In addition, the transmitter laser source is pulsed continually at the bit rate instead of being intermittently pulsed by the bit sequence.

The error probability is simply the probability that one Poisson variate containing signal plus noise energy does not exceed another Poisson variate containing noise energy alone. Hence

$$PE = \sum_{k_1=0}^{\infty} \sum_{k_2=k_1}^{\infty} Pos(k_1, K_s + K_b) \, Pos(k_2, K_b) \gamma_{k_1 k_2} \qquad (7.4.11)$$

where K_s and K_b are signal and noise energies over $T/2$ sec, and again $\gamma_{k_1 k_2} = \frac{1}{2}$ for $k_1 = k_2$ and 1 elsewhere. These error probabilities are again in the form of a discrete summation, and therefore amenable to digital computation, provided the infinite summations are properly truncated (enough terms must be included such that a negligible change in PE is generated when the remaining terms are discarded). Some exemplary plots computed in this way are shown in Figure 7.7. Pratt [3] has shown that

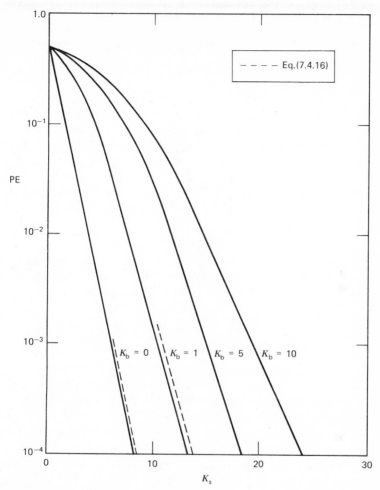

Figure 7.7. Bit error probability, PPM binary system: K_s = average signal count, K_b = average noise count per pulse interval.

the error probability in (7.4.11) can be written in terms of the Marcum Q function:

$$Q(a, b) \triangleq \int_b^\infty \exp\left[-\frac{(a^2 + x^2)}{2} \right] I_0(ax) \, dx \tag{7.4.12}$$

where $I_0(x)$ is the imaginary Bessel function. This relation follows from the identity

$$Q(a, b) = \sum_{k=0}^\infty \frac{\exp(-b^2/2)(b^2/2)^k}{k!} \sum_{j=0}^{k-1} \frac{\exp(a^2/2)(a^2/2)^j}{j!} \tag{7.4.13}$$

which, when substituted into (7.4.11), yields

$$PE = Q(\sqrt{2m_0}, \sqrt{2m_1}) - \tfrac{1}{2} \exp[-(m_1 + m_0)] I_0(2\sqrt{m_1 m_0}) \quad (7.4.14)$$

where $\quad m_1 = K_s + K_b$

$\qquad m_0 = K_b$

Equation 7.4.14 is useful since the Q function has been tabulated [4] and recursive computational methods have been developed for its evaluation [5]. It can also be shown that (7.4.14) can be manipulated into the form (see Problem 7.10)

$$PE = \exp[-(m_1 + m_0)]\left\{ \sum_{n=1}^{\infty} \left(\frac{m_0}{m_1}\right)^{n/2} I_n(2\sqrt{m_1 m_0}) + \frac{1}{2} I_0(2\sqrt{m_1 m_0}) \right\}$$

$$(7.4.15)$$

where $I_n(x)$ is the Bessel function of order n. Equation 7.4.15 is convenient since it involves only a single summation of the tabulated Bessel functions. Hubbard [6] has further noted that since $I_n(x) \le I_0(x)$, a useful bound follows if I_n is replaced by I_0 in (7.4.15), which then simplifies to

$$PE \le \left[\frac{\sqrt{m_0}}{\sqrt{m_1} - \sqrt{m_0}} + \frac{1}{2}\right] \exp[-(m_1 + m_0)] I_0(2\sqrt{m_1 m_0}) \quad (7.4.16)$$

The right side also serves as an accurate and simple approximation at PE values less than 10^{-3} (Figure 7.7). It is tempting to superimpose the PPM curves on the OOK curves, but care must be used in making a direct comparison. The OOK system uses pulses twice as long as PPM and therefore has a higher noise count K_b. If the systems are compared at equal signal energy (same K_s), the PPM always shows the better performance. If the systems are compared at equal signal power (same n_s), the OOK has twice the signal energy and therefore better performance, as evident from Figures 7.6 and 7.7.

It is important to note that the expressions for PE in (7.4.5) and (7.4.8) depend on both the signal and noise count energies K_s and K_b, and not simply on their ratio. This fact is emphasized in Figure 7.8 in which PE in (7.4.8) is plotted as a function of K_b for two fixed ratios of K_s/K_b. Notice that PE can vary orders of magnitude at the same ratio, depending on the value of K_b. This is an important point, since the ratio K_s/K_b is often used as an indication of system quality and, as we see, can often be misleading in digital systems. It is precisely this dependence on both signal and noise energy that distinguishes the optical Poisson detection problem from the analogous RF detection problem; that is, coherent detection in an additive Gaussian noise channel [10, 11].

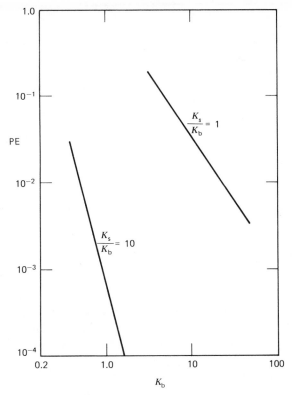

Figure 7.8. Variation of PPM PE with fixed K_s/K_b. K_s = signal count, K_b = noise count.

Another parameter often quoted is the detection signal-to-noise ratio ρ_d, defined as the ratio of the square of the average decisioning signal with no noise to its variance when noise is present. For Poisson counting in PPM, this ratio becoming $\rho_d = K_s^2/K_s + K_b$. The behavior of PE with K_b for a fixed ρ_d is illustrated in Figure 7.9. The results again indicate the ambiguity in even using ρ_d as a design criterion. For large values of K_b, PE is asymptotic to $\text{Erfc}(\sqrt{\rho_d/2})$ where

$$\text{Erfc}(x) = \frac{1}{\sqrt{2\pi}} \int_x^\infty \exp\left[-\frac{z^2}{2}\right] dz \qquad (7.4.17)$$

We note that even if the background noise is negligible, that is, $K_b = 0$, PE in (7.4.8) does not go to zero, but rather approaches the value $\exp(-K_s)/2$. This can be explained as follows. With no noise, no count will occur in the nonsignaling half-interval. Thus a correct bit decision is always made as

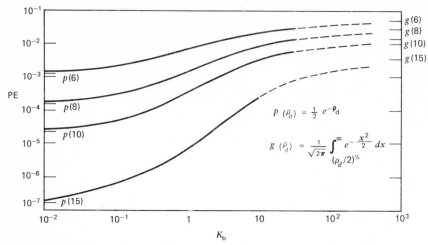

Figure 7.9. Variation of PPM PE with fixed ρ_d. $\rho_d = K_s^2/K_s + K_b$.

long as at least one count occurs in the correct half-interval. The only way an error is made is if the signaling half-interval produces no counts, in which case a random choice is made with a probability of $\frac{1}{2}$ of erring. Since the probability of no count in the signaling half-interval is $\exp(-K_s)$, the error probability limit follows as $(\frac{1}{2})\exp(-K_s)$. The latter is often called the PPM quantum limited bound on PE.

Antipodal Signaling (Phase Shift Keying). In this format binary information is sent by using a signal or its negative to intensity modulate an optical carrier over $(0, T)$. Decoding is accomplished by determining which signal polarity is being received. Such signal sets are called *antipodal*, or opposite polarity, signals. For an arbitrary antipodal signal set, the intensity waveforms become

$$n_1(t) = n_s[1 + s(t)],$$
$$n_0(t) = n_s[1 - s(t)] \qquad 0 \le t \le T \qquad (7.4.18)$$

where $s(t) \le 1$. The most common example of an antipodal signal set is a sine wave transmitted with or without a 180° phase shift during each bit interval to represent the binary symbol. Since a 180° phase shift of a sine wave is equivalent to multiplying by -1, antipodal signaling and phase shift keying (PSK) are synonomous. PSK has the advantage of requiring a relatively simple encoder (a free running oscillator, phase shifted according to the bit stream, followed by an intensity modulator of the light source, as shown in Figure 7.5c). The MAP decoder for log correlation in two separate channels can be implemented by a log amplification of a receiver oscillator running at the same frequency and phase as the two possible transmitted signals.

Antipodal error probability is more difficult to determine than in the preceding cases. Consider the integral log likelihood function of (7.3.9). If we assume a small counting interval τ_s, the output process approaches a delta function whenever a count occurs. Labeling these points t_j, (7.3.9) becomes

$$
\tilde{\Lambda}_1 = \int_0^T x(t) \log\left[1 + \frac{n_s}{n_b} [1 + s(t)] \right] dt
$$

$$
= \sum_{j=1}^k \log\left[1 + \frac{n_s}{n_b} [1 + s(t_j)] \right] \tag{7.4.19}
$$

where $k = k(0, T)$ is the total count over $(0, T)$. $\tilde{\Lambda}_0$ is the same with $s(t_j)$ replaced by $-s(t_j)$. Note that the decoder uses both the total electron count k and their location in time t_j to evaluate each $\tilde{\Lambda}_i$. If $s(t)$ has zero average value over $(0, T)$, then the MAP test is simply a comparison of $\tilde{\Lambda}_1 \gtrless \tilde{\Lambda}_0$, which is identical to the test

$$
Y \triangleq \sum_{j=1}^k \log\left[\frac{n_b + n_s + n_s s(t_j)}{n_b + n_s - n_s s(t_j)} \right] \gtrless 0 \tag{7.4.20}
$$

The average error probability is obtained from the probability that Y is negative when $n_1(t)$ is sent. (See Problem 9.15) We note that Y is no longer a Poisson variable as in the previous cases. However, we can write

$$
PE = \int_{-\infty}^0 p_{Y_1}(y)\, dy \tag{7.4.21}
$$

where $p_{Y_1}(y)$ is the probability density of Y given that $n_1(t)$ was transmitted. To evaluate (7.4.21), it is more convenient to deal with characteristic functions, since (7.4.20) involves a sum of independent variables. Recalling the joint density of the shot noise occurrence times $\{t_j\}$ in (4.2.8), the characteristic function of any term in (7.4.20), given intensity $n_1(t)$ is sent, is

$$
\Psi_1(\omega) = E\left[\exp\left(j\omega \log\left[\frac{1 + \varepsilon s(t_j)}{1 - \varepsilon s(t_j)} \right] \right) \right]
$$

$$
= \frac{1}{T} \int_0^T \left[\frac{1 + \varepsilon s(t)}{1 - \varepsilon s(t)} \right]^{j\omega} [1 + \varepsilon s(t)]\, dt
$$

$$
= \frac{1}{T} \int_0^T \frac{[1 + \varepsilon s(t)]^{j\omega + 1}}{[1 - \varepsilon s(t)]^{j\omega}}\, dt \tag{7.4.22}
$$

where $\varepsilon = n_s/(n_s + n_b)$. The conditional characteristic function of Y_1 is then

$$\psi_{Y_1|k,}(\omega) = [\Psi_1(\omega)]^k \qquad (7.4.23)$$

Averaging over the Poisson count variable k occurring over the bit observation interval $(0, T)$ yields

$$\psi_{Y_1}(\omega) = \sum_{k=0}^{\infty} \psi_{Y_1|k}(\omega) \, \text{Pos}(k, m_T)$$

$$= \exp[[\Psi_1(\omega) - 1]m_T] \qquad (7.4.24)$$

where $m_T = (n_s + n_b)T$. To determine the desired PE we must transform (7.4.24) and evaluate (7.4.21). Hence,

$$\text{PE} = \int_{-\infty}^{0} \frac{1}{2\pi} \int_{-\infty}^{\infty} \psi_{Y_1}(\omega) e^{-j\omega y} \, d\omega \, dy$$

$$= \frac{1}{2\pi} \int_{-\infty}^{0} \int_{-\infty}^{\infty} \exp[(\Psi_1(\omega) - 1)(n_s + n_b)T - j\omega y] \, d\omega \, dy \qquad (7.4.25)$$

This represents the bit error probability for an arbitrary antipodal signal set used for binary intensity modulated signaling. An exact solution for any signal set $\pm s(t)$ requires evaluation of (7.4.22), followed by the integration in (7.4.25). Herrmann [7] has derived approximations and bounds to PE in (7.4.25) for the PSK (sinusoidal) case using a similar development with Chernov bounds [8], suggesting the approximation

$$-\log \text{PE} \gtrsim \{(1 - \Psi_1(\hat{\omega})) \log_{10} e\}(n_s + n_b)T + \log_{10} 2 \qquad (7.4.26)$$

for PSK systems. Here $\hat{\omega}$ is the value of ω that minimizes $\Psi_1(\omega)$. For the sinusoidal case, $\Psi_1(\omega)$ is minimized for $j\omega = -\frac{1}{2}$, for which

$$\Psi_1(\hat{\omega})|_{j\hat{\omega}=-1/2} = \frac{1}{T} \int_0^T \frac{(1 + \varepsilon \sin \omega_s t)^{1/2}}{(1 - \varepsilon \sin \omega_s t)^{-1/2}} \, dt$$

$$= \frac{1}{2\pi} \int_0^{2\pi} [1 - \varepsilon^2 \sin^2 \theta]^{1/2} \, d\theta \qquad (7.4.27)$$

when T is assumed an integer multiple of the sine wave period $2\pi/\omega_s$. Figure 7.10 displays some error probabilities evaluated using (7.4.27) in (7.4.26).

As an alternative, the sine waves can be replaced by antipodal square waves. That is, we use a positive or negative square wave to intensity modulate the source when sending binary data. Although requiring more bandwidth, the square wave intensity has the advantage of maximizing the optical

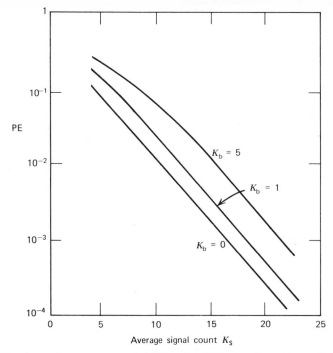

Figure 7.10. Approximate error probabilities for PSK signaling, $K_s = n_s T$, $K_b = n_b T$ (Equation (7.4.26)).

pulse energy during each bit under an average power constraint (Problem 7.5). The corresponding intensity set is then

$$n_1(t) = n_s[1 + \text{sq}(t)], \quad 0 \le t \le T \qquad (7.4.28)$$
$$n_0(t) = n_s[1 - \text{sq}(t)],$$

where $\text{sq}(t)$ is a unit level, symmetric, square wave of period T_s, $T_s \le T$. For detection, the discrete optimal decoder simplifies to the following test. The counting interval $(0, T)$ is subdivided into subintervals, each $T_s/2$ sec long. For intensity $n_1(t)$ in (7.4.11), the received intensity is $2n_s$ over every subinterval where $\text{sq}(t) = +1$, and is zero elsewhere. For the other intensity the opposite is true. Hence, the MAP test is then

$$\sum_{j \in T^+} k_j \gtrless \sum_{j \in T^-} k_j \qquad (7.4.29)$$

where k_j is the jth subinterval count, and T^+ (T^-) is the collection of subintervals where $\text{sq}(t) = +1$ (-1). The decoder therefore processes the photodetector output by collecting energy over time intervals corresponding to positive and negative values of sq(t), and comparing the results. We now have the left and right side of (7.4.29) as independent Poisson count variables, with average value $K_b = n_b T$ when noise alone is present, and $K_s + K_b$, where $K_s = 2n_s(T/2) = n_s T$, when signal plus noise is present. The corresponding error probability is therefore identical to (7.4.8) with these values of K_s and K_b. Note that the resulting PE does not depend on T_s, the period of the square wave, but only on the total energy involved. Also, we see that antipodal square wave signaling performs identically to a PPM system having the same value of K_s and K_b. (In fact, when $T_s = T$ the square wave system is a PPM system.) The antipodal square wave system is advantageous when pulse amplitudes and pulse widths are fixed, since the square waves spread signal power over a longer time period and thereby generate more signal energy for detection.

Noncoherent FSK. The principal difficulty with PSK intensity modulation is that any arbitrary phase angle θ of the unmodulated subcarrier must be known exactly at the decoder, relative to the bit period, in order to achieve MAP decoding. This generally requires some type of auxiliary phase referencing (discussed in Chapter 10) in order to maintain this phase angle over many bit intervals. To avoid phase referencing in PSK, however, one can use instead frequency shift keying (FSK) in which two separate frequencies are used for the binary signals. The encoder therefore generates a burst of a sine wave of one of the two frequencies during each bit period, and intensity modulates the source as shown in Figure 7.5d. Since detection is based on different frequencies, the phase angle of each can be considered a random variable with respect to the decoding operation. The MAP decoder, however, becomes more complicated since it must account for this random phase in its operation. Mathematically, the decoder must generate the generalized likelihood functions of (7.2.7) by averaging over the phase angle θ. The FSK intensities are

$$n_1(t, \theta) = n_s[1 + \sin(\omega_1 t + \theta)],$$
$$n_0(t, \theta) = n_s[1 + \sin(\omega_0 t + \theta)], \qquad 0 \le t \le T \qquad (7.4.30)$$

where ω_1 and ω_0 are the signaling frequencies. If θ is assumed a uniform phase angle for each bit, the averaged likelihood function is

$$\Lambda_i = E_\theta[P(\mathbf{k}|i, \theta)]$$

$$= \frac{1}{2\pi} \int_0^{2\pi} \prod_{j=1}^{2B_n T} \text{Pos}[k_j, \mu_{ij}(\theta) + \mu_b] \, d\theta \qquad (7.4.31)$$

where the dependence of μ_{ij} on θ has been emphasized. Using (7.3.3) and passing to the integral form of the exponent, the above becomes

$$\Lambda_i = C \int_0^{2\pi} \exp\left[\int_0^T x(t) \log\left(1 + \frac{n_i(t, \theta)}{n_b}\right) dt\right] d\theta \qquad (7.4.32)$$

where $x(t)$ is the detector shot noise and the constant C includes all terms not depending on θ. Since the log term in the exponent is periodic over $(0, T)$ with period $2\pi/\omega_i$, it admits a Fourier expansion, which allows its phase shifted version to be written as

$$\log\left(1 + \frac{n_i(t, \theta)}{n_b}\right) = \sum_{q=1}^{\infty} a_q \sin[q\omega_i t + \psi_q - q\theta] \qquad (7.4.33)$$

where $\{a_q, \psi_q\}$ are the harmonic amplitudes and phases of the log function. Substituting into (7.4.32) and manipulating trigonometrically, yields

$$\Lambda_i = C \int_0^{2\pi} \exp\left[\sum_{q=1}^{\infty} X_{qi} \cos q\theta + Y_{qi} \sin q\theta\right] d\theta$$

$$= C \int_0^{2\pi} \exp\left[\sum_{q=1}^{\infty} \mathscr{E}_{qi} \cos(q\theta + \phi_{qi})\right] d\theta \qquad (7.4.34)$$

where

$$X_{qi} = a_q \int_0^T x(t) \cos[q\omega_i t + \psi_q] dt \qquad (7.4.35a)$$

$$Y_{qi} = a_q \int_0^T x(t) \sin[q\omega_i t + \psi_q] dt \qquad (7.4.35b)$$

$$\mathscr{E}_{qi}^2 = X_{qi}^2 + Y_{qi}^2 \qquad (7.4.35c)$$

$$\phi_{qi} = \tan^{-1}\left[\frac{Y_{qi}}{X_{qi}}\right] \qquad (7.4.35d)$$

Here (X_q, Y_q) are cosinusoidal and sinusoidal harmonic correlations of the shot noise process, and $(\mathscr{E}_{qi}, \phi_{qi})$ are the corresponding envelope and phase variables. Thus the generalized likelihood Λ_i becomes a rather complicated integral involving all harmonics of the log intensity variations. Although it is possible to reduce (7.4.34) somewhat (see Problem 7.14), the resulting Λ_i still remains a rather complicated function. Some simplification occurs if we neglect all harmonics beyond the first in (7.4.33), that is, let $a_q = 0$, $q \geq 2$. In this case (7.4.34) reduces to

$$\Lambda_i = C \int_0^{2\pi} \exp[\mathscr{E}_{1i} \cos(\theta + \phi_{1i}) d\theta$$

$$= 2\pi C I_0(\mathscr{E}_{1i}) \qquad (7.4.36)$$

where $I_0(\mathscr{E})$ is the imaginary Bessel function. The decoder test, $\Lambda_1 \gtrless \Lambda_0$, is therefore the same as the test $I_0(\mathscr{E}_{11}) \gtrless I_0(\mathscr{E}_{10})$. However, since $I_0(\mathscr{E})$ is itself a monotonic function, an equivalent test is simply

$$\mathscr{E}_{11} \gtrless \mathscr{E}_{10} \qquad (7.4.37)$$

That is, decoding is achieved by computing $\mathscr{E}_{11}$ and $\mathscr{E}_{10}$ in (7.4.35c), using the correlator in (7.4.35a and b) and selecting the largest. The correlations involve only the harmonic sine waves at the bit frequencies. We emphasize that (7.4.37) is in general a suboptimal decoding procedure since the remaining harmonic frequencies of each bit intensity have been neglected. The single harmonic decoder is depicted in Figure 7.11. Note the more complicated processing being performed to account for the random phase. This is generally the price paid for the lack of knowledge of system parameters.

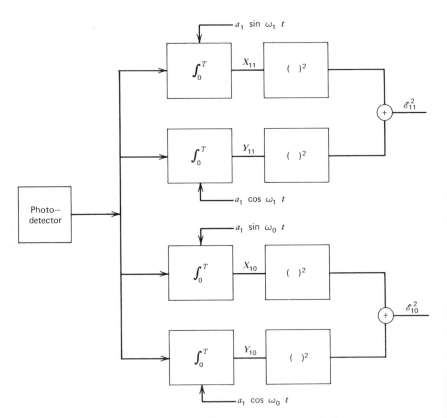

Figure 7.11. Noncoherent bit decoder, FSK signaling, first harmonic.

Spatial Keying. We have been assuming that information transmission is achieved by modulating the optical field in the time domain, while collecting received field power over all available spatial modes. Theoretically, however, binary data can be transmitted by making use of different spatial modes at the receiver. Modulation is achieved by transmitting optical energy in one of two possible spatial modes, or group of modes, as shown in Figure 7.5e. This corresponds to use of the intensity set

$$
\begin{aligned}
n_1(t, \mathbf{r}) &= n_{\mathrm{s}}, & \mathbf{r} \in \mathscr{A}_1 \\
n_0(t, \mathbf{r}) &= n_{\mathrm{s}}, & \mathbf{r} \in \mathscr{A}_0
\end{aligned}
\tag{7.4.38}
$$

where $\mathscr{A}_1$ and $\mathscr{A}_0$ are the spatial areas in the focal plane spanned by the signaling modes. Since the received field is now spatially distributed according to the data bit, two separate photodetectors must be used. Generation of $\tilde{\Lambda}_i$ requires collecting counts over spatial areas $\mathscr{A}_i$ simultaneously, which are then compared. Note that bit disjointness is achieved spatially rather than in amplitude, phase, or frequency encoding, as in previous examples. Since distinct spatial modes produce independent counts just as in distinct time intervals, the resulting error probability becomes identical with that for PPM in (7.4.3), except that signal and noise energy are collected over the full bit interval from each binary set of modes. Under a transmitter power constraint, this doubles the signal energy over that of PPM, but also doubles the noise energy. The primary examples of modal encoding is polarization modulation [3, Chap. 2], where the "spatial" modes are considered to be the polarization states of the optical field.

7.5 DIVERGENCE

Although an exact formulation for the MAP decoder has been obtained, a general expression for PE is not always possible. The examples showed several special cases in which PE was in fact computable as an infinite summation, while in others we had to bound or approximate PE. This inability to determine an exact PE in the general case prevents an accurate comparison of design alternatives. In particular, it hinders the system designer from relating properties of the binary format used for signaling with the eventual performance. To overcome this, several alternative design criteria are used. One concentrates solely on the average values of the normalized likelihood functions $\{\tilde{\Lambda}_i\}$, rather than attempting to determine their actual probability density. We would expect that if the difference of the mean values of the correct and incorrect $\tilde{\Lambda}_i$ can be increased, the resulting PE will also be improved. In essence, the "further apart" the $\tilde{\Lambda}_i$ are during decisioning, the less the probability overlap, and the less the chance for detection errors. This criterion, based on "separation" of likelihood mean values, is called a *divergence* criterion [9]. For the general binary MAP decoder the mean distance, or divergence, becomes

$$\text{DIV} \triangleq \tfrac{1}{2}[E(\tilde{\Lambda}_1 - \tilde{\Lambda}_0) \text{ given a one sent}]$$
$$+ \tfrac{1}{2}[E(\tilde{\Lambda}_0 - \tilde{\Lambda}_1) \text{ given a zero sent}] \tag{7.5.1}$$

where the average is taken over the counting statistics. (We assume no random parameters are involved.) Since each $\tilde{\Lambda}_i$ in (7.3.6) involves a log correlation with the detector count vector, the mean values above can be easily computed. Using the discrete model of (7.3.6) in (7.5.1) yields

$$\text{DIV} = \frac{1}{2} \sum_j E(k_j | 1) \log\left(1 + \frac{\mu_{1j}}{\mu_b}\right)$$

$$- \frac{1}{2} \sum_j E(k_j | 1) \log\left(1 + \frac{\mu_{0j}}{\mu_b}\right)$$

$$+ \frac{1}{2} \sum_j E(k_j | 0) \log\left(1 + \frac{\mu_{0j}}{\mu_b}\right)$$

$$- \frac{1}{2} \sum_j E(k_j | 0) \log\left(1 + \frac{\mu_{1j}}{\mu_b}\right) \tag{7.5.2}$$

when $E(k_j | i)$ is the average of k_j conditioned upon bit i. For multimode detection $E(k_j | 1) = \mu_{1j} + \mu_b$ and $E(k_j | 0) = \mu_{0j} + \mu_b$. The DIV simplifies to

$$\text{DIV} = \frac{1}{2} \sum_j \left[\mu_{1j} \log\left(1 + \frac{\mu_{1j}}{\mu_b}\right) + \mu_{0j} \log\left(1 + \frac{\mu_{0j}}{\mu_b}\right)\right]$$

$$- \frac{1}{2} \sum_j \left[\mu_{1j} \log\left(1 + \frac{\mu_{0j}}{\mu_b}\right) + \mu_{0j} \log\left(1 + \frac{\mu_{1j}}{\mu_b}\right)\right] \tag{7.5.3}$$

Thus, in Poisson detection, the divergence depends only on the intensity vectors used, and DIV can be used for a direct comparison of arbitrary intensity pairs.

We may immediately inquire if (7.5.3) indicates that any one intensity set is preferable over all others, from the point of view of maximizing divergence. Let us first consider DIV under the condition that the average transmitter power be constrained. That is, we require $n_i(t) = n_s$ for $i = 0, 1$. Since the intensities must be nonnegative, we see that we can immediately write the following inequalities for DIV:

$$DIV \le \frac{1}{2}\sum_j \left[\mu_{1j} \log\left(1 + \frac{\mu_{1j}}{\mu_b}\right) + \mu_{0j} \log\left(1 + \frac{\mu_{0j}}{\mu_b}\right) \right]$$

$$\le n_s T \log\left(1 + \frac{n_s}{n_b}\right) \qquad (7.5.4)$$

The first inequality holds only if the two negative terms in (7.5.3) are zero, which occurs only if at every j one or the other of the intensities must be zero whenever the other is nonzero, and the intensity vectors are non-overlapping. The second inequality holds if the full available power is used for all intensity components when nonzero, and if the average energy over both intensities is $n_s T$. It can be easily seen that all these conditions will occur if the on-off signal set in (7.4.1) is used. Hence, DIV is maximized under a power constraint by an OOK system.

Consider now an average energy constraint of K_s on both signals, as in (7.1.7). Again we find that DIV can be bounded, this time by

$$DIV \le \left(\frac{K_s}{2}\right)\left[\max_j \log\left(1 + \frac{\mu_{1j}}{\mu_b}\right) + \max_j \log\left(1 + \frac{\mu_{0j}}{\mu_b}\right) \right]$$

$$= K_s\left[\log\left(1 + \frac{K_s}{n_b \tau_s}\right) \right] \qquad (7.5.5)$$

where τ_s is the counting subinterval. The equality holds if the intensities are nonoverlapping, and each contains count energy K_s over one τ_s sec interval. Hence, the signaling takes the form of an intensity set using the full available energy in two different counting intervals. This corresponds to a PPM format in which optical energy is placed in one of two separate subintervals for bit encoding. Note that the time interval τ_s appears explicitly in (7.5.5), and the value of DIV will increase as τ_s is made smaller. That is, the PPM signaling intervals should be as small as possible, as permitted by the intensity bandwidth B_n (recall $\tau_s \cong 1/2B_n$). The smallest value of τ_s is $1/2B_0$ where B_0 is the optical detector bandwidth. However, if τ_s is reduced to this value, the number of time modes in τ_s sec reduces to one, and the multimode (Poisson) assumption may be invalid. It is interesting that although we allowed T sec (T/τ_s counting intervals) to represent a bit, maximum divergence occurs if only two of the intervals are used, and theoretically the bit time T can be reduced to $2\tau_s$ sec. Of course, this assumes that the available signal energy K_s can be generated in each τ_s sec interval, which implies high peak power values for small τ_s.

7.6 SINGLE MODE BINARY DETECTION—LAGUERRE DETECTION

In the analysis of the preceding sections we have tacitly assumed multimode reception, which justified the use of Poisson counting models. As stated, this is almost always a valid assumption in practical system analysis. However, we have also seen that under these conditions the performance can be improved (sometimes by several orders of magnitude) by reducing the amount of background noise collected, the latter directly related to the number of time-space modes observed by the detector. This immediately suggests that system design should be directed toward reducing the number of observed modes in order to improve PE. Mathematically, however, a significant reduction of the number of detector modes may violate our multimode condition, and make our Poisson assumption somewhat suspect. In this case, we must return to the actual Laguerre counting statistics to describe the interval counts. We then must question whether the previously computed decoders are still optimal, and whether the previous PE curves are still usable. Equivalently, the question is one of determining how small D, the number of modes, can be before the multimode analysis is invalid. In this section we address this problem, referring to the nonmultimode case as single mode detection, or simply Laguerre detection. The implication here is that Laguerre probabilities must be used to describe the interval counts.

Consider an intensity modulated system using τ_s sec counting subintervals to transmit the T sec binary intensity vector $\mu_i = (\mu_{i1}, \mu_{i2}, \ldots)$. Let the detected field during each subinterval contain the identical number of modes, D, with the interest here in the case when D is small ($1 \leq D \leq 20$). The formulation above generates a Laguerre distributed count k_j during each subinterval. A decoder is to perform a MAP test on the count sequence $\mathbf{k} = \{k_j\}$ by computing the likelihood function in (7.2.4). This becomes

$$\Lambda_i = P(\mathbf{k}|i) = \prod_{j=1}^{2B_nT} \frac{\mu_{b0}^{k_j}}{(1 + \mu_{b0})^{D+k_j}} \exp\left[-\frac{\mu_{ij}}{1 + \mu_{b0}}\right] L_{k_j}^{D-1}\left(\frac{-\mu_{ij}}{\mu_{b0}(1 + \mu_{b0})}\right)$$

$$(7.6.1)$$

where $\mu_{b0} \triangleq \alpha N_{0b}$. Canceling common terms in the test $\Lambda_1 \gtrless \Lambda_0$ yields the equivalent test

$$\prod_{j=1}^{2B_nT} L_{k_j}^{D-1}(\hat{\mu}_{1j}) \exp\left[-\frac{\mu_{1j}}{1 + \mu_{b0}}\right] \gtrless \prod_{j=1}^{2B_nT} L_{k_j}^{D-1}(\hat{\mu}_{0j}) \exp\left[-\frac{\mu_{0j}}{1 + \mu_{b0}}\right] \quad (7.6.2)$$

where $\hat{\mu}_{ij} = -\mu_{ij}/\mu_{b0}(1 + \mu_{b0})$. Thus the test performed on the count sequence requires a comparison of products of Laguerre polynomials. We see that the MAP test is not a simple linear count correlation with the observed count vector as in the multimode case, but rather involves the nonlinear processing indicated. A simplification occurs if we assume the counting intervals are small enough such that only a count of one or zero is expected in each mode. If k_j approaches zero or one, the Laguerre terms in (7.6.2) behave as

$$L_{k_j}^{D-1}(\hat{\mu}_{ij}) = 1, \qquad \text{if } k_j = 0$$

$$= D + \frac{n_i(t_j)\tau}{\mu_{b0}}, \qquad \text{if } k_j = 1 \qquad (7.6.3)$$

The equivalent to each term in (7.3.5) then becomes

$$\tilde{\Lambda}_i = \log C + \sum_{j=1}^{T/\tau_s} \left\{ k_j \log\left[1 + \frac{n_i(t_j)\tau_s}{n_b\tau_s} \right] - \frac{\mu_{ij}}{1 + \mu_{b0}} \right\}$$

$$= \log C + \int_0^T x(t) \log\left(1 + \frac{n_i(t)}{n_b} \right) dt - \frac{K_i}{1 + \mu_{b0}} \qquad (7.6.4)$$

where C represents all terms not dependent on i, and $n_b = \alpha N_{0b} 2B_0 D_s$ is again the noise count rate. Thus the single mode MAP decoder, when observing the detector counting process over small intervals, performs the same log intensity correlation as in the multimode case. The only difference is that the noise count is due to a relatively few modes in the Laguerre case.

Consider a binary PPM system using the intensity set in (7.4.9). We assume the counting over each bit subinterval is made in the Laguerre case; that is, D is small. The MAP decoder in (7.6.4), performing the test $\tilde{\Lambda}_1 \gtrless \tilde{\Lambda}_0$, reduces to a simple count comparison over each subinterval, just as in (7.4.10) for multimode detection. When this count comparison test is implemented, the single mode PPM error probability is then

$$\text{PE}_\text{L} = \frac{A}{2} \sum_{k_1=0}^{\infty} B^k L_k^{D-1}(K_s') \sum_{k_2=k_1}^{\infty} \gamma_{k_1 k_2} B^{k_2} \binom{D + k_2}{k_2} \qquad (7.6.5)$$

Here k_1 and k_2 represent the count variable over each bit subinterval,

$$A = (1/1 + \mu_{b0})^{2D+1} \exp[-K_s/1 + \mu_{b0}], \quad B = \mu_{b0}/1 + \mu_{b0},$$

and

$$K_s' = -K_s/\mu_{b0}(1 + \mu_{b0}).$$

The term $\mu_{b0} \triangleq \alpha N_{0b}$ is the noise count per time-space mode, whereas the total noise in the counting interval is $D\mu_{b0}$. The L subscript on PE is used to denote that Laguerre counting was used. The probability in (7.6.5) has been digitally computed for several values of the parameters K_s, μ_{b0}, and D. Some typical plots are shown in Figure 7.12 as a function of signal

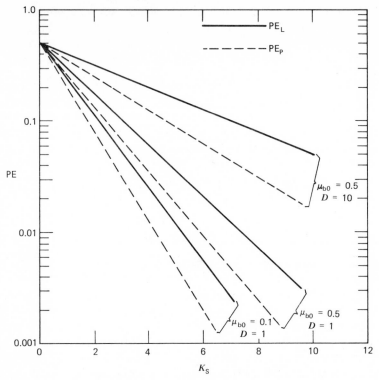

Figure 7.12. Bit error probabilities for Laguerre counting (PE$_L$) and Poisson counting (PE$_P$): μ_{b0} = noise count per mode, D = number of modes, K_s = signal count.

count K_s, and cross plotted in Figure 7.13 as a function of the counting dimension D. This latter curve is particularly useful for determining the advantage of reducing the number of optical modes (reducing D while keeping K_s fixed). The reduction in PE$_L$ does indeed reflect the fact that it is advantageous to reduce the number of modes. That is, the system should encounter a few spatial noise modes as possible, while the signal energy should occupy the full available optical bandwidth (conversely, the optical bandwidth should theoretically be reduced to that of the intensity modulation). Interestingly, the advantage gained depends on the background noise level, and may be relatively small at low noise levels. This may preclude an extensive effort to reduce the number of modes.

If Poisson counting is assumed, the corresponding PE is given in Figure 7.7. For a comparison with the Laguerre case, some Poisson results (denoted PE$_P$) are superimposed in Figures 7.12 and 7.13. These illustrate the difference

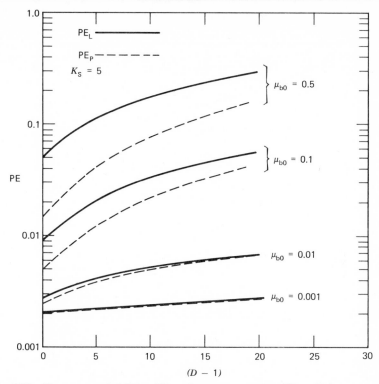

Figure 7.13. Bit error probabilities: PE_L = Laguerre counting, PE_P = Poisson counting. μ_{b0} = noise count per mode, D = number of modes, K_s = signal count.

between single mode error probabilities (Laguerre) and multimode error probabilities (Poisson). Two primary conclusions are immediately evident: (1) The Poisson error probabilities are universally lower than the corresponding Laguerre probabilities, and (2) when $\mu_{b0} \ll 1$, PE_P yields a fairly accurate approximation to PE_L, even if the dimension D is not particularly large. This latter fact appears to indicate that the differences between the two probability densities have little effect on error probability when the noise per mode is small. This is further emphasized in Figure 7.14, in which K_s and $D\mu_{b0}$ are held fixed while PE_L is plotted as a function of D. The PE_L curve approaches PE_P asymptotically from above as D increases, and the corresponding μ_{b0} decreases. In other words, for fixed signal and noise energies, PE_L is accurately predicted by PE_P if the noise energy is produced from a relatively low μ_{b0} value.

The magnitude of the difference between PE_L and PE_P depends on the actual values of D, K_s, and μ_{b0}. This can be seen by investigating the behavior

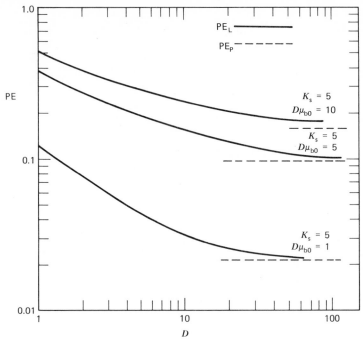

Figure 7.14. Bit error probabilities: PE_L = Laguerre counting, PE_P = Poisson counting. μ_{b0} = noise count per mode, D = number of modes, K_s = signal count.

of the two functions at $D = 1$, which is the point at which the largest difference occurs. When $D = 1$, and $\mu_{b0} \ll 1$, PE_P has the quantum limited bound

$$PE_P \cong \tfrac{1}{2}e^{-K_s} \tag{7.6.6}$$

The corresponding Laguerre limit can be determined by noting that for $D = 1$, (7.6.5) is

$$PE_L = \left(\frac{1}{1 + \mu_{b0}}\right)^2\left(\frac{1 + 2\mu_{b0}}{2}\right)\exp\left[-\frac{K_s}{1 + \mu_{b0}}\right]\sum_{k=0}^{\infty} B^{2k}L_k(K_s') \tag{7.6.7}$$

By applying the Laguerre identity (C.6.1), of Appendix C, and manipulating algebraically, the above becomes

$$PE_L = \frac{1}{2}\exp\left[-\frac{K_s}{1 + 2\mu_{b0}}\right] \tag{7.6.8}$$

The ratio of (7.6.8) to (7.6.6) is then

$$\left.\frac{PE_L}{PE_P}\right|_{D=0} = \exp\left[\frac{\mu_{b0}K_s}{1 + 2\mu_{b0}}\right] \tag{7.6.9}$$

This shows that the ratio of the two error probabilities depends explicitly on the bracket above. When expressed in this manner, one can append earlier statements and conclude that PE_P and PE_L are fairly close over the range of all $D \geq 1$ if $2\mu_{b0}K_s/(1 + 2\mu_{b0}) \ll 1$, or if $\mu_{b0} \ll 1$ *and* $\mu_{b0}K_s \ll 1$.

The Laguerre error probabilities PE_L can also be related mathematically to the Poisson error probabilities PE_P by recalling that Laguerre probabilities are averages of Poisson probabilities. That is,

$$\text{Lag}(k, K_s, N_{0b}) = \int_0^\infty \text{Pos}(k, x)p(x)\, dx \tag{7.6.10}$$

where x is the integrated intensity of a Gaussian field over the counting interval and $p(x)$ is its exponential probability density given in (2.4.12). Recall that the Laguerre probability in (7.6.5) is simply the probability that a Laguerre count in the nonsignaling interval exceeds the count in the signaling interval. Using (7.6.10) this becomes

$$PE_L = \frac{1}{2}\sum_{k_1=0}^\infty \sum_{k_2=k_1}^\infty \gamma_{k_1k_2} \int_0^\infty \int_0^\infty \text{Pos}(k_1, x_1)\, \text{Pos}(k_2, x_2)p(x_1, x_2)\, dx_1\, dx_2 \tag{7.6.11}$$

where $p(x_1, x_2)$ is the joint energy density of the Gaussian fields in the two intervals. Summing inside the integrals simplifies this to

$$PE_L = \int_0^\infty \int_0^\infty PE_P(x_1 - x_2, x_2)p(x_1, x_2)\, dx_1\, dx_2 \tag{7.6.12}$$

where $PE_P(x_1 - x_2, x_2)$ is the Poisson error probability in (7.4.8) with signal energy $x_1 - x_2$ and noise energy x_2. Equation 7.6.12 is simply the conditional error probability jointly averaged over the received energy during each counting interval. When conditioned upon x_1 and x_2, this conditional error probability is simply the Poisson error probability with x_1 and x_2 as average energies of the first and second interval, respectively. Thus Laguerre error probabilities are in fact averages of Poisson error probabilities.

7.7 TIMING ERROR EFFECTS

In deriving our expressions for PE we have been assuming that the bit timing is perfect, and the decoder counts photoelectrons (i.e., integrates the photodetector shot noise) exactly over the proper intervals. If timing offsets occur during a bit period, because of inaccurate time synchronization of the transmitter and receiver, counting occurs over offset intervals, leading to system degradation. In this section we consider this effect for PPM and OOK operation under a multimode (Poisson) counting assumption.

Consider a PPM system with a Δ sec offset. That is, the decoder starts and stops counting over a $T/2$ sec interval that is displaced by Δ sec from that containing the bit information, as shown in Figure 7.15. As a result only a portion of the true signal energy is included in the signal count, while some signal energy may contribute to the count in the adjacent interval, causing intersymbol interference in the form of energy spillover. The effect of this

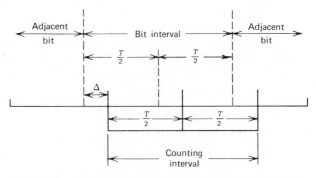

Figure 7.15. Offset counting intervals during PPM bit detection.

interference depends on the form of the adjacent bit, that is, whether or not it contains signal energy. Assuming a positive timing offset $(0 < \Delta < T/2)$, the various effects on the counting statistics are summarized in Table 7.1 where n_s is the average signal count rate. If we let K_s be the average count over $T/2$ due to signal energy, and assume equiprobable bits, the error probability for a positive timing error Δ, averaging over all possibilities given in Table 7.1, is then

$$
\text{PE} | \Delta = \frac{1}{2} \sum_{k_1=0}^{\infty} \sum_{k_2=k_1}^{\infty} \{ \gamma_{k_1 k_2} \, \text{Pos}[k_1, K_s(1 - \varepsilon) + K_b] \, \text{Pos}[k_2, K_s \varepsilon + K_b]
$$

$$
+ \frac{1}{2} \gamma_{k_1 k_2} \, \text{Pos}[k_1, K_s(1 - \varepsilon) + K_b] \, \text{Pos}[k_2, K_b] \}
$$

$$
+ \frac{1}{4} \sum_{k_2=0}^{\infty} \sum_{k_1=k_2}^{\infty} \gamma_{k_1 k_2} \, \text{Pos}[k_1, K_s \varepsilon + K_b] \, \text{Pos}[k_2, K_s + K_b]
$$

$$
(7.7.1)
$$

Table 7.1 Count Probabilities in PPM Detection[a]

Transmitted bit	Subsequent bit	$\text{Prob } k\left(0, \dfrac{T}{2}\right) = k_1$	$\text{Prob } k\left(\dfrac{T}{2}, T\right) = k_2$
1	0	$\text{Pos}\left[k_1, n_s\left(\dfrac{T}{2} - \Delta\right) + K_b\right]$	$\text{Pos}[k_2, K_b]$
1	1	$\text{Pos}\left[k_1, n_s\left(\dfrac{T}{2} - \Delta\right) + K_b\right]$	$\text{Pos}[k_2, K_b + n_s\,\Delta]$
0	1	$\text{Pos}[k_1, n_s\,\Delta + K_b]$	$\text{Pos}\left[k_2, n_s\dfrac{T}{2} + K_b\right]$
0	0	$\text{Pos}[k_1, n_s\,\Delta + K_b]$	$\text{Pos}\left[k_2, n_s\left(\dfrac{T}{2} - \Delta\right) + K_b\right]$

[a] T = bit period; $T/2$ = pulse interval; Δ = timing offset; n_s = signal count rate; K_b = noise count in $T/2$ sec.

where $\varepsilon = 2\Delta/T$ is the fractional timing error. The error probability for negative time shifts is identical to the preceding, when all possibilities are considered, if we interpret $\varepsilon = 2|\Delta|/T$ when $\Delta < 0$. Note that if each of the double sum terms in (7.7.1) is compared to our earlier result (7.4.11), which assumed perfect timing, we can rewrite

$$PE|\Delta = \tfrac{1}{2}PE(K_s', K_b') + \tfrac{1}{4}PE(K_s'', K_b) + \tfrac{1}{4}PE(K_s'', K_b') \qquad (7.7.2)$$

where $PE(K_s, K_b)$ is the PPM error probability in (7.4.11) and

$$
\begin{aligned}
K_s' &= K_s(1 - 2\varepsilon) \\
K_s'' &= K_s(1 - \varepsilon) \qquad\qquad (7.7.3) \\
K_b' &= K_b + K_s\varepsilon
\end{aligned}
$$

Thus timing errors in PPM can be accounted for by merely reinterpreting the effective signal and noise count per $T/2$ interval while assuming perfect timing. Note that the timing errors always act to reduce the effective signal energy, while increasing the effective noise, the overall result degrading the error probability. It is important to realize that the fact that the spilled over signal energy appears as effective noise energy is intrinsic in the Poisson assumption and is valid as long as the multimode assumption is true.

A plot of (7.7.2), obtained by digital computation, is shown in Figure 7.16 for positive or negative timing errors. The results show a relatively fast increase in PE (system degradation) as the offset is increased. The system is

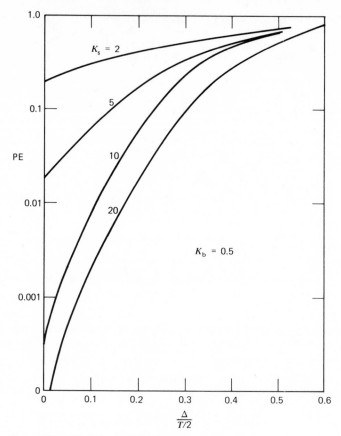

Figure 7.16. Variation of PPM PE with timing error Δ.

essentially ruined (PE $\approx$ 0.5) when $\varepsilon \approx$ 0.5 or when $|\Delta| \approx \frac{1}{2}(T/2)$. These results are of considerable interest since they show that even with increasing signal energy, K_s, a relatively sharp system degradation can still be expected.

A fairly accurate approximation to system performance can be obtained by using the fact that at high noise energy levels, the Poisson error probability are closely related to the Erfc function in (7.4.17), as evident from Figure 7.9. For the higher values of K_s, the signal energy spill over into the nonsignaling intervals produces a high noise condition. Thus for $K_s \gg 1$, we can approximate

$$\text{PE}|\Delta \approx \text{Erfc}\left[\frac{1}{2}\frac{K_s^2(1-\varepsilon)^2}{K_s(1-\varepsilon)+K_s\varepsilon}\right]^{1/2} = \text{Erfc}[\tfrac{1}{2}K_s(1-\varepsilon)^2]^{1/2} \quad (7.7.4)$$

This approximation clearly shows the system degradation with ε, even with large values of K_s. This can again be attributed to the fact that timing errors cause a portion of the signal energy to appear as noise energy. Therefore even though a high signal energy is available, there is consequently a high noise energy as well, when $|\Delta| \neq 0$. The overall result produces the degradation predicted by (7.7.4).

When on-off keyed data bits are transmitted and threshold tests are used for bit decisions at the decoder, the effect of timing errors can be determined by a procedure similar to the PPM case. The actual bit decisions are influenced by the adjacent bit (the subsequent bit when $\Delta > 0$, the preceding bit when $\Delta < 0$), just as in the previous case. If we consider the four possible combinations of transmitted and adjacent bits, and the associated error probability for each the total error probability when a threshold k_T is used and an offset Δ occurs is then

$$
\text{PE}|\Delta = \frac{1}{4} \underbrace{\sum_{k=0}^{k_T} \gamma_{kk_T} \, \text{Pos}(k, K_s + K_b)}_{11} + \frac{1}{4} \underbrace{\sum_{k=k_T}^{\infty} \gamma_{kk_T} \, \text{Pos}(k, K_b)}_{00}
$$

$$
+ \frac{1}{4} \underbrace{\sum_{k=0}^{k_T} \gamma_{kk_T} \, \text{Pos}(k, K_s(1 - \varepsilon) + K_b)}_{10} + \frac{1}{4} \underbrace{\sum_{k=k_T}^{\infty} \gamma_{kk_T} \, \text{Pos}(k, \varepsilon K_s + K_b)}_{01}
$$

$$(7.7.5)$$

where now $\varepsilon = |\Delta|/T$ and K_s, K_b are the received signal and noise counts over $(0, T)$, respectively. The symbols below each sum represent the combination of data bits causing the corresponding error probability, with the left-hand bit the transmitted bit and the other the adjacent bit. Comparison of (7.7.5) with our earlier result (7.4.5) allows us to write

$$
\text{PE}|\Delta = \tfrac{1}{2}\text{PE}(K_s, K_b) + \tfrac{1}{2}\text{PE}(K_s', K_b') \tag{7.7.6}
$$

where the terms on the right are error probabilities with perfect timing and K_s' and K_b' are defined in (7.7.3). We again observe that timing error effects can be interpreted as degradations in signal energy and increases in noise energy in a perfectly timed system. Note that timing errors are exhibited only in the second term in (7.7.6), and can be attributed to the last two terms in (7.7.5), where the adjacent bit is opposite from the true bit. These error probabilities depend on the choice of threshold k_T used for decisioning. When the threshold in (7.4.3) is used, PE in (7.7.6) is plotted as in Figure 7.17 as a function of timing offset for several values of K_s and K_b. The curves manifest similar behavior as in PPM, except the degradation is faster, and the curves exhibit crossovers. That is, at small offsets increasing K_s decreases

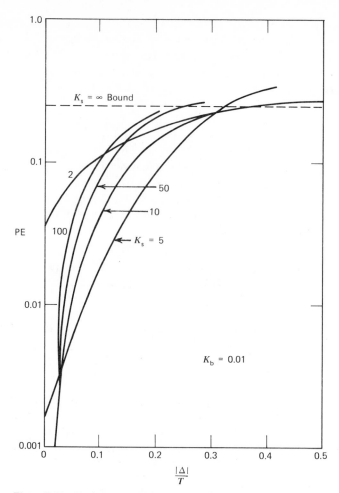

Figure 7.17. Variation of OOK PE with timing error Δ.

error probability, but at larger offsets the opposite is true. An examination of the sums of (7.7.5) reveals that for $K_b \ll 1$ and $K_s \geq 1$, the first three terms tend to zero and the resulting PE is directly attributable to the last term, that is, the error probability when a zero is sent and the adjacent bit is a one. In the limits as $K_s \to \infty$, it follows that even though $k_T \to \infty$, this latter probability becomes exactly unity for any $\varepsilon \neq 0$. The overall PE$|\Delta$ therefore becomes 0.25, and the result is plotted as the $K_s = \infty$ curve in Figure 7.17. The behavior of all these curves can be directly attributed to the fact that optimal OOK requires proper threshold selection, and timing offsets cause changes in effective signal and noise counts, and hence suboptimal operation. As these effective counts become widely different from the design values, the resulting system performance is severely degraded. Synchronization subsystems used to maintain the required timing are discussed again in Chapter 10.

7.8 THERMAL NOISE AND PHOTOMULTIPLIERS IN POISSON DETECTION

So far in our analysis we have assumed a shot noise limited receiver. In particular, we have neglected the effect of circuit noise following photodetection, which adds a thermal random process to the shot noise output. This tends to weaken our original assertion that the decoder counts exactly the number of electrons released in a given time interval. Recall that the subsequent MAP decoder processing depended strongly on our ability to count accurately in each interval. For example, a voltage generated across a 50 ohm resistor resulting from the flow of a single detector electron during a time interval 10^{-6} sec is $(e \cdot 50 \cdot 10^6) = 8 \times 10^{-12}$ V. Now the same resistor operating at room temperature would contribute an integrated thermal noise output voltage whose root square (rms) value is approximately 28×10^{-9} V (Boltzmann's constant $\cdot$ temperature $\cdot$ resistance $\cdot 10^6$). Thus the rms noise is about 3500 times greater than the voltage of a single electron, and it is clear that the count contributed by this single electron could not be easily made by observing this integrated voltage. For this reason photomultiplication is used, which effectively amplifies the current effect of each electron, resulting in a larger electron voltage at the integrator output. We would like to examine the effect on our previous results when the photomultiplier and noise are included.

The actual electronic circuitry following the photodetector is most accurately modeled as in Figure 7.18. The current from the detector passes

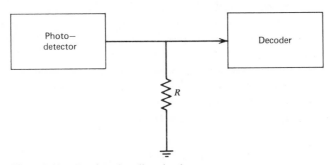

Figure 7.18. Receiver decoding circuitry.

through a resistor R, which generates a voltage that serves as the input to the decoder integrator (counter). In Section 1.8 it was shown that an ideal photomultiplier effectively increases the electron charge e by the gain G. Each electron forming in the photodetector therefore contributes a voltage value GeR/τ to the output of a τ sec integrator. If k electrons are generated in the τ sec counting interval, then the integrator voltage value is

$$y = k\left(\frac{GeR}{\tau}\right) \tag{7.8.1}$$

With Poisson statistics for the count, the probability density of y is then

$$p(y) = \sum_{j=0}^{\infty} \frac{m_\tau^j}{j!} \exp(-m_\tau)\,\delta\left(y - j\,\frac{GeR}{\tau}\right) \tag{7.8.2}$$

where $\delta(x)$ is the delta function and m_τ is the average count over the τ sec counting interval and detector area. Thermal noise generated in the resistor adds directly to the photodetector current. This thermal noise voltage is modeled as Gaussian white noise with spectral level

$$N_{oc} = 4\kappa \mathcal{T} R$$

where κ is Boltzmann's constant and $\mathcal{T}$ is the resistor temperature. This noise voltage is integrated by the decoder counter and adds to the voltage sample in (7.8.1) a noise variable that is Gaussian distributed with zero mean and variance N_{oc}/τ. Thus the combined integrator sample value z after τ sec of counting has a probability density obtained by convolving the Gaussian density with the discrete density in (7.8.2), yielding

$$p_z(z, m_\tau) = \sum_{j=0}^{\infty} \frac{m_\tau^j}{j!} \exp(-m_\tau)\,\text{Gsn}\left(z; j\,\frac{GeR}{\tau}, \frac{N_{oc}}{\tau}\right) \tag{7.8.3}$$

where $\text{Gsn}(z; b, c)$ denotes a Gaussian density in the variable z with mean b and variance c. Observe that the voltage sample probability densities are now continuous densities rather than discrete. Let us now reexamine the PPM case with T sec bit periods and $T/2$ sec pulse intervals. The average count m_τ is $(K_s + K_b)$ when signal is present, and is K_b when signal is absent. Therefore, the probability of incorrect decision is simply the probability that the observable voltage z after the correct interval is less than the observable z after the incorrect intervals. Hence

$$\text{PE} = \int_{-\infty}^{\infty} dz\, p_z(z, K_s + K_b) \int_{z}^{\infty} p_z(y, K_b)\, dy \tag{7.8.4}$$

This can be written more compactly as

$$PE = \sum_{j=0}^{\infty} \frac{(K_s + K_b)^j}{j!} \exp[-(K_s + K_b)] \cdot \int_{-\infty}^{\infty} dz \; Gsn(z; \mu, \sigma^2)\Phi(z)$$

(7.8.5)

where

$$\Phi(z) = \frac{1}{2}\left\{ 1 + \sum_{j=0}^{\infty} \left[\frac{K_b^j}{j!} e^{-K_b} \right] \text{Erf}\left(\frac{z - j\mu}{\sqrt{2\sigma^2}} \right) \right\}$$

$$\text{Erf}(x) = \frac{1}{\sqrt{2\pi}} \int_0^x Gsn(y; 0, 1) \, dy$$

$$\mu = \frac{GeR}{T/2}$$

$$\sigma^2 = \frac{N_{oc}}{T/2}$$

Equation 7.8.5 has been evaluated for several values of K_s and K_b and is shown in Figure 7.19 with $T/2 = 10^{-9}$ sec, $R = 50$ ohms, and N_{oc} corresponding to a noise temperature of 300°K. The asymptotic values at large photomultiplier gains are precisely the values obtained earlier in (7.4.11).

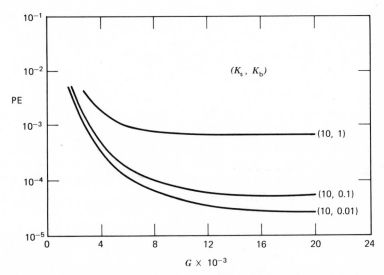

Figure 7.19. PPM Bit error probability versus photomultiplier gain G. Pulse interval = 10^{-9} s, load resistance = 50 Ω, receiver temperature = 300° Kelvin.

At low gains, however, the thermal noise becomes the dominating source of error, and the probability of error increases rapidly. Note that to overcome the thermal environment a photomultiplier gain of about 10^4 is necessary for the operating conditions above. As a rough rule of thumb we can note that the Gaussian noise effect is basically eliminated as the Gaussian density in (7.8.5) approaches a "sharp" delta function about the mean μ. This requires $\mu \gg \sigma$ in (7.8.5) or $2GeR/T \gg (2N_{0c}/T)^{1/2}$. Thus the ideal photomultiplier essentially overcomes the thermal environment if its gain satisfies

$$G \gg (2 \times 10^7)\left(\frac{T\mathscr{T}}{R}\right)^{1/2} \qquad (7.8.6)$$

This gives a rough quantitative measure for determining gain values under which the Poisson detection error probabilities are valid. Note the required gain increases with bit period T and temperature $\mathscr{T}$, but depends inversely on the load resistance R. Thus gain requirements are reduced by using high load impedance photodetectors.

In the preceding discussion we have assumed ideal photomultipliers in which the gain G was the same for every released electron. A nonideal multiplier generally has a gain factor that is itself a random variable and different for each electron, as discussed in Section 4.10. In practice, this gain is often modeled as a Gaussian random variable with a variance or "spread" usually taken as a percentage of the mean gain. We would now like to recompute PE under this situation. If we let G_i be the electron gain of the ith released photoelectron, then the integrator sample value for k electrons is now

$$y = \sum_{i=1}^{k} G_i\left(\frac{eR}{T/2}\right) \qquad (7.8.7)$$

instead of (7.8.1). Here each G_i is an independent random variable with probability density $p_G(x)$. The probability density of y is then obtained from

$$p(y) = \sum_{k=0}^{\infty} p(y|k)\, \text{Pos}(k, m) \qquad (7.8.8)$$

where m is the Poisson level and $p(y|k)$ is the conditional density of y, given k photoelectrons. From (7.8.7)

$$p(y|k) = \frac{1}{(2eR/T)^k}\left[p_G\left(\frac{y}{2eR/T}\right) \underset{k-1}{\otimes} p_G\left(\frac{y}{2eR/T}\right)\right] \qquad (7.8.9)$$

where $\underset{k}{\otimes}$ denotes k-fold convolution. Let $p_G(x)$ be a Gaussian density with mean G and standard deviation $\zeta G/2$, where $0 \leq \zeta \leq 1$. Then y is a Gaussian random variable with mean $2kGeR/T$ and variance $k[(\zeta G/2)(2eR/T)]^2$ so that (7.8.8) becomes

$$p(y) = \sum_{j=0}^{\infty} Gsn\left[y; j\left(\frac{2eGR}{T}\right), j\left(\frac{\zeta GeR}{T}\right)^2\right]\left[\frac{m^j}{j!} e^{-m}\right] \qquad (7.8.10)$$

If one adds the voltage contribution due to the thermal noise, then the integrator output z has a probability density

$$p_z(z, m) = \sum_{j=0}^{\infty} Pos(j, m)\, Gsn\left[z; j\left(\frac{2GeR}{T}\right), j\left(\frac{\zeta GeR}{T}\right)^2 + \frac{2N_{0c}}{T}\right] \qquad (7.8.11)$$

Note that (7.8.11) is identical to (7.8.3), except for the variance terms in the Gaussian densities and is identical to it for $\zeta = 0$. Hence the probability of error is given exactly by (7.8.5), with the variance σ^2 replaced by this new variance. The resulting error probabilities are shown in Figure 7.20 as a function of the spreading parameter ζ, using the parameters of Figure 7.19.

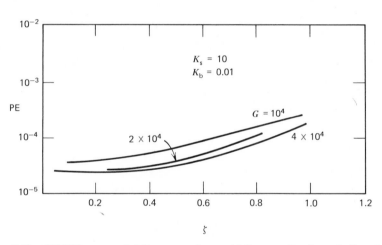

Figure 7.20. PPM Bit error probability versus photomultiplier spreading factor ζ. G = photomultiplier gain.

Although the results vary somewhat as a function of the signal and noise counts, one observes that for mean gains between 10^4 and 10^5 the error probabilities obtained earlier (Figure 7.19) are valid with gain spreads as high as 30 to 40%. Even with spreads as high as 70%, the results indicate only about a factor of 2 increase in PE. Thus, with suitably high gain non-ideal photomultipliers, system error probabilities can be basically divorced from additive thermal noise effects, and the error probability curves plotted earlier based on ideal counting do represent the overall system probabilities. These results also imply that the multiplier spread parameter ζ should be as small as possible for best operation.

In summary, then, we have seen that the effect of circuit thermal noise is to degrade system performance by increasing error probabilities over those previously calculated using Poisson detection models. The latter considered only the anomalous effects of background radiation and detector shot noise. The circuit noise effects can be eliminated, however, by using high gain photomultipliers in the detector stages. These multipliers basically "lift" the detected shot noise processes out of the thermal noise, so that the system is shot noise dominated, and the Poisson detection curves are valid. Randomness in the multiplier gain weakens their advantages, and essentially adds more "noise" to the detection procedure.

7.9 HETERODYNED BINARY SYSTEMS

The alternative to intensity modulation of the bits is to digitally modulate the phase or frequency of the laser source, using heterodyning or homodyning at the receiver to recover the data. A typical system block diagram is shown in Figure 7.21. Such systems are referred to as (spatial) coherent binary systems, and suffer from the usual problem of having to maintain spatial coherence during transmission. (Recall from our discussions in Chapter 6 that loss of coherence reduces the recovered power levels, thereby degrading the heterodyning operation.) Binary modulation formats can be in terms of phase shifting (PSK) or frequency shifting (FSK) the optical carrier. Under spatial coherent conditions, a strong receiver laser oscillator can homodyne (for PSK) or heterodyne (for FSK). Bit decoders following the photodetector then recover the bits by signal processing. In PSK the optical field is homodyned and processed at baseband (binary waveform)

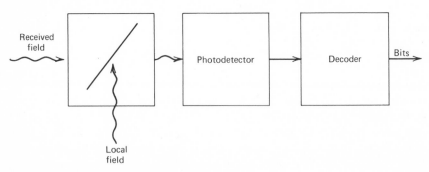

Figure 7.21. The heterodyned binary digital receiver.

frequencies. In FSK the field is heterodyned to some intermediate frequency, and decoding is made by frequency detection. The primary advantage of coherent systems is that a strong receiver oscillator can cause shot noise limited operation and overcome the circuit noise effects discussed in the preceding section. The disadvantage of a heterodyned digital system is the requirement to maintain not only spatial coherence but phase and frequency coherence as well. In PSK, extraneous phase shifts of the optical carrier during transmission mask the true modulated phase, and extraneous frequency shifts degrade FSK performance.

The use of a strong oscillator during heterodyne reception also causes the photodetector output to be accurately modeled as an additive Gaussian noise process. Analytically, this has the advantage of allowing us to resort to known results for the Gaussian noise channel for decoder design and performance [10]. For a PSK system, the detector output after homodyning with a strong source is approximately

$$x(t) \cong (Ge\alpha A_r)a_L a_c \cos[\phi(t)] + b(t) \tag{7.9.1}$$

where G is the detector gain, a_L is the local oscillator field amplitude, a_c is the receiver laser field amplitude, and $\phi(t)$ is the phase modulation of the optical carrier (either 0 or 180°, depending on the bit). The process $b(t)$ is the combined noise effect of the background noise and detector shot noise, and is modeled as a Gaussian process with zero mean and spectral level equal to the shot noise level of the detector output, as developed in Section 6.1. This level is given by $N_0 \triangleq (Ge)^2[\alpha P_L + \alpha^2 P_L N_{ob}]$. The MAP decoder for this model is known to be a single integrator, operating over each bit interval, followed by a bit decision based on the sign of the integrator output. The mean integrator value is $+ Ge\alpha A_r a_L a_c$ for one binary symbol and minus this value for the other. The integrated effect of $b(t)$ in (7.9.1) causes decision errors. The probability of a bit error is simply the probability that the incorrect sign is detected. Hence,

$$\begin{aligned} PE = \frac{1}{2} \int_{-\infty}^{0} &\text{Gsn}\left(x; Ge\alpha A_r a_L a_c, \frac{N_0}{T}\right) dx \\ &+ \frac{1}{2} \int_{0}^{\infty} \text{Gsn}\left(x; -Ge\alpha A_r a_L a_c, \frac{N_0}{T}\right) dx \end{aligned} \tag{7.9.2}$$

The symmetry of the Gaussian density reduces this to

$$PE = \text{Erfc}(\text{SNR}_p)^{1/2} \tag{7.9.3}$$

where the Erfc function is defined in (7.4.17) and

$$\text{SNR}_p = \frac{\alpha A_r a_c^2 T}{1 + \alpha N_{ob}} = \frac{n_s T}{1 + \alpha N_{ob}} \tag{7.9.4}$$

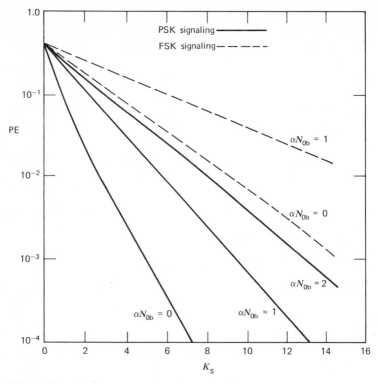

Figure 7.22. Bit error probability, heterodyne receiver.

where $n_s = \alpha(A_r a_c^2)$ is the signal count rate due to the carrier amplitude a_c. Equation 7.9.3 is plotted in Figure 7.22 for several noise counts, αN_{0b}, as a function of $K_s = n_s T$. It is interesting to observe that for the case when the background temperature $\mathcal{T}$ is very high, or if the carrier frequency f is in the microwave or RF range, so that $\kappa \mathcal{T}/hf \gg 1$, then $N_{0b} \approx \kappa \mathcal{T}$ and (7.9.4) becomes

$$\text{SNR}_p \cong \frac{\alpha(A_r a_c^2)T}{\alpha(\kappa \mathcal{T})} = \frac{(A_r a_c^2/2)T}{(\kappa \mathcal{T}/2)} \tag{7.9.5}$$

The numerator is recognized as the receiver carrier energy in the bit time T, while the denominator is the two-sided thermal noise spectral level. Hence (7.9.5) is the familiar "bit energy to noise level" ratio that determines bit error probability in microwave systems [11].

In FSK, the laser frequency is shifted to one of two frequencies during each bit. Receiver heterodyning converts this to one of two intermediate frequencies. The decoder must decide which of the two frequencies is being received. The detector output now takes the form

$$x(t) \approx (Ge\alpha A_r)a_L a_c \sin(\omega t + \theta) + b(t) \tag{7.9.6}$$

where ω is either ω_1 or ω_0 depending on the bit, and θ is a random phase. The MAP decoder, when $b(t)$ is a Gaussian noise process, is known to be a pair of bandpass envelope detectors tuned to each frequency [10]. The envelope detector outputs are sampled and compared at the end of each bit interval. If the bandpass filters preceding the envelope detectors are properly "matched," and have a bandwidth of approximately $1/T$ about each ω_i, the resulting error probability is [11]

$$PE = \frac{1}{2} \exp\left[-\frac{SNR_p}{4} \right] \tag{7.9.7}$$

where now

$$SNR_p = \frac{\alpha A_r a_c^2 T/2}{1 + \alpha N_{0b}} = \frac{1}{2}\left(\frac{n_s T}{1 + \alpha N_{0b}} \right) \tag{7.9.8}$$

Equation 7.9.7 is included in Figure 7.22.

REFERENCES

[1] Abramowitz, M. and Stegun, I., *Handbook of Mathematical Functions*, National Bureau of Standards Appl. Math. Series, Washington, D.C., June 1964.

[2] Ross, M., *Laser Receivers*, John Wiley and Sons, New York, 1966, Appendix A.

[3] Pratt, W., *Laser Communication Systems*, John Wiley and Sons, New York, 1969, Chap. 11.

[4] *Tables of the Q-Function*, Rand Corp. Research Memo RM 399, January 1950.

[5] Brennan, L. and Reed, I., "A Recursive Method of Computing the Q-Function," *IEEE Trans. Inf. Theory*, **IT-11**, No. 1, 312 (April 1965).

[6] Hubbard, W., "Binary Detection in Optical Twin Channel Receivers," *IEEE Trans. Commun. Tech.*, **COM-19**, 221 (April 1971).

[7] Herrmann, G., "Optimum versus Suboptimum Detection under the Poisson Regime," *IEEE Trans. Communications*, **COM-21**, 800 (July 1973).

[8] Gallager, R., *Information Theory and Reliable Communications*, John Wiley and Sons, New York, 1969, Appendix 5A.

[9] Kailath, T., "The Divergence and Bhattacharya Distance Measures in Signal Selection," *IEEE Trans. Commun. Tech.*, **COM-15**, 52–60 (February 1967).

[10] Van Trees, H., *Detection, Estimation, and Modulation Theory*, John Wiley and Sons, New York, 1969, Chap. 4.

[11] Viterbi, A., *Principles of Coherent Communications*, McGraw-Hill Book Co., New York, 1969, Chap. 7.

PROBLEMS

1. Show that if any additive or multiplicative terms are common to each of the likelihood functions Λ_i in a binary test, they can be canceled in performing the ML comparison.

2. Let $n_1(t)$, $n_0(t)$, $0 \le t \le T$, be two intensities used for binary signaling in a direct detection optical system. Show that if $n_1(t)$ and $n_0(t)$ are identical over some interval τ in $(0, T)$, then the photodetector count need not be observed during the interval τ in making ML detection.

3. Show that if the signal intensities $n_i(t)$ in digital signaling contain a random parameter θ with density $p(\theta)$, then the ML binary test in (7.2.2) is equivalent to the MAP test in (7.2.5) if the $\{\Lambda_i\}$ are defined as in (7.2.7).

4. Prove that MAP decoding [using the test defined in (7.2.2)] will yield the minimum value of PE in (7.2.8) over all possible tests. [*Hint*: Write PE|1 as

$$\text{PE}|1 = \sum_{\text{all } \mathbf{k}} P(\mathbf{k}|1)P(0|\mathbf{k})$$

where $P(i|\mathbf{k})$ is the probability of decoding bit i given $\mathbf{k}$, and then show PE is minimized if $P(i|\mathbf{k})$ is selected according to the MAP procedure.]

5. Given the optical intensity $n(t) = n_s[1 + s(t)]$ where $|s(t)| \le 1$, prove that if the optical field is constrained in peak power, the maximum pulse energy in a time interval $(0, T)$ occurs when $s(t) = \pm 1$ for all t. that is, a square wave of arbitrary period.

6. Prove that the MAP decoder for a PPM system performs the test in (7.4.10).

7. Show that if an OOK threshold test is used for bit decisioning in a single mode detection with Laguerre counting in each mode, the resulting PE can be written as

$$\text{PE} = \frac{1}{2} \left\{ \exp\left(\frac{-K_s}{1 + \mu_{bo}}\right) \left(\frac{1}{1 + \mu_{bo}}\right) \sum_{k=0}^{k_T} \left(\frac{\mu_{bo}}{1 + \mu_{bo}}\right)^k L_k\left(-\frac{K_s}{\mu_{bo}(1 + \mu_{bo})}\right) \right.$$

$$\left. + \frac{1}{1 + \mu_{bo}} \sum_{k=k_T}^{\infty} \left(\frac{\mu_{bo}}{1 + \mu_{bo}}\right)^k \right\}$$

where k_T is the threshold, K_s is the signal energy, and μ_{bo} is noise per mode.

8. Using the identities relating the chi-square Q function [1, pp. 941] to the Poisson sum:

$$Q(2m|2c) = \sum_{k=0}^{c-1} \frac{m^k}{k!} e^{-m}$$

and to the gamma function:

$$Q(b|a) = \frac{1}{(a-1)!} \int_b^\infty e^{-t} t^{a-1} \, dt \triangleq \frac{\Gamma(a,b)}{(a-1)!}$$

show that the OOK PE in (7.4.5) can be written as in (7.4.8).

9. Show that the threshold in (7.4.3) minimizes the error probability PE in (7.4.5). [*Hint*: Differentiate to find the minimizing k_T.]

10. Given the identities:

$$I_0(x) \triangleq \sum_{j=0}^{\infty} \frac{x^{2j}}{j!j!}$$

$$Q(a, b) = \exp\left[-\frac{a^2 + b^2}{2} \right] \sum_{n=0}^{\infty} \left(\frac{a}{b}\right)^n I_n(ab)$$

show that (7.4.14) can be written as in (7.4.15).

11. Calculate DIV for the following intensity sets operating under (a) a power constraint and (b) an energy constraint:
(1) On-off keying
(2) PPM
(3) PSK (square wave)
(4) FSK (square wave).

12. Consider a photomultiplier modeled as a branching process, as in Problem 2.11.
(a) Assuming large gain per stage, determine how large the first stage gain should be in order to maintain a spreading factor less than 40%.
(b) If all stage gains are identical, determine how much the first stage gain should be increased to account for the remaining stages while producing the same spreading factor.

13. Consider the function Erfc(x), defined as

$$\text{Erfc}(x) = \frac{1}{\sqrt{2\pi}} \int_x^\infty e^{-y^2/2} \, dy$$

Integrate by parts to establish the bounds

$$\text{Erfc}(x) < \frac{1}{\sqrt{2\pi}\, x} \exp\left[-\frac{x^2}{2} \right], \qquad x > 0$$

$$\text{Erfc}(x) > \frac{1}{\sqrt{2\pi}\, x} \left(1 - \frac{1}{x^2} \right) \exp\left[-\frac{x^2}{2} \right], \qquad x > 0$$

14. By expanding $\exp(\alpha \cos \beta)$ into sums of Bessel functions, show that (7.4.34) can be written as

$$\Lambda = C \prod_{\mathbf{m}(0)} \left[I|_{m_i}|(\mathscr{E}_i) \cos \left(\sum_{i=0}^{\infty} \cos m_i \varphi_i \right) \right]$$

where $\mathbf{m}(0)$ is the set of integer vectors $\mathbf{m} = (m_1, m_2, \cdots)$ whose components satisfy $\sum_{i=0}^{\infty} im_i = 0$.

15. (a) Prove that for antipodal intensity modulation with the intensity set in (7.4.18), the probability of making an error in (7.4.20) when $n_1(t)$ is sent is equal to that when $n_0(t)$ is used. That is, show that

$$\int_{-\infty}^{0} p_{Y_1}(y)\, dy = \int_{0}^{\infty} p_{Y_0}(y)\, dy$$

where Y_1 is the conditional value of Y in (7.4.20). [*Hint*: Show that $\Psi_1(\omega) = \Psi_0(-\omega)$ in (7.4.22). Then prove the above using $\psi_{Y_1}(\omega)$ and $\psi_{Y_0}(\omega)$, as in (7.4.24).]

(b) Use the result of (a) to show that the antipodal average PE is then given by (7.4.21).

8 Digital Communications—
Block Coded Signaling

In binary systems bits are transmitted one at a time, and receiver detection of each bit is made independently of others when timing is accurate. The alternative to binary transmission is to encode blocks of bits, and decode the whole block simultaneously at the receiver. Such systems are called block encoded systems, and have received much attention in modern communication theory and practice. Block encoding generally leads to savings in power and improved performance, but usually at the cost of added system complexity and processing time. The advent of high speed processors and microelectronic circuitry has aided in overcoming these drawbacks, and has inspired digital system designers to devote more emphasis to block coding techniques. In this chapter we study block encoding principles as applied to our optical digital system.

8.1 THE BLOCK CODED OPTICAL SYSTEM

In block coding the transmitter encodes blocks of bits in succession for transmission. Let l represent the block length, that is, the number of bits in a block to be encoded. The block is encoded into a particular waveform of finite time length, say T sec. The waveform is then transmitted to represent the block. Since a distinct waveform must be available for each possible block pattern, and since any block of l bits can take on any one of 2^l different patterns, a block encoded system must have $M = 2^l$ waveforms available for each block. The system therefore transmits $l = \log_2 M$ bits every T sec, or at a rate of $\log_2 M/T$ bits/sec. Note that each waveform now corresponds to more than one bit, and is referred to as a digital word. Thus a block coded system transmits words rather than individual bits. Of course, when $l = 1$, $M = 2$ and the system reduces to the binary system of the preceding chapter.

At the receiver a decoder must decide which of the M possible words is being received during each T sec interval. If an error is made, the incorrect pattern is decoded, which can correspond to one, several, or all of the bits in

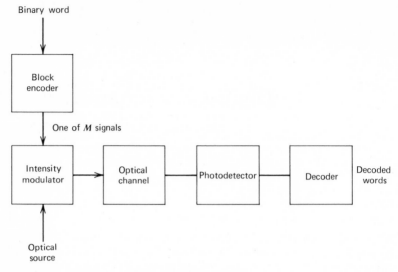

Figure 8.1. The block coded direct detection system.

the block being incorrect. If correct decoding occurs, all of the bits are detected correctly.

A block encoding model for an optical system is shown in Figure 8.1, obtained by extending the binary model of Chapter 7. The diagram shows a direct detection system using a single photodetector at the receiver. The encoded waveforms are used to intensity modulate the optical source. At the receiver the field is photodetected and the resulting output process used for decoding. The detector output can be modeled as a shot noise process or as a count vector. The counts can be generated from either multimode (Poisson) or single mode (Laguerre) models.

The MAP decoder decides the block bit pattern corresponding to the most probable intensity, based on its observation over each T sec block interval. Timing for the intervals is provided by an appropriate synchronization subsystem. Thus the decoder must compute $P(i|\mathbf{k})$ for $i = 1, 2, \ldots, M$, and determine the largest. For equally likely words, this will again reduce to a ML test in which the decoder computes

$$\Lambda_i = P(\mathbf{k}|q) \tag{8.1.1}$$

where $P(\mathbf{k}|q)$ is again the likelihood for the qth intensity $n_q(t)$. The decoder then decides with the test:

$$\text{decide intensity (block pattern) } j \text{ if } \Lambda_j = \max \Lambda_q \tag{8.1.2}$$

Since each Λ_q must be computed simultaneously to perform this test, the decoder will be performing more complicated processing than for the binary case. The general MAP decoder therefore takes the form of a parallel bank of M channels, each computing one of the Λ_q (or log Λ_q), followed by a comparison for the largest. The overall decoder model is shown in Figure 8.2. If the observable is composed of counts over τ sec intervals, then the processing in each channel depends on the counting statistics. For multimode counting, each channel computes

$$\tilde{\Lambda}_q = \sum_{j=1}^{2B_nT} k_j \log\left(1 + \frac{\mu_{qj}}{\mu_b}\right) - K_q \qquad (8.1.3)$$

where k_j are the observed counts, μ_b is the noise count per counting interval, K_q is the average signal field count over T sec, B_n is the intensity bandwidth, and

$$\mu_{qj} = \int_{t_j-\tau}^{t_j} n_q(t)\, dt \qquad (8.1.4)$$

For Laguerre counting with D modes, each channel must compute instead

$$\tilde{\Lambda}_q = \sum_{j=1}^{2B_nT} \log L_{k_j}^{D-1}(\hat{\mu}_{qj}) - \frac{K_q}{1 + \mu_{b0}} \qquad (8.1.5)$$

where now $\hat{\mu}_{qj} = [-\mu_{qj}/\mu_{b0}(1 + \mu_{b0})]$ and μ_{b0} is the noise count per mode.

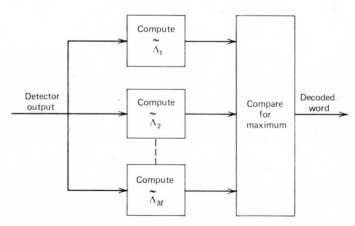

Figure 8.2. The block coded word decoder.

Performance of block coded systems is usually assessed in terms of the average probability of making a word error, PWE. This is defined as

$$\text{PWE} = \frac{1}{M} \sum_{q=1}^{M} \text{PWE}|q \qquad (8.1.6)$$

where $\text{PWE}|q$ is the probability of deciding the incorrect word when the qth word is sent. Exact calculation for PWE is difficult to determine, just as in the binary case, and we must often settle for approximation, simulation, or bounding methods. A simple bound often applied in block encoding analysis is the "union bound" [1]. This involves use of the fact that if $x_1, \ldots, x_M$ are M random numbers, the probability that x_1 is less than the $M - 1$ remaining numbers is bounded from above by the union (sum) of probabilities that x_1 is less than each x_j individually (Problem 8.2). If the numbers $\{x_i\}$ correspond to the $\{\tilde{\Lambda}_i\}$ where a particular word is sent, this union rule implies

$$\text{PWE}|q \le \sum_{\substack{j=1 \\ j \ne q}}^{M} \text{Prob}[\tilde{\Lambda}_q \le \tilde{\Lambda}_j | q] \qquad (8.1.7)$$

where the probability is conditioned on the qth word. Note we do not disturb the inequality by assuming all likelihood equalities are interpreted as errors. The average word error probability then satisfies

$$\text{PWE} \le \frac{1}{M} \sum_{q=1}^{M} \sum_{\substack{j=1 \\ j \ne q}}^{M} \text{Prob}[\tilde{\Lambda}_q \le \tilde{\Lambda}_j | q] \qquad (8.1.8)$$

By properly pairing terms, we can then write (8.1.8) as

$$\text{PWE} \le \frac{1}{M} \sum_{q=1}^{M} \sum_{\substack{j=1 \\ j \ne q}}^{M} \{ \tfrac{1}{2} \text{Prob}[\tilde{\Lambda}_q \le \tilde{\Lambda}_j | q] + \tfrac{1}{2} \text{Prob}(\tilde{\Lambda}_j \le \tilde{\Lambda}_q | j) \} \qquad (8.1.9)$$

The term in the braces is the probability of erring in attempting to decide between a pair of the $\tilde{\Lambda}_i$, and is therefore the error probability associated with a binary test. Often the latter can be evaluated and (8.1.9) is a useful upper bound to PWE. Its value is in the fact that we are guaranteed that performance must be at least this good. Since (8.1.9) involves a binary error probability many of our results in the preceding chapter are immediately applicable for establishing this bound.

It is often desirable to convert word error probability to an equivalent bit error probability. This can be accomplished by determining the probability that a given bit of the word will be incorrect after incorrect decoding. If the incorrectly decoded word is equally likely to be any of the remaining

$M - 1$ words, then a given bit will be decoded as any of the bits in the same position of each other word. In M equally likely patterns a given bit will be a one or a zero $M/2$ times. The chance of it being the incorrect bit in the $(M - 1)$ incorrect patterns is then $(M/2)/(M - 1)$. This means the probability of a given bit being in error is then this probability times the probability that the word was in error. Hence the equivalent bit error probability PE is related to the word error probability PWE by

$$PE = \frac{1}{2}\left(\frac{M}{M - 1}\right)PWE \qquad (8.1.10)$$

Thus bit and word error probabilities can be related by the use of (8.1.10).

Formats for the M-ary block encoded intensity waveform set used for signaling are generally derived as extensions of the binary case. In a pulse position (PPM) format the T sec interval is divided into M time slots, and an optical pulse is placed in one of the slots to represent each block word. In a pulse amplitude modulation (PAM) format, an optical pulse is sent with its intensity selected as one of M distinct levels. Another popular method is to use multilevel phase shift keying (PSK) or frequency shift keying (FSK). In the former, a subcarrier oscillator signal with one of M different phases is used to intensity modulate the optical carrier. In the latter, one of M different subcarrier frequencies is used. Another example of FSK is to use multiple shifts of the laser frequency. If the frequency shifts are not outside the optical bandwidth of the detector, heterodyning followed by frequency detection must be used. If the shifts are outside the optical bandwidth (e.g., using a different color for each word), direct detection can be used with multiple detectors, one for each color. Spatial encoding uses one of M groups of spatial modes for each word but such encoding is difficult to generate. A form of spatial encoding is polarization encoding, in which one of M polarization states is used. The basic objective in format selection is to find waveform sets that are easy to distinguish (less chance of error) yet easy to generate and decode. The PPM system, spatial encoded system, and color coded system are all examples of systems in which detection is made with counts from "orthogonal" modes (i.e., counts from a sequence of nonoverlapping modes, each of which corresponds to a different word). Such systems are often called orthogonal coded systems, and their performance is identical when parameters are properly equated.

8.2 MULTIMODE PPM (ORTHOGONAL) SIGNALING

PPM signaling is a possible technique to use for block encoding, requiring only the time delaying of an optical pulse. In addition, large values of M can theoretically be accommodated by a relatively simple tapped optical delay line. The optical pulse is generated, fed into the line, and delayed until the proper interval, then extracted. Let us consider the optical pulse to occupy a time interval τ sec wide, where again $\tau = 1/B_n$, and B_n is the intensity bandwidth. In T sec there will be $M = T/\tau = 2B_n T$ possible pulse positions, or words. After photodetection the block decoder observes the count vector $\mathbf{k} = \{k_j\}$, where k_j is the count over the jth interval. In a multimode counting model the decoder processing is described by (8.1.3). If the qth intensity is sent, only the qth counting interval contains signal energy K_s so that $\mu_{qj} = K_s$, $j = q$, and $\mu_{qj} = 0$, $j \neq q$. Equation 8.1.3 therefore involves only a count over the qth interval. Thus PPM decoding reduces to a simple count comparison over each interval, selecting the word corresponding to the maximum count. A block diagram of the PPM block decoder is shown in Figure 8.3. This decoder model can in fact be extended to any type of orthogonal signaling in which counts are generated from independent modes. The channels then would correspond to counters operating over each of the possible signaling modes.

The probability of making a word error can be derived in exact form for the multimode PPM, or any orthogonal signaling system. The count k_j is a Poisson variable with mean $(K_s + K_b)$ if the jth intensity is sent, and with

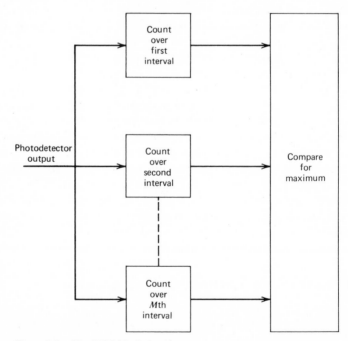

Figure 8.3. The PPM block decoder.

mean K_b otherwise. If the qth intensity is sent, a correct decision is made with probability $1/(r + 1)$ if k_q equals r other k's while exceeding the remaining $M - 1 - r$. Therefore, upon considering all possibilities,

$$\text{PWE} = 1 - \frac{\exp[-(K_s + MK_b)]}{M}$$

$$- \sum_{r=0}^{M-1} \sum_{k=1}^{\infty} \text{Pos}(k, K_s + K_b) \left[\sum_{j=0}^{k-1} \text{Pos}(j, K_b) \right]^{M-1-r}$$

$$\times [\text{Pos}(k, K_b)]^r \frac{(M-1)!}{j!(M-1-r)!r+1} \qquad (8.2.1)$$

The first negative term accounts for the case when a zero count occurs in all intervals. The $r = 0$ negative term assumes no equalities occur, and the remaining negative terms consider all possibilities involving count equalities. By applying the identity

$$\sum_{r=0}^{M-1} \frac{(M-1)!}{(r+1)!(M-1-r)!} A^{M-1-r} B^r = \frac{A^{M-1}}{M(B/A)} \left[\left(1 + \frac{B}{A}\right)^M - 1 \right] \qquad (8.2.2)$$

we can rewrite the error probability as

$$\text{PWE} = 1 - \frac{\exp[-(K_s + MK_b)]}{M} - \sum_{k=1}^{\infty} \text{Pos}(k, K_s + K_b)$$

$$\times \left[\sum_{j=0}^{k-1} \text{Pos}(j, K_b) \right]^{M-1} \left(\frac{1}{Ma}\right) [(1 + a)^M - 1] \qquad (8.2.3)$$

where

$$a \triangleq \frac{K_b{}^k}{k! \sum_{t=0}^{k} K_b{}^t/t!}$$

The PWE just given has been digitally computed for various values of K_s, K_b, and M. An exemplary plot is shown in Figure 8.4 in which PWE has been plotted for various M as a function of K_s, with $K_b = 3$. The same data has been cross plotted in Figure 8.5a, b, and c, for various K_s and several K_b, as a function of M. Note that PWE increases (system becomes poorer) as M is increased with fixed K_s and K_b. This can be directly attributed to the fact that decoder decisioning involves more alternatives, and chance of error will invariably increase.

Although the curves show poorer performance as M increases, the use of these curves to compare M-ary systems at different values of M is somewhat

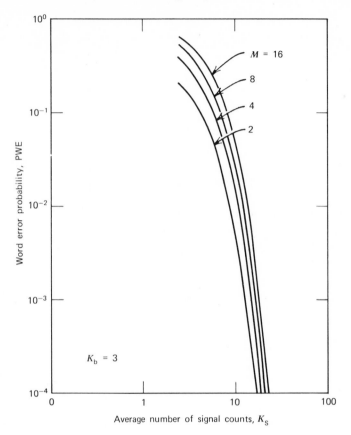

Figure 8.4. Word error probability for block coded PPM. M = number of signals, K_b = average noise count per pulse.

misleading. This is because as M increases, more information is transmitted in each word, and larger error probabilities may in fact be tolerable. In other words, if one system transmits more data than another system, it perhaps can afford to make more errors in transmission. Therefore, comparison of M-ary digital systems should be made only on an individual bit error basis and only after normalization to account for the difference in rates. An M-ary block coded system with τ sec counting intervals transmits $\log_2 M$ bits of information in $M\tau$ sec. It therefore transmits at a rate

$$\mathcal{R} \triangleq \frac{\log_2 M}{T} = \frac{\log_2 M}{M\tau} \text{ bits/sec} \qquad (8.2.4)$$

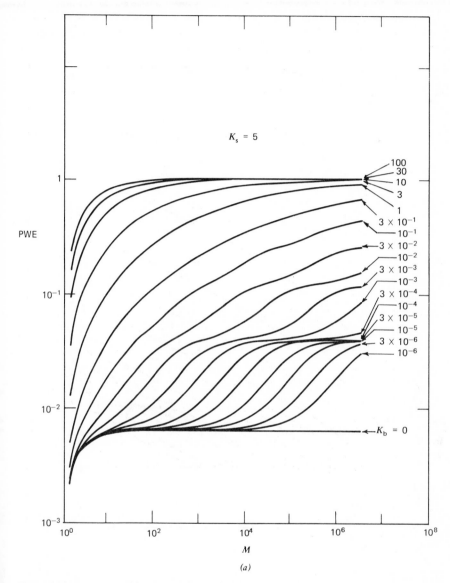

PWE

$K_s = 5$

100
30
10
3
1
3×10^{-1}
10^{-1}
3×10^{-2}
10^{-2}
3×10^{-3}
10^{-3}
3×10^{-4}
10^{-4}
3×10^{-5}
10^{-5}
3×10^{-6}
10^{-6}

$K_b = 0$

M

(a)

Figure 8.5. Error probabilities as a function of M, fixed K_s, K_b. (a) $K_s = 5$, (b) $K_s = 20$, (c) $K_s = 40$.

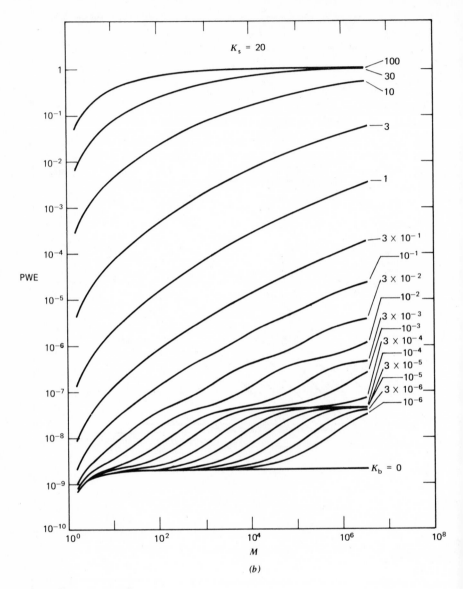

Figure 8.5. (*Continued*)

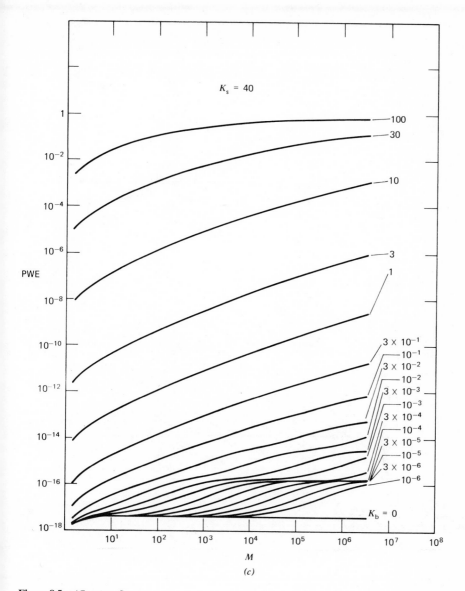

Figure 8.5. (*Continued*)

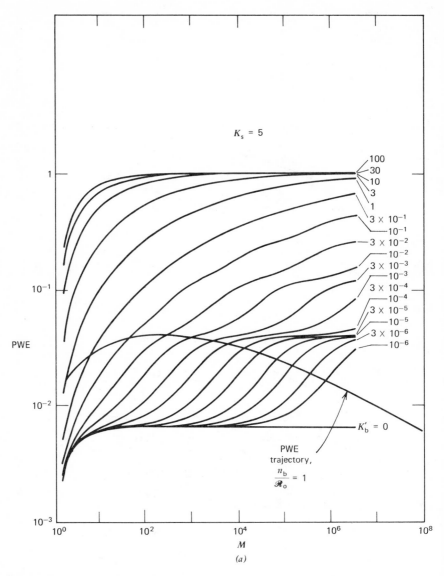

Figure 8.6. PWE trajectories for normalized bit rates. $K'_b = n_b \log_2 M / \mathscr{R}_0 M$.

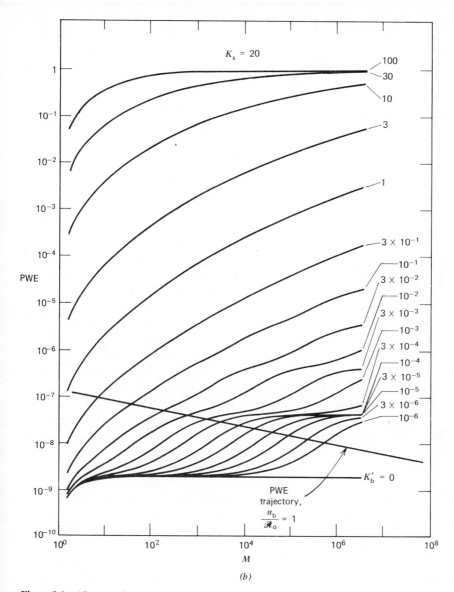

Figure 8.6. (*Continued*)

If the transmission rate $\mathcal{R}$ is normalized for each M, τ must be readjusted to maintain a fixed rate, say $\mathcal{R} = \mathcal{R}_0$. Thus, for each M, τ must be decreased as

$$\tau = \frac{\log_2 M}{M\mathcal{R}_0} \tag{8.2.5}$$

The corresponding noise count per counting interval τ is then

$$K'_b = n_b\tau = \left(\frac{n_b}{\mathcal{R}_0}\right)\left(\frac{\log_2 M}{M}\right) \tag{8.2.6}$$

where n_b is the noise count rate and $n_b/\mathcal{R}_0$ is the noise count per bit time. The normalized noise count therefore decreases with increasing M if bit rate is normalized. Thus, for comparison of different block length systems operating with fixed signal count K_s and equal bit rates $\mathcal{R}_0$, we should compare at values of K'_b given by (8.2.6) and convert to bit error probabilities using (8.1.10). This adjustment in Figure 8.5 corresponds to "jumping" from one K_b curve to a lower K_b curve as M increases along the abscissa. Several trajectories of this type are shown superimposed in Figure 8.6 for different values of $n_b/\mathcal{R}_0$. Note that word error probabilities may actually initially increase, then decrease for the higher M values (Figure 8.6a). If K_s is large enough, PWE continually decreases with M, as for example in Figure 8.6b. The conversion of these normalized PWE to bit error probabilities using (8.1.10) leads to typical curves shown in Figure 8.7 for several values of M. Note the crossover at lower values of K_s, beyond which a distinct advantage over binary transmission occurs as block size is increased. Of course, we emphasize that this improvement is obtained at the expense of intensity bandwidth, transmitter peak power, and decoder complexity. As τ is decreased in (8.2.5), both B_n and pulse power must increase, the latter to maintain fixed count K_s per signaling interval. In addition, as M is increased the number of decoder channels in Figure 8.2 must be increased. These disadvantages may cause the study of block coded systems at large values of M to be somewhat academic; nevertheless, the communication advantages are available for exploitation.

A curve corresponding to $M = \infty$ can be derived as a limiting condition, but care must be used in interpreting its physical implication. This limit can be obtained by replacing K_b in (8.2.3) by K'_b in (8.2.6) and noting that as $M \to \infty$

$$\exp[-(K_s + MK'_b)] \xrightarrow[M \to \infty]{} 0$$

$$\left[\sum_{j=0}^{k-1} \frac{(K'_b)^j}{j!} e^{-K'_b}\right]^{M-1}\left[\frac{(1+a)^M - 1}{M}\right] \xrightarrow[M \to \infty]{} \begin{cases} 0, & k = 1 \\ 1, & k > 1 \end{cases} \tag{8.2.7}$$

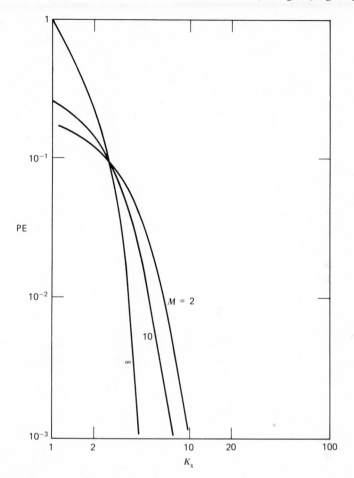

Figure 8.7. Equivalent bit error probability with block coded orthogonal signaling: $K_s =$ signal count per word, $K_b = n_b/\mathscr{R}_0 = 1$.

Using the above we then have

$$\text{PWE} \xrightarrow[M \to \infty]{} 1 - \sum_{k=2}^{\infty} \frac{K_s^k e^{-k_s}}{k!}(e^{-K_s})$$

$$= 1 - (1 - K_s e^{-K_s} - e^{-K_s})$$

$$= (1 + K_s)e^{-K_s} \qquad (8.2.8)$$

Equation 8.2.8 is plotted as $M = \infty$ in Figure 8.7. It is noteworthy that it is precisely the probability of a count of zero or one occurring in a noiseless counting interval of signal energy K_s. This has the following interpretation. As $M \to \infty$, the number of intervals becomes infinite, but the normalized noise energy per interval K_b' approaches zero. The probability that more than one count will occur in any one of $M - 1$ independent nonsignaling intervals having noise energy K_b' approaches zero as $M \to \infty$, indicating that counts of only zero or one will occur in every such interval with probability one. Furthermore, there will be an infinite number of intervals with a zero count and with a one count. Therefore, as $M \to \infty$, an error occurs (with probability approaching one) whenever the signaling interval has a count of zero or one, and an error never occurs when the latter interval has a count greater than one. Hence, (8.2.8) follows.

Although the $M = \infty$ curve serves as an ultimate limit its performance is actually not achievable. This is because the optical bandwidth places a lower limit to values that τ can achieve in (8.2.5). Since $\tau \le 1/2B_0$, the maximum value of M is that for which

$$\frac{M}{\log_2 M} \ge \frac{B_0}{2\mathscr{R}_0} \qquad (8.2.9)$$

For example, if $B_0 = 10^{12}$ Hz and $\mathscr{R}_0 = 1$ gigabit, then $M_{max} \approx 8 \times 10^3$. This value is still large enough to achieve significant coding advantage, although we must again question the validity of the Poisson assumption when $2B_0\tau \approx 1$. For this reason limiting (large M) behavior should be investigated as a single mode condition. This we do in the next section.

8.3 SINGLE MODE PPM (ORTHOGONAL) SIGNALING

Let us consider again a PPM signaling format except that a single mode assumption is made for each counting interval. This means we have reduced the spatial and temporal modes to their limit. The corresponding count probabilities are Laguerre distributed, and the MAP decoding operation is given in (8.1.5). For PPM, $\mu_{qj} = K_s$ for $q = j$, and 0 otherwise, and the decoder computation simplifies to

$$\tilde{\Lambda}_q = L_{k_q}(-\hat{K}_s) \qquad (8.3.1)$$

where $\hat{K}_s = K_s/\mu_{b0}(1 + \mu_{b0})$ and terms not dependent on q have been dropped. However, the Laguerre polynomials with negative arguments are monotonic increasing in their index for any D [Appendix C, (C.16)]. Hence, the MAP comparison of the $\{\tilde{\Lambda}_q\}$ again resolves to an equivalent comparison of the count components. That is, even in single mode detection, the counts over each interval (pulse position) must be compared for the maximum in MAP decoding. Thus the single mode MAP decoder is identical to the multimode count decoder in Figure 8.2.

The resulting word error probability can be derived similarly to (8.2.1) with Laguerre probabilities inserted. Intervals containing optical energy have probabilities given by

$$\text{Lag}(k, K_s, \mu_{b0}) \triangleq \frac{\mu_{b0}{}^k}{(1 + \mu_{b0})^{k+1}} \exp\left[-\frac{K_s}{1 + \mu_{b0}}\right] L_k(-\hat{K}_s) \quad (8.3.2)$$

Intervals containing only noise have Bosé–Einstein counts

$$\text{Bos}(k, \mu_{b0}) \triangleq \frac{\mu_{b0}{}^k}{(1 + \mu_{b0})^{k+1}} \quad (8.3.3)$$

Thus,

$$\text{PWE} = \sum_{r=0}^{M-1} \left(\frac{1}{1+r}\right)\binom{M}{r} \sum_{k=0}^{\infty} \text{Lag}(k, K_s, \mu_{b0})\left[\sum_{j=0}^{k-1} \text{Bos}(j, \mu_{b0})\right]^{M-1-r}$$

$$\cdot [\text{Bos}(k, \mu_{b0})]^r = \frac{e^{-a_2}}{(1 + \mu_{b0})^M} \sum_{r=0}^{M-1}\left(\frac{1}{1+r}\right)$$

$$\cdot \binom{M}{r} \sum_{k=0}^{\infty} L_k(-\hat{K}_s) a_1{}^{k(r+1)}\left[\sum_{j=0}^{k-1} a_1{}^j\right]^{M-r-1} \quad (8.3.4)$$

where $a_1 \triangleq \mu_{b0}/1 + \mu_{b0}$, $a_2 \triangleq -K_s/1 + \mu_{b0}$, and $\hat{K}_s = K_s/\mu_{b0}(1 + \mu_{b0})$. Using the identity

$$\sum_{j=0}^{k-1} a_1{}^j = \frac{1 - a_1{}^k}{1 - a_1} \quad (8.3.5)$$

will reduce this to

$$\text{PWE} = \frac{e^{-a_2}}{(1 + \mu_{b0})^M} \sum_{r=0}^{M-1} \binom{M-1}{r} \frac{1}{1+r} \sum_{k=0}^{\infty} L_k(-\hat{K}_s)$$

$$\times [(1 - a_1{}^k)^{M-1-r} a_1{}^{k(r+1)}] \quad (8.3.6)$$

Performing a binomial expansion in the brackets, and summing over k with identity (C. 6.1) of Appendix C then yields

$$\text{PWE} = \frac{e^{-a_2}}{(1 + \mu_{b0})^M} \sum_{r=0}^{M-1} \binom{M}{r} \frac{1}{1+r} \sum_{j=0}^{M-1-r} \frac{\binom{M-1-r}{j}}{[(1 - a_1{}^{j+1+r})]}$$

$$\cdot \exp\left[\frac{(-1)^{j+1}\hat{K}_s \, a_1{}^{j+r+1}}{a_1{}^{j+r+1} - 1}\right] \quad (8.3.7)$$

This contains no infinite summations, and involves only the two finite sums, depending on M. The form is convenient for desk calculation and some exemplary curves are shown in Figure 8.8. The results closely resemble the PPM curves of the preceding section.

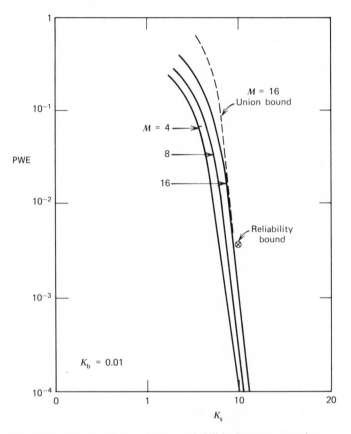

Figure 8.8. Block coded word error probabilities Laguerre counting.

A convenient upper bound can be derived by using the union bound in (8.1.9). Since the signaling involves disjoint counts the probability that the correct count is less than a particular incorrect count is the same for each of the incorrect counts. Hence, (8.1.9) reduces to

$$\text{PWE} \leq (M - 1) \sum_{k=0}^{\infty} \text{Lag}(k, K_s, \mu_{b0}) \sum_{j=k}^{\infty} \text{Bos}(j, \mu_{b0}) \qquad (8.3.8)$$

Noting

$$\sum_{j=k}^{\infty} \text{Bos}(j, \mu_{b0}) = \left(\frac{\mu_{b0}}{1 + \mu_{b0}}\right)^{k-1} \tag{8.3.9}$$

we reduce (8.3.8) to

$$\text{PWE} \leq \frac{M - 1}{1 + 2\mu_{b0}} \exp\left[-\frac{K_s}{1 + 2\mu_{b0}}\right] \tag{8.3.10}$$

This is superimposed in Figure 8.8, and we see it serves as an accurate approximation for small PWE. More accurate approximations have been suggested. A common procedure is to find approximations of the form

$$\text{PWE} \cong \exp[-\log_2 M\mathscr{E}(K_s, \mu_{b0})] \tag{8.3.11}$$

where $\mathscr{E}(K_s, \mu_{b0})$ is called the system *reliability function*. Forms for this reliability have been derived by Liu [2] and Kennedy [3, 4] for the orthogonal optical system. A typical reliability function is shown in Figure 8.9 as a function of $K_s/\log_2 M$ (average signal count per bit) and K_s/μ_{b0}. Values of PWE generated in this way are also superimposed in Figure 8.8.

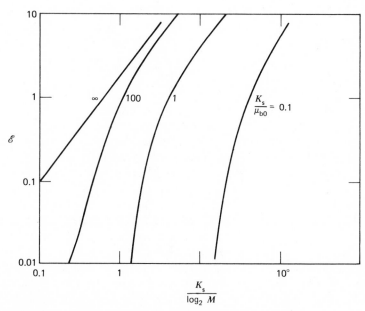

Figure 8.9. Word error probability reliability function. (From [3]).

Let us use the PWE bound in (8.3.10) to convert to an equivalent bit error probability bound using (8.1.10). Hence,

$$\begin{aligned} \text{PE} &\leq \frac{(M/2)}{1 + 2\mu_{b0}} \exp\left[-\frac{K_s}{1 + \mu_{b0}} \right] \\ &= \frac{1/2}{1 + 2\mu_{b0}} \exp\left[-\frac{K_s}{1 + \mu_{b0}} + (\ln 2)\log_2 M \right] \end{aligned} \qquad (8.3.12)$$

This bound has an interesting interpretation. First let us define $n_s = K_s/T$ as the average signal count rate per word length. Then we note

$$\frac{K_s}{\log_2 M} = \frac{K_s}{\mathcal{R}_0 T} = \frac{n_s}{\mathcal{R}_0} \qquad (8.3.13)$$

and

$$\text{PE} < \frac{1/2}{1 + 2\mu_{b0}} \exp\left[-(\log_2 M)\left(\frac{n_s}{\mathcal{R}_0(1 + \mu_{b0})} - \ln 2 \right) \right] \qquad (8.3.14)$$

Now consider this result as $M \to \infty$. The bit error probability must necessarily go to zero (since the bound above goes to zero) as long as

$$\frac{n_s}{\mathcal{R}_0(1 + \mu_{b0})} - \ln 2 > 0 \qquad (8.3.15)$$

or equivalently,

$$\frac{n_s}{\mathcal{R}_0} > (1 + \mu_{b0}) \ln 2 \qquad (8.3.16)$$

Thus a block coded system with single mode operation has the capability of approaching perfect transmission (zero bit errors) while maintaining a fixed bit rate $\mathcal{R}_0$, provided that $n_s/\mathcal{R}_0$ remains constant and satisfies (8.3.16). We see from (8.3.13) that the parameter $n_s/\mathcal{R}_0$ is the average count per bit of the system. Hence perfect transmission is possible with infinite block sizes provided that the count per bit is large enough. Note that to maintain $n_s/\mathcal{R}_0$ constant for all M it is necessary that K_s, the count per pulse, increase as $\log M$. In other words, as $M \to \infty$ a stronger pulse must be transmitted so as to maintain a constant signal count per bit, which must also satisfy (8.3.16). In essence, this is saying that even though we transmit blocks of bits, the effective signal count being allowed each bit must be large enough in order to obtain a coding advantage. We point out that the time T to send a word is also related to M through $T = M\tau$, where τ is the signalling pulse time. Thus if T is to remain fixed, then τ must decrease with increasing M, implying an increase in the system bandwidth. Thus as $M \to \infty$ the zero error condition for a fixed word time will only occur with an infinite system bandwidth. On the other hand, if we maintain constant counting intervals τ at the limit allowed by the optical bandwidth, then T, the word interval, must increase with M. This means that in the limit of zero bit error probability, the word length becomes infinitely long, and an infinite delay occurs in decoding with a finite bandwidth.

We could also interpret (8.3.15) as a condition on $\mathcal{R}_0$, which requires that

$$\mathcal{R}_0 < \frac{n_s}{(1 + 2\mu_{b0}) \ln 2} \tag{8.3.17}$$

The right-hand side places an upper bound on the transmission rate that can achieve zero bit error probability. This bound is dependent on the available count rate n_s (i.e., the transmitter average power) and the noise count μ_{b0}. This bound is often called the *channel capacity* of the optical digital system [5, 6].

8.4 DIVERGENCE

Earlier we showed that in binary communications some design directions could be generated by examining divergence, as an alternative to error probability. In the block coded case, we can extend our earlier definition of binary divergence to an equivalent M-ary average divergence. We define this by

$$\text{Div} \triangleq \frac{1}{M^2} \sum_{j=1}^{M} \sum_{q=1}^{M} \{E[\tilde{\Lambda}_q - \tilde{\Lambda}_j] | q\text{th intensity sent}$$

$$+ E[\tilde{\Lambda}_j - \tilde{\Lambda}_q] | j\text{th intensity sent}\} \tag{8.4.1}$$

The braces represent the average distance between two likelihoods, and the average is then taken over all likelihood pairs. If we take a multimode counting vector model for the observable, while defining equal signal energy intensity vectors $\boldsymbol{\mu}_i = (\mu_{i1}, \mu_{i2}, \ldots, \mu_{i2B_nT})$, (8.4.1) reduces to

$$\text{DIV} = \frac{1}{M^2} \sum_{j=1}^{M} \sum_{q=1}^{M} \sum_{i=1}^{2B_nT} (\mu_{qi} - \mu_{ji}) \left[\log\left(1 + \frac{\mu_{qi}}{\mu_b}\right) - \log\left(1 + \frac{\mu_{ji}}{\mu_b}\right) \right]$$

$$= \sum_{i=1}^{2B_nT} \left[\frac{2}{M} \sum_{q=1}^{M} \mu_{qi} \log\left(1 + \frac{\mu_{qi}}{\mu_b}\right) - \frac{2}{M^2} \sum_{\substack{j \\ j \neq q}}^{M} \sum_{q}^{M} \mu_{qi} \log\left(1 + \frac{\mu_{ji}}{\mu_b}\right) \right] \tag{8.4.2}$$

For an arbitrary intensity vector set, the above can be easily evaluated (see Problems 8.6 and 8.7). For our design objective, we might again attempt to determine conditions under which DIV is maximized for fixed count energy K_s. This is achieved by writing the following inequalities:

$$\text{DIV} \leq \frac{2}{M} \sum_{i=1}^{2B_n T} \sum_{q=1}^{M} \mu_{qi} \log\left(1 + \frac{\mu_{qi}}{\mu_b}\right)$$

$$\leq \frac{2}{M} \left[\max_{q,i} \log\left(1 + \frac{\mu_{qi}}{\mu_b}\right)\right] \sum_{i=1}^{M} \sum_{q=1}^{M} \mu_{qi}$$

$$\leq 2K_s \log\left(1 + \frac{K_s}{\mu_b}\right) \tag{8.4.3}$$

where the last term serves as a true upper bound. The first equality holds if the second term in (8.4.2) is zero, requiring μ_{qi} to be zero for all i at which μ_{ji} is nonzero, $j \neq q$. That is, the M intensities must be mutually disjoint over the counting intervals, which implies $2B_n T \geq M$. The second and third equalities hold only if $\mu_{qi} = K_s$ for one i and is zero for all other i. That is, the qth intensity must have all its available energy concentrated in one counting interval. Thus the upper bound to DIV occurs only if the intensities of the transmitter set are disjoint and wholly concentrated in one of the counting intervals. This of course is satisfied with a set of PPM intensities in which pulses occupy one interval. It is significant that any disjoint intensity set, no matter how many intervals are used, yields the bound of the first inequality, but only the PPM pulsed set yields the last upper bound. Thus, of all disjoint intensity sets, only the pulsed set maximizes DIV. Note that this immediately implies that only M counting intervals are needed for transmitting M intensities. Conversely, we can state that M should be increased to the value $2B_n T$ to take full advantage of PPM signaling. This same conclusion was derived by Bar–David [7, 8] from an information theory point of view.

For single mode counting with $D = 1$, the likelihood is given in (8.1.5), and the averages in (8.4.1) must be taken over Laguerre probabilities. Since the counts are independent, we must evaluate averages of the form

$$E[L_{k_i}(\hat{\mu}_{ji})] = \sum_{k=0}^{\infty} \text{Lag}(k, \mu_{qi}, \mu_{b0}) L_k(\hat{\mu}_{ji}) \tag{8.4.4}$$

Using identity (C.6.4) of Appendix C, (8.4.4) evaluates to

$$E[L_{k_i}(\hat{\mu}_{ji})] = \exp\left[\frac{\mu_{ji}}{1 + \mu_{b0}}\right] I_0[2(\hat{\mu}_{qi}\hat{\mu}_{ji}\mu_{b0})^{1/2}] \tag{8.4.5}$$

where I_0 is the imaginary Bessel function of zero order. Averaging a likelihood function therefore yields

$$E[\Lambda_j | q \text{ sent}] = \exp\left[\frac{2K_s}{1 + \mu_{b0}}\right] \prod_{i=1}^{2B_n T} I_0[2(\hat{\mu}_{qi}\hat{\mu}_{ji}\mu_{b0})^{1/2}] \tag{8.4.6}$$

The divergence between the qth and jth words is then of the form

$$
(\text{DIV})_{jq} = C\left\{ \prod_{i=1}^{2B_nT} I_0[\hat{\mu}_{ji}2\sqrt{\mu_{b0}}] + \prod_{i=1}^{2B_nT} I_0[\hat{\mu}_{qi}2\sqrt{\mu_{b0}}] \right.
$$
$$
\left. - 2\prod_{i=1}^{2B_nT} I_0[2(\hat{\mu}_{qi}\hat{\mu}_{ji}\mu_{b0})^{1/2}] \right\} \tag{8.4.7}
$$

where C represents common factors involving the total count energy of the two intensities. We see that the last term is minimized if the Bessel function argument is made zero for every i. This requires either $\hat{\mu}_{ji} = 0$ or $\hat{\mu}_{qi} = 0$ for each i, which means at least one of the jth and qth intensities has zero energy in each counting interval. Since this must be true for all j and q, $j \neq q$, this again suggests a disjoint intensity set for the M words. Furthermore, it can be shown (Problem 8.8) that each of the first two terms is maximized with a pulsed intensity set. The simultaneous minimization of the negative term and maximization of the positive terms produces a maximum of $(\text{DIV})_{jq}$. Hence, we again establish that, based on a divergence criterion, an M-ary PPM pulsed intensity set is the optimal signaling format for intensity modulated optical communications.

8.5 MULTILEVEL PAM, PSK, AND FSK ENCODING

Although PPM optical signaling is most favorable in terms of divergence, as well as having analytical convenience, it also has certain disadvantages. For example, it requires pulsed operation, uses relatively wide intensity bandwidths, requires accurate delay lines for generation, and is not a historically common signaling technique. For this reason there is interest in using more familiar intensity signaling techniques that circumvent these disadvantages while giving acceptable performance. Examples of such systems are multilevel amplitude modulation (MPAM), multilevel phase shift keying (MPSK) and multilevel frequency shift keying (MFSK).

In MPAM, each block word is transmitted as an optical pulse with a different intensity level. After counting over each block interval a decision is made as to which level is being received. If the intensity levels selected are such that $(K_1, K_2, \ldots, K_m)$ are the corresponding average counts, the MAP decoder for multimode counting computes

$$
\tilde{\Lambda}_q = k \log\left(1 + \frac{K_q}{K_b}\right) - K_q \tag{8.5.1}
$$

for each q, where $k = k(0, T)$ is the total count observed over the $(0, T)$ and K_b is the noise count. Word j is selected if $\tilde{\Lambda}_j = \max(\tilde{\Lambda}_q)$. Noting that each $\tilde{\Lambda}_q$ in (8.5.1) is linear in k, regions wherein each $\tilde{\Lambda}_q$ is maximum can be easily

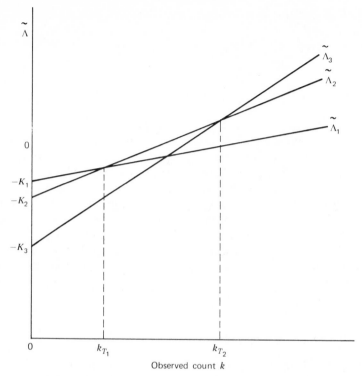

Figure 8.10. Likelihood functions versus observed count k in MPAM. k_{T_i} = decision thresholds.

ascertained, as shown in Figure 8.10. If the $\{K_q\}$ are indexed in ascending order, the MAP decoder will decide $\tilde{\Lambda}_j$ if

$$\frac{K_j - K_{j-1}}{\log[(K_j + K_b)/(K_{j-1} + K_b)]} < k < \frac{K_{j+1} - K_j}{\log[(K_{j+1} + K_b)/(K_j + K_b)]} \quad (8.5.2)$$

Thus a sequence of $M - 1$ thresholds is set, and the appropriate word is selected based upon the threshold values between which the observed count falls.

The probability of word error is then

$$\text{PWE} = \frac{1}{M} \sum_{j=1}^{M} \{\gamma_{kk_{T_{j-1}}}[\text{Prob}(k \leq k_{T_{j-1}})|\text{intensity } j \text{ sent}]$$

$$+ \gamma_{kk_{T_j}}[\text{Prob}(k \geq k_{T_j})|\text{intensity } j \text{ sent}]\} \quad (8.5.3)$$

where $k_{T_{j-1}}$, k_{T_j} are the lower and upper thresholds in (8.5.2), with $k_{T_0} = 0$ and $k_{T_M} = \infty$, and $\gamma_{kk_{T_j}} = \frac{1}{2}$ for $k = k_{T_j}$ and one otherwise. For given intensity levels $\{K_i\}$, the above can be easily evaluated for either Poisson or Laguerre counting. The intensity levels can be either spaced uniformly or nonuniformly. If K_M is the upper intensity level based upon energy restrictions, and the remaining levels are selected on a uniform spaced basis, then

$$K_j = \left(\frac{j}{M}\right)K_M \tag{8.5.4}$$

On the other hand, the variance of counts varies proportionally to the average intensity so that the count spread is larger for the higher intensities, suggesting a nonuniform spacing. An example would be one whose intensity levels were selected as

$$K_j = \left(\frac{j}{M}\right)^2 K_M \tag{8.5.5}$$

Word error probabilities in (8.5.3) have been determined by Kinsel [9] for both the uniform spacing in (8.5.4) and nonuniform spacing in (8.5.5), assuming multimode counting. The results are shown in Figure 8.11 for

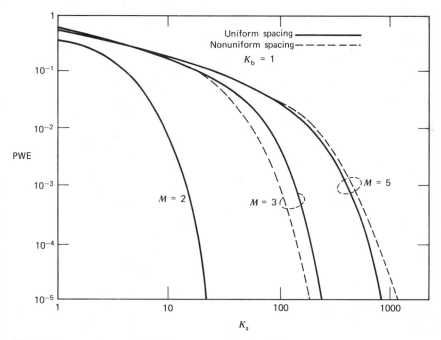

Figure 8.11. Word error probabilities for MPAM. (From [9]).

several values of M. An improvement occurs in nonuniform spacing for $M = 3$, but a degradation occurs for $M \geq 5$. This can be related to the fact that the nonuniform spacing in (8.5.5) tends to squeeze down the lower levels causing more errors at these intensities. As the background noise is decreased, the advantage of graded spacing extends to larger values of M. Optimum spacing in fact can be determined, but is a complicated function of maximum intensity and background (Problem 8.9).

In MPSK, the intensity is modulated by a sine wave having one of M different phases during each word interval. The system has the advantage of utilizing a familiar encoding procedure, and requires less intensity bandwidth than other multilevel schemes. Decoding is accomplished by correlating with log intensities of each possible phase state. The resulting system PWE is extremely difficult to evaluate, even for low values of M, since the signaling is not orthogonal. Hence, one must resort to bounding performance or use of simulation. Some indication of performance is obtained by examining the signal divergence between two particular words. The intensities used for these words are, say,

$$n_q(t) = n_s[1 + \sin(\omega_s t + \theta_q)]$$
$$n_j(t) = n_s[1 + \sin(\omega_s t + \theta_j)] \tag{8.5.6}$$

where θ_q and θ_j are the phase angles, and ω_s is the sine wave frequency, assumed here to be a multiple of the word interval j, that is, $\omega_s = i2\pi/T$ for some integer $i \geq 1$. The divergence between the two words, in integral form, is then

$$(\text{DIV})_{qj} = \int_0^T n_q(t) \log\left(1 + \frac{n_q(t)}{n_b}\right) dt + \int_0^T n_j(t) \log\left(1 + \frac{n_j(t)}{n_b}\right) dt$$
$$- \int_0^T n_q(t) \log\left(1 + \frac{n_j(t)}{n_b}\right) - \int_0^T n_j(t) \log\left(1 + \frac{n_q(t)}{b_b}\right) dt \tag{8.5.7}$$

If we again expand the periodic log intensities into their Fourier series, as in (7.4.27), (8.5.7) reduces to a series of integrals of the form

$$\int_0^T \sin(\omega_s t + \theta_q) \sin(i\omega_s t + \theta_i) \, dt = \begin{cases} 0, & i \neq 1 \\ \dfrac{T}{2} \cos(\theta_q - \theta_1), & i = 1 \end{cases} \tag{8.5.8}$$

where θ_1 is either θ_q or θ_j, depending on the integral. When (8.5.8) is used in (8.5.7), the divergence becomes

$$(\text{DIV})_{qj} = s_1(n_s T)[1 - \cos(\theta_q - \theta_j)] \tag{8.5.9}$$

with s_1 the amplitude of the first harmonic of the log intensity. Thus the divergence between two words of the PSK set is related to the phase angle between them. Clearly, the minimum divergence occurs for two adjacent phase states. If we assume the phase states are symmetrically located around the unit circle, adjacent phases will be separated by $2\pi/M$ rad. This immediately indicates a predominate disadvantage of PSK signaling. When M is large, the divergence between adjacent phases in (8.5.9) becomes approximately

$$(\text{DIV})_{q+1,q} \cong s_1(n_s T)\left(\frac{2\pi}{M}\right)^2 \qquad (8.5.10)$$

and the minimum divergence decreases as the square of M. This means transmitter power (n_s) must increase as the square of M in order to compensate. When the minimum divergence is small we would expect a fairly poor error probability for this system. For this reason multilevel PSK systems are usually restricted to low values of M (e.g., $M = 2, 4, 6,$ or 8). Although these values of M may not produce significant improvement in bit error probability, MPSK systems do allow increased bit rate without significant increase of the intensity bandwidth. This is because the required intensity bandwidth is essentially not dependent on M, and is approximately equal to the reciprocal of the word time for any M.

In MFSK, one of M different frequencies is used to intensity modulate for word signaling. The system requires only the frequency shifting of an oscillator to form the words used to modulate the optical source. If the frequencies are harmonically related, the required intensity bandwidth increases linearly with M much as in the PPM case. Let us restrict our analysis here to an FSK system using square wave signaling (which can be considered an approximation to the sinusoidal case). We assume the square waves are phase locked to a known reference phase, and we let the lowest frequency square wave have period T (the word interval). The remaining ones are harmonically related, so that the jth harmonic has period $T/2^{j-1}$, as shown in Figure 8.12. Denote the qth square wave by $s_q(t)$. Since $\log[1 + (s_q(t)/K_b)] = \log(1 + 1/K_b)s_q(t)$, the MAP decoder need only compute

$$\tilde{\Lambda}_q = \sum_{\{T_q\}} k_j \qquad (8.5.11)$$

where $\{T_q\}$ is the collection of subintervals in $(0, T)$ where $s_q(t) = 1$. The MAP decoder therefore collects counts according to the rule given above, and can be implemented by a single counter operating over subsequent intervals of width equal to half the smallest period, followed by sum and store logic which computes $\{\tilde{\Lambda}_q\}$ in (8.5.11). In attempting to decide between any two harmonics, common intervals in (8.5.11) do not contribute to the

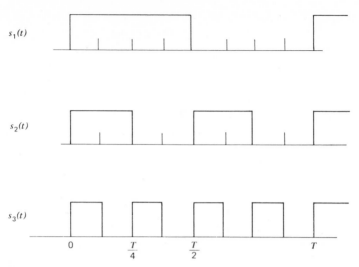

Figure 8.12. Harmonic square waves for MFSK signaling.

resulting decisioning. It follows that the test between any intensity pair $\tilde{\Lambda}_q \gtrless \tilde{\Lambda}_i$ corresponds to the test $\sum_j k_j [s_q(t) - s_i(t)] \gtrless 0$, which in turn is equal to the test

$$\sum_{T_{qi}^+} k_j \gtrless \sum_{T_{qi}^-} k_j \qquad (8.5.12)$$

where T_{qi}^+ is the collection of time intervals in $(0, T)$ where $s_q(t) - s_i(t) = 1$, and T_{qi}^- is the collection where $s_q(t) - s_i(t) = -1$. By referring to Figure 8.12 we can easily establish that T_{qi}^+ and T_{qi}^- are mutually disjoint time intervals, and each involves a total time of $T/4$ sec. If intensity q is sent, then the set T_{qi}^- contains no signal energy, while the collection T_{qi}^+ contains $T/4$ sec of signal energy. Similarly, if intensity i is sent, only T_{qi}^- contains signal energy. This is true for any q and i, $q \neq i$. Thus the binary test between any correct harmonic and any other incorrect harmonic always corresponds to an orthogonal counting (PPM) test with $T/4$ sec of counts with signal energy plus noise energy being compared to counts with $T/4$ sec of noise alone. The corresponding block coded PWE is therefore given by (8.2.1) with $K_s = n_s T/4$ and $K_b = n_b T/4$, where n_s and n_b are the signal and noise count rates. Note that no matter which harmonic (word) is sent the FSK system always transmits a total word energy of $n_s T/2$ during each block interval, but only one-half of this energy is used for detection, because of harmonic overlap. The FSK system is advantageous when peak power (n_s) is limited since it allows signal power to be spread over a larger time period to generate more signal energy. That is, if n_s is fixed, T can be increased to improve detection performance.

8.6 TIMING ERRORS IN PPM BLOCK CODED SYSTEMS

We have already shown that a serious disadvantage with narrow pulse operation, such as in PPM, is that timing of the system becomes difficult to maintain. This causes counting, or integration, to occur over offset intervals, leading to false count values which degrade performance. This effect may be particularly costly in a block coded system where the timing degradation can in fact cancel out the inherent coding advantage of using large size blocks.

Let us consider again the PPM block coded signaling format using T sec word intervals, τ sec pulse intervals, and $M = T/\tau$ encoding words. Let a Δ sec timing offset occur during a word interval. This means each τ sec counting interval is offset by Δ sec during T. In a Poisson counting model, the effect is to cause a decrease in signal energy in the correct interval, while introducing additional energy in the adjacent interval. The corresponding PWE, considering all intervals, must then be obtained by recomputing (8.2.1) with this adjustment. Let k_1 be the correct interval count and let k_2 be the adjacent interval containing the signal energy spill over, when a Δ sec offset occurs, as shown in Figure 8.13. The remaining interval counts, which we denote as simply k_r, contain only background noise. We neglect the fact that the end intervals have a slightly different effect, since they may contain energy spill over from adjacent words. (When M is fairly large, $M \geq 5$, eliminating end effects can be justified, see Problem 8.10.) Neglecting end effects allows us to assume that all M intervals exhibit identical degradation when they are the correct interval. The word error probability with a timing offset Δ, PWE$|\Delta$, is obtained by rederiving (8.2.1). This requires us to consider the correct

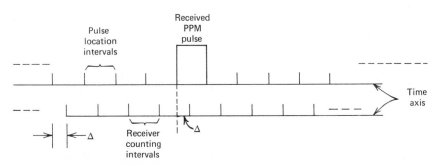

Figure 8.13. Timing offset in MPPM counting.

interval with energy $K_s(1 - \varepsilon) + K_b$, one incorrect interval with energy $\varepsilon K_s + K_b$, and $M - 2$ remaining intervals with noise energy K_b, where $\varepsilon = |\Delta|/\tau$. We must then sum over the probabilities of all situations in which an error can occur, including count equalities.

A simplification producing an accurate approximation and lower bound can be obtained by considering count equalities to always produce correct decisions. Under this assumption,

$$\text{PWE}|\Delta \gtrsim 1 - \text{Prob}[k_1 \geq (k_3, k_r)] \tag{8.6.1}$$

For any ε, $0 \leq \varepsilon \leq 1$, we then have

$$\text{Prob}[k_1 \geq (k_2, k_r)] = \sum_{k=0}^{\infty} \text{Pos}(k, K_s(1 - \varepsilon) + K_b).$$

$$\cdot \left[\sum_{j=0}^{k} \text{Pos}(j, K_s\varepsilon + K_b) \right] \left[\sum_{j=0}^{k} \text{Pos}(j, K_b) \right]^{M-2} \tag{8.6.2}$$

Equation (8.6.2) was computed and used in (8.6.1) to derive the lower bound curves shown in Figure 8.14. (The curves represent an accurate approximation when the signal energy K_s is much larger than the noise energy K_b and count equalities have negligible effect.) Again we see a fairly rapid degradation in system performance as the offset delay becomes a significant portion of the pulse interval. This means that the advantage gained in block coding (when PWE is converted to bit error probabilities at normalized rates) is almost eliminated as Δ increases. The result is to place a restriction on PPM block coded capabilities when timing effects are included. For example, suppose we can only achieve accurate timing to within Δ_0 sec. If we attempt to increase the intensity bandwidth B_n in order to reduce pulse size τ, and therefore increase M, we see that we must maintain $\tau \geq 2\Delta_0$, or equivalently,

$$B_n \leq \frac{1}{2\Delta_0} \tag{8.6.3}$$

Thus the intensity bandwidth can never exceed one-half the "frequency" of the timing accuracy. When converted to block size, M is bounded by

$$M = 2B_n T \leq \frac{T}{\Delta_0} \tag{8.6.4}$$

For example, if $\Delta_0 = 10^{-8}$ sec and T corresponds to a 10^6 word/sec rate, then $B_n \leq 50$ MHz and $M \leq 100$. From our analysis in Section 8.2, we see

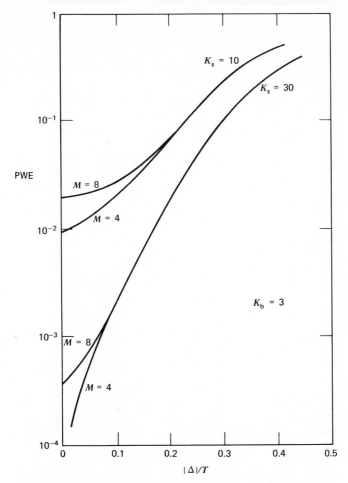

Figure 8.14. PWE degradation due to timing errors in MPPM systems.

that no coding advantage may exist for this range of parameter values. This can be seen by examining specifically the normalized trajectories in Figure 8.6(a) and (b). For $M \leq 100$, the significant improvement in PWE is never realized. If we attempt to increase M, reducing τ accordingly, then the timing offset, ε, increases, and the system degrades as in Figure 8.14. Hence timing errors severely restrict the theoretical advantages of narrow pulsed, MPPM operation. This implies that there is a need to further develop other optical intensity encoding schemes that make better use of the available optical bandwidth than simply transmitting extremely narrow light pulses.

REFERENCES

[1] Wozencraft, J. and Jacobs, I., *Principles of Communication Engineering*, John Wiley and Sons, New York, 1965, Chap. 4.

[2] Liu, J., "Reliability of Quantum Mechanical Communication Systems," *IEEE Trans. Inf. Theory*, **IT-16**, 319 (May 1970).

[3] Kennedy, R., "Communication Through Optical Scattering Channels," *Proc. IEEE*, **58-10**, 1651 (October 1970).

[4] ———, *Fading Dispersive Communication Channels*, John Wiley and Sons, New York, 1969, Chap. 5.

[5] Gallager, R., *Information Theory and Reliable Communication*, John Wiley and Sons, New York, 1968, p. 18.

[6] Helstrom, C., "Quantum Mechanical Communication Theory," *Proc. IEEE*, **58-10**, 1591 (October 1970).

[7] Bar-David, I., "Communication under the Poisson Regime," *IEEE Trans. Inf. Theory*, **IT-15**, 31–37 (January 1969).

[8] ———, "Information in Time of Arrival of Photons," *J. Opt. Soc.*, **63**, 166 (July 1973).

[9] Kinsel, T., "Wideband Optical Communication Systems," *Proc. IEEE*, **58-10**, 1666 (October 1970).

PROBLEMS

1. Show that in MAP decoding of block coded systems with equally likely words;

$$\max_q [P(q \mid \mathbf{k})] = \max_q [P(\mathbf{k} \mid q)]$$

where q is the event that the qth word is transmitted.

2. Prove formally the union bound inequality. That is, prove if $x_1, x_2, \ldots, x_m$ are random numbers, then

$$\text{Prob}[x_1 \leq (x_2, x_3, \ldots, x_m)] \leq \sum_{i=2}^{m} \text{Prob}[x_1 \leq x_i]$$

3. Consider the problem of MAP detecting one of M events from observations $\mathbf{k}$. Let PWD $= 1 - $ PWE be the word detection probability.

(a) Show that the conditional probability PWD$|q$ can be written

$$\text{PWD}|q = \sum_{r=0}^{M-1} \left(\frac{1}{r+1} \right) \sum_{J_{qr}} P(\mathbf{k} \mid q)$$

where J_{qr} is the set of $\mathbf{k}$ such that $\tilde{\Lambda}_q = \max \tilde{\Lambda}_i = r$ other Λ_i. [*Hint*: Over J_{qr}, $P(\mathbf{k} \mid q) = \max_i P(\mathbf{k} \mid i) = r$ other $P(\mathbf{k} \mid i)$.]

(b) Noting $\{J_{qr}\}$ are disjoint over r, but not over q, show that

$$\sum_{q=1}^{M} \sum_{J_{qr}} \frac{P(\mathbf{k} \mid q)}{r+1} = \sum_{\cup_q J_{qr}} \max_i P(\mathbf{k} \mid i)$$

(c) Use (b) to show

$$\text{PWD} = \frac{1}{M} \sum_J \max_i P(\mathbf{k}|i)$$

where J is the space of all possible $\mathbf{k}$.

4. Using Problem 8.3, write a general expression for detection probability under Poisson counting, using the optical model of Section 8.1.

5. Show that if we neglect likelihood equalities (assume their probability of occurring is zero), the M-ary Laguerre word error probability becomes

$$\text{PWE} = \sum_{i=1}^{M-1} \left(\frac{1}{i+1}\right)\binom{M-1}{i}\left\{\frac{1}{1+2\mu_{b0}} \exp\left(\frac{-K_s}{1+2\mu_{b0}}\right)\right\}^i$$

$$\times \left\{1 - \exp\left(-\frac{K_s}{1+2\mu_{b0}}\right)\left(\frac{1+\mu_{b0}}{1+2\mu_{b0}}\right)\right\}^{M-1-i}$$

6. Show that for an M-level PPM system using signal count $K_s = n_s T$ and noise count rate n_b, the DIV in Section 8.4 becomes

$$\text{DIV} = K_s \log\left(1 + \frac{n_s}{n_b}\right)$$

7. Determine the average DIV for a quadraphase PSK system with signal count rate n_s. [Hint: Use (8.5.9).]

8. Show that $\prod_{i=1}^{M} I_0(x_i)$ is maximized under an energy constraint, $\sum_i x_i = E$, with a pulsed intensity set $(x_1 = E, x_i = 0, i \neq 1)$. [Hint: Consider a product of two and iterate the result to a general product.]

9. Derive the equation that must be solved for the optimal interval spacing that minimizes PWE for an M-level PAM system in (8.5.3).

10. Consider an M-level PPM system with timing errors Δ. Show that if end effects are taken into account (i.e., if we consider all possible pulse locations in two consecutive word intervals, $\text{PWE}|\Delta$ is given by

$$\text{PWE}|\Delta = \left(\frac{M^2 - 2M + 2}{M^2}\right)P[k_1, k_2, (M-2)k_3]$$

$$+ \left(\frac{M-1}{M^2}\right)P[k_1, 2k_2, (M-2)k_3]$$

$$+ \left(\frac{M-1}{M^2}\right)P[k_1, 0, (M-1)k_3]$$

where $P[k_1, bk_2, ck_3]$ is the probability of error with the correct interval count k_1, b intervals of count k_2, and c intervals of count k_3. Here $k_1, k_2,$ and k_3 are Poisson counts with energy $(K_s(1 - \varepsilon) + K_s), (\varepsilon K_s + K_b),$ and K_b, respectively.

9 Parameter Estimation in Optical Communications

In certain operations, the basic role of a communication receiver is to derive or extract pertinent parameters from the received field. This operation can be equivalently viewed as one of estimating the parameter from an observation of the received field. When stated in this way, the design of suitable receivers can be cast into a framework of parameter estimation. By this means, one analytically strives to construct optimal devices for performing the necessary estimation. This approach has been successfully applied to microwave communication systems [1–3], and in this chapter we examine similar applications to the optical communication channel.

Parameter estimation, as applied to communication system design, deals with the derivation of optimum procedures for determining parameter values immersed within an observable process. To accomplish this, the meaning of optimum must be clearly defined, and the statistical relationship between the parameter to be estimated and the observable must be accurately described. Thus formulation of a valid system model and choice of meaningful measures of optimality are important prerequisites. In applying estimation procedures to accomplish system design, use is made of the fact that the mathematical system models performing the estimation are derived. In fact, in certain cases the form of the system is more readily obtainable than a closed form solution for the estimate itself. It is the hope of the system designer that the analytical model will generate, or at least suggest, practical implementations of the optimal estimation procedure.

Most of the tractable results in estimation theory occur when analysis is confined to the single parameter case, that is, when the entity to be estimated is a random parameter, fixed over the observation interval. The extension to vector estimators is basically straightforward, but generally with significant increase in complexity. Since one can often model waveform estimation in terms of changing random parameters (the entity to be estimated is a random parameter over a given observation interval but may change from one observation interval to the next), the results of parameter estimation are often directly applicable to more general cases. For this reason, we concentrate only on the case of parameter estimation, with cursory reference to the other types. The reader is referred to more rigorous discussion in References 1 to 5. In the following section we review some basic definitions, procedures, and results from classical parameter estimation, and spend the remainder of the chapter applying these to the optical communication system model.

9.1 REVIEW OF PARAMETER ESTIMATION

The basic format of an estimation problem is shown in Figure 9.1. A random parameter θ is embedded within an observable y. We temporarily ignore exact specification of the structure of y, but in eventual applications it represents a time process or vector expansion of the noisy photodetector output. A system H is to operate upon y and produce at its output an optimum estimate $\hat{\theta}$ of the parameter θ. This estimator is described by its functional operation on the input, which we denote $H(y)$. An optimum estimate is defined in terms of some particular performance criterion, usually one relating $\hat{\theta}$ to θ. When y is a real time process, the estimator H may take on either of two forms. A *recursive* estimator will produce continuously in time the best estimate $\hat{\theta}$ based on the observable y up to that time. A nonreal time estimator processes the input y for a fixed period of time and produces a single estimate after all processing is completed.

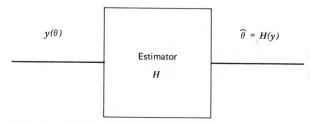

$$y(\theta) \qquad\qquad \hat{\theta} = H(y)$$

Estimator

H

Figure 9.1. The basic estimator model.

There are many well-accepted criteria for defining optimum estimates. We confine our attention to only the two more popular in system analysis. These are the maximum a posteriori (MAP) estimate, $\hat{\theta}_m$, and the conditional mean (CM) estimate, $\hat{\theta}_c$. Both estimators are derived directly from the conditional a posteriori density of the parameter θ given the observable y, which we denote $p(\theta|y)$. The MAP estimate is that value of θ for which $p(\theta|y)$ is maximum. That is,

$$p(\theta_m|y) = \max_{\theta} p(\theta|y) \qquad\qquad (9.1.1)$$

In words, $\hat{\theta}_m$ is the "most likely" value of θ for the given observation y. A MAP estimator is the device $H(y)$ that operates on y to produce the maximizing value $\hat{\theta}_m$. The conditional mean estimate $\hat{\theta}_c$ is the mean of the a posteriori density $p(\theta|y)$:

$$\hat{\theta}_c = \int_{-\infty}^{\infty} \theta p(\theta|y) \, d\theta \qquad (9.1.2)$$

Thus $\hat{\theta}_c$ is the expected value of θ given the observable y, and a CM estimator is a device that produces $\hat{\theta}_c$ from y. The CM estimate is often called a minimum mean squared estimate, or a quadratic cost estimate, since (9.1.2) is also the value of θ_c minimizing the mean squared (quadratic) error $E(\theta - \hat{\theta}_c)^2$. Both estimates (9.1.1) and (9.1.2) depend on y, and although both are physically intuitive, each has immediately obvious disadvantages. Mathematically, $\hat{\theta}_m$ requires finding a maximum while $\hat{\theta}_c$ requires an integration. Both estimates will be different, in general, except when the maximum of $p(\theta|y)$ occurs at its mean value. A system designer is primarily interested in the system (estimator) H that must be implemented. Physically, H must produce the desired estimate, and its underlying operation is to first compute the density $p(\theta|y)$ from the observable y. In many cases, however, the subsequent operation (maximization or averaging) may simplify the computation by eliminating certain portions unnecessary to the final solution.

The generation of $p(\theta|y)$ is most easily carried out by noting

$$p(\theta|y) = \frac{P(y|\theta)p(\theta)}{P(y)} \qquad (9.1.3)$$

Here, $P(y|\theta)$ is the probability of y conditioned on θ and is generally known from the system model, while $p(\theta)$ is the a priori density on θ, and $P(y)$ is the probability of the observable y occurring.† Note that the right-hand side is a probability density in θ, but is a function of y. Since the MAP estimate maximizes $p(\theta|y)$ over θ, and since only the numerator in (9.1.3) involves θ, $\hat{\theta}_m$ equivalently satisfies

$$\hat{\theta}_m = \theta \text{ such that } P(y|\theta)p(\theta) \text{ is maximum} \qquad (9.1.4)$$

Furthermore, since the maximum value of a function is also the maximum value of any monotonic increasing functional of that function, we can seek instead the maximum of the logarithm of $P(y|\theta)p(\theta)$. Thus we can replace (9.1.4) by

$$\hat{\theta}_m = \theta \text{ such that } [\log P(y|\theta) + \log p(\theta)] \text{ is maximum} \qquad (9.1.5)$$

---

† The form of (9.1.3) is convenient when y is a discrete random variable. When y has a continuous probability density, $P(y|\theta)$ and $P(y)$ are replaced by their probability densities $p(y|\theta)$ and $p(y)$, respectively.

If a solution for $\hat{\theta}_m$ exists, it must satisfy the extremal condition

$$\frac{\partial}{\partial \theta} [\log P(y|\theta) + \log p(\theta)]|_{\theta = \hat{\theta}_m} = 0 \qquad (9.1.6)$$

when the derivatives exist. Equations 9.1.4 to 9.1.6 each define the same estimates and each may be used for determining $\hat{\theta}_m$. Equation 9.1.6 is useful since it defines an equation that the optimal estimate must satisfy providing the derivatives exist.

The CM estimate in (9.1.2) can be rewritten using (9.1.3)

$$\hat{\theta}_c = \frac{1}{P(y)} \int_{-\infty}^{\infty} \theta P(y|\theta) p(\theta) \, d\theta \qquad (9.1.7)$$

The integration must be performed as shown, and no simplification using logarithms can be invoked. In addition, the denominator $P(y)$, which did not affect $\hat{\theta}_m$, plays the role of a normalizing factor in (9.1.7) and cannot be neglected. [Note that for a given y, $P(y)$ is simply a number.] We stress that although the optimal estimates can be defined rather compactly as in the equations above, the actual solution in a particular communication application is often not attainable. From an engineering design viewpoint it is hoped that in these cases the form of the estimator, or at least a close approximation, is suggested.

Once a suitable estimator has been found, it is generally of interest to determine its performance. To measure how well a particular estimator $H(y)$ will perform we often examine its bias and variance, defined as

$$\text{bias } H(y) \triangleq E(\theta - H(y))$$
$$\text{variance } H(y) = E(\theta - H(y))^2 - [\text{bias } H(y)]^2 \qquad (9.1.8)$$

where the average is taken over the joint statistics of θ and y. The bias indicates the average error in estimating θ, while the variance indicates the spread about this average. If the bias is zero, then the variance of $H(y)$ is also the mean squared error in estimation. To determine bias and variance, the estimator $H(y)$ must be found explicitly as a function of y, from which (9.1.8) can be computed. However, it can be shown [1] that the variance of any unbiased estimator is always lower bounded by the Cramer–Rao bound (CRB) given by

$$\text{CRB} = \left\{ -E\left[\frac{\partial^2 \log[P(y|\theta)p(\theta)]}{\partial^2 \theta} \right] \right\}^{-1} \qquad (9.1.9)$$

The CRB lower bounds the smallest variance (mean squared error) that can be obtained in unbiased estimations. An estimator that achieves the CRB is said to be an *efficient* estimator. Not all estimators are efficient but for most operating conditions the MAP and CM estimators perform reasonably close. Generally, an estimator is useful if it is *asymptotically efficient* (becomes efficient under some limiting condition on the observable). The primary usefulness of the CRB is that it is reasonably straightforward to compute, can be determined without explicit solution of $H(y)$, and often indicates the key system parameters directly affecting estimation performance.

9.2 MAP PARAMETER ESTIMATION IN MULTIMODE OPTICAL RECEIVERS

We now apply the procedures of the preceding section to an optical communication model operating under multimode conditions. We wish to estimate a random parameter of the transmitted signal field. We assume the signal field is deterministic, except for the parameter value. The observable is the output of the photodetector of the received optical field, as shown in Figure 9.2. We

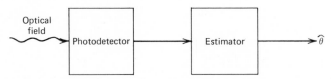

Figure 9.2. The direct detection optical estimator model.

assume the optical field is discretized into a finite number of time intervals, and the photodetector produces an output count vector for the estimator observable. We first consider an idealized situation in which the true electron count is observed, and temporarily postpone the effects of circuit noise and random photomultiplication. The optimum estimator follows the photodetector and indicates the postdetection processing that should be performed for best parameter estimation. In this section we concentrate on MAP estimation only. The observable y is the discretized vector of photodetector counts $\mathbf{k} = (k_1, k_2, \ldots, k_M)$ corresponding to the count sequence collected over a time interval $(0, T)$ and over the detector area. The counts are produced by the reception of an optical field containing signal intensity $n(t, \theta)$, where $0 \le t \le T$, and θ is a real random parameter to be estimated at the receiver. The latter parameter may correspond to the frequency, phase, amplitude, delay, and so on, of the transmitted signal waveform. The parameter θ has associated with it an a priori density $p(\theta)$. The receiver estimator is to generate the MAP estimate of θ based on the vector of counts $\mathbf{k}$ that has been observed.

For multimode detection, the counts $\{k_i\}$ associated with each time interval $(t_i - \tau, t_i)$ are independent and Poisson distributed when conditioned on the intensity. Hence, we write

$$P(\mathbf{k}|\theta) = \prod_{i=1}^{M} P(k_i|\theta)$$

$$= \prod_{i=1}^{M} \frac{[\mu_i(\theta) + \mu_b]^{k_i}}{k_i!} \exp\{-[\mu_i(\theta) + \mu_b]\} \qquad (9.2.1)$$

where again $\mu_b = n_b\tau$ is the average background noise count per interval τ and

$$\mu_i(\theta) = \int_{t_i-\tau}^{t_i} n(t, \theta)\, dt \qquad (9.2.2)$$

The MAP estimate of the parameter θ is defined by (9.1.5) or (9.1.6). Substituting (9.2.1) into (9.1.5) requires determination of

$$\max_{\theta} \left\{ \sum_{i=1}^{M} k_i \log(\mu_i(\theta) + \mu_b) - \int_0^T n(t, \theta)\, dt + \log p(\theta) \right\} \qquad (9.2.3)$$

where all terms not involving θ, and therefore not affecting the maximization, have been dropped. We see that the only operation to be performed with the count observable prior to maximization is the summation, or correlation of the count sequence with the log of the intensity sequence, over each interval. The specific form of a practical implementation of this correlation depends on the particular intensity involved. When derivatives exist, (9.1.6) may be used instead to define the MAP estimate. Using primes to denote differentiation with respect to θ, this becomes

$$\frac{p'(\hat{\theta}_m)}{p(\hat{\theta}_m)} + \sum_{i=1}^{M} k_i\left(\frac{\mu_i'(\hat{\theta}_m)}{\mu_i(\hat{\theta}_m) + \mu_b}\right) - \int_0^T n'(t, \hat{\theta}_m)\, dt = 0 \qquad (9.2.4)$$

Again the sum can be interpreted as a correlation, except the correlating function is differentiated. We point out that not every solution to (9.2.4) is a MAP estimate, since false solutions (maxima) can occur. In most applications, however (9.2.4) is useful primarily for suggesting system implementation rather than deriving specific solutions.

The CRB for the estimator variance can be obtained by using (9.2.1) in (9.1.9) and expanding. The joint expectation can be obtained by averaging over $\mathbf{k}$ first, conditioned on θ, then averaging over θ. Thus,

$$\text{CRB} = \left\{ -E_\theta\left[\frac{\partial^2 \log p(\theta)}{\partial^2\theta}\right] + E_\mathbf{k} \sum_{i=1}^{M} \left[k_i\left[\frac{\partial^2 \log(\mu_i(\theta) + \mu_b)}{\partial\theta^2}\right] - \frac{\partial^2 \mu_i(\theta)}{\partial\theta^2}\right]\right\}^{-1}$$

$$(9.2.5)$$

Noting that $E(k_i|\theta) = \mu_i(\theta) + K_b$, taking the derivatives, and simplifying yields the general result

$$\text{CRB} = \left\{ E_\theta \left[-\frac{\partial^2 \log p(\theta)}{\partial \theta^2} + \sum_i \frac{[\mu_i'(\theta)]}{\mu_i(\theta) + K_b} \right] \right\}^{-1} \tag{9.2.6}$$

We apply these equations to several examples.

Example 9.1

Consider the problem of estimating the intensity level of a monochromatic optical field in background noise following multimode photodetection. In this case

$$n(t, \theta) = \theta n_s \tag{9.2.7}$$

where n_s is the average count rate produced by the signal field and $\mu_i(\theta) = \theta n_s \tau$. Let θ have an exponential probability density so that

$$p(\theta) = \frac{1}{2\sigma^2} \exp\left[-\frac{\theta}{2\sigma^2} \right], \qquad \theta \geq 0 \tag{9.2.8}$$

Since $d \log p(\theta)/d\theta = -1/2\sigma^2$, the MAP solution equation in (9.2.4) becomes

$$-\frac{1}{2\sigma^2} + \frac{n_s \tau}{\hat{\theta}_m n_s \tau + n_b \tau} \sum_{i=1}^M k_i - n_s T = 0 \tag{9.2.9}$$

The solution is then

$$\hat{\theta}_m = \left(\frac{1}{K_s + 1/2\sigma^2} \right) \sum_{i=1}^M k_i - \frac{K_b}{K_s} \tag{9.2.10}$$

where $K_s = n_s T$ and $K_b = n_b T$. Thus the MAP estimator simply collects (integrates) the counts over the total observable interval $(0, T)$ and computes (9.2.10). The estimate has an average value given by

$$E(\hat{\theta}_m) = \left\{ E\left[\frac{(\sum k_i)}{K_s + 1/2\sigma^2} \right] - \frac{K_b}{K_s} \right\}$$

$$= \frac{E(\theta)K_s + K_b}{(K_s + 1/2\sigma^2)} - \frac{K_b}{K_s} \tag{9.2.11}$$

Thus the bias in (9.1.8), in general, is not zero, and the estimate is not unbiased. However, we note that if $K_s \gg 1/2\sigma^2$, then $E(\hat{\theta}_m) \approx E(\theta)$, and the MAP estimator approaches an unbiased estimator. The condition that K_s be large is equivalent to the condition that the observation time T be long, since

$K_s = n_s T$. The estimate variance follows as

$$\text{variance }(\hat{\theta}_m) = E[\hat{\theta}_m - E(\hat{\theta}_m)]^2$$

$$= \frac{E(\theta)K_s + K_b}{(K_s + 1/2\sigma^2)^2}$$

$$= \frac{2\sigma^2 K_s + K_b}{(K_s + 1/2\sigma^2)^2} \qquad (9.2.12)$$

As the observation time is increased, we note that

$$\text{variance }(\hat{\theta}_m) \approx \frac{2\sigma^2}{K_s} \qquad (9.2.13)$$

and the variance of the MAP estimator varies inversely with the signal count energy K_s, while depending directly on σ^2. The latter can be considered the a priori uncertainty in the intensity value. This means the ability to estimate intensity can be improved by observing as much of the optical energy as possible, and by reducing the range of a priori uncertainty in the parameter. As $K_s \to \infty$, the estimator variance approaches zero, and the estimate approaches the true parameter value.

The intensity estimation example above can be considered as a model for the more general problem of estimating intensity modulation of a received optical field. If we assume the observation time is much smaller than the rate at which the intensity modulation changes, then the MAP intensity demodulator (estimator) therefore corresponds to a device that computes (9.2.10) over each subsequent T interval. The estimator therefore attempts to follow the intensity variation from one T sec interval to the next by computing its MAP estimate over that interval. Note that the intensity demodulator generated in this way involves only a count collection, or integration, over each interval. In other words, the MAP demodulator simply integrates the detector counts over intervals shorter than the intensity variations. This result may be contrasted with our analysis in Section 5.3 where we considered the received intensity to be demodulated by a simple low pass filter. We see now that if the bandwidth of the filter is large enough relative to the bandwidth of the modulating intensity, the filter in fact approximates a MAP demodulator.

It is often desirable to derive an integral version for the MAP equations. This can be accomplished similar to our procedure in Section 7.3. We replace the summation of sample values by its integral equivalent. Thus we write

$$\max_\theta \left\{ \sum_{j-1}^{T/\tau} k(t_j - \tau, t_j)\log\left[n(t_j, \theta) + n_b\right] - \int_0^T n(t, \theta)\, dt + \log p(\theta) \right\} \qquad (9.2.14)$$

where terms not involving θ have been dropped, and n_b is the noise count rate

$$n_b = (\alpha N_{0b})2B_0 D_s \qquad (9.2.15)$$

The integral version follows by substituting with the detector output process $x(t)$ in (7.1.3). Equation (9.2.14) then becomes

$$\max_\theta \left\{ \int_0^T x(t) \log(n(t, \theta) + n_b)\, dt - \int_0^T n(t, \theta)\, dt + \log p(\theta) \right\} \qquad (9.2.16)$$

The discrete correlation derived in (9.2.3) is now replaced by an integral correlation. In a similar manner the integral version of (9.2.4) becomes

$$\frac{p'(\hat{\theta}_m)}{p(\hat{\theta}_m)} - \int_0^T x(t) \left[\frac{n'(t, \hat{\theta}_m)}{n(t, \hat{\theta}_m) + n_b} \right] dt - \int_0^T n'(t, \hat{\theta}_m)\, dt = 0 \qquad (9.2.17)$$

These integral equations use as their input (observable) the detector output rather than the count vector. The value of $\hat{\theta}_m$ satisfying the maximization in (9.2.16), or the equality in (9.2.17), is the MAP estimate, assuming the intensity derivatives exist.

The integral version of the CRB in (9.2.6) has the form

$$\text{CRB} = \left\{ E_\theta \left[-\frac{\partial^2 \log p(\theta)}{\partial^2 \theta} + \int_0^T \frac{[n'(t, \theta)]^2}{n(t, \theta) + n_b}\, dt \right] \right\}^{-1} \qquad (9.2.18)$$

Again the only difference is that (9.2.18) uses the intensity function $n(t, \theta)$ itself while the discrete CRB uses the interval counts $\mu_i(\theta)$.

Example 9.2

Consider the problem of estimating a Gaussian phase angle of a sinusoidal intensity. Let

$$n(t, \theta) = n_s[1 + b \cos(\omega_s t + \theta)], \qquad 0 \le t \le T \qquad (9.2.19)$$

$$\omega_s \gg \frac{1}{T}, \qquad b \le 1$$

where b is the intensity modulation index, and n_s and ω_s are constants. The condition $\omega_s \gg 1/T$ implies that there are many frequency cycles in T sec. The phase angle θ has density

$$p(\theta) = \frac{1}{\sqrt{2\pi}\sigma} \exp\left[-\frac{\theta^2}{2\sigma^2} \right] \qquad (9.2.20)$$

and is to be estimated at the receiver in the presence of noise of count rate n_b.

Equation 9.2.17 becomes

$$\frac{\hat{\theta}_m}{\sigma^2} = \int_0^T x(t) \left[\frac{n_s b \sin(\omega_s t + \hat{\theta}_m)}{(n_s + n_b) + n_s b \cos(\omega_s t + \hat{\theta}_m)} \right] dt$$

$$+ \int_0^T n_s [b \sin(\omega_s t + \hat{\theta}_m)] \, dt \qquad (9.2.21)$$

Since the intensity frequency ω_s is assumed large compared to $1/T$, the second integral integrates to zero for any $\hat{\theta}_m$. Hence, the MAP estimate $\hat{\theta}_m$ is that satisfying

$$\hat{\theta}_m = \sigma^2 \int_0^T x(t) \left[\frac{n_s b \sin(\omega_s t + \hat{\theta}_m)}{(n_s + n_b) + n_s b \cos(\omega_s t + \hat{\theta}_m)} \right] dt \qquad (9.2.22)$$

The desired parameter $\hat{\theta}_m$ appears on both sides, and an explicit solution for $\hat{\theta}_m$ is not immediately available. However, we can interpret the right side as a time correlation of the detector output with the time function in the bracket. Hence, the MAP estimate of $\hat{\theta}_m$ can be interpreted as that value of θ which forces the time cross correlation of the detector output and the bracketed function to equal $\hat{\theta}_m$. This suggests an estimator similar to that shown in Figure 9.3, employing a feedback loop to generate the proper $\hat{\theta}_m$ for forcing the loop to lock in (when $\hat{\theta}_m$ is correct, the output of the correlator is that necessary to maintain the steady state of the loop). Feedback loops of this type are common in phase estimating systems [3, 6]. System implementation requires a signal generator that can produce the correlating function in (9.2.22). If $n_b \gg n_s$, or if $b \ll 1$, the correlating function approaches a sinusoidal signal that can be generated from a simple voltage oscillator. The resulting correlating loop is called a phase lock loop. If $n_b = 0$ and $b = 1$, the correlating function is simplified to $\tan(\omega_s t + \hat{\theta}_m/2)$, and the corresponding loop is referred to as a tan-lock loop.

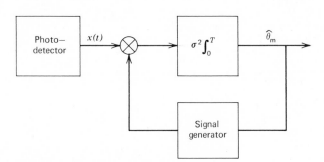

Figure 9.3. The feedback MAP phase estimator.

The CRB for the phase estimator can be evaluated from (9.2.18). We consider the case for $b = 1$ and $n_b = 0$, making use of the high frequency assumption on ω_s. The CRB is

$$\text{CRB} = \left\{ \frac{1}{\sigma^2} + E_\theta \left[\int_0^T n_s \left[\frac{\sin^2(\omega_s t + \theta)}{1 + \cos(\omega_s t + \theta)} \right] dt \right] \right\}^{-1}$$

$$= \left\{ \frac{1}{\sigma^2} + E_\theta \left[n_s \int_0^T [1 - \cos(\omega_s t + \theta)] \, dt \right] \right\}^{-1}$$

$$= \frac{\sigma^2}{1 + \sigma^2 K_s} \qquad (9.2.23)$$

where $K_s = n_s T$ is again the collected count energy over the observation time. Again we see the importance of collecting as much field energy as possible for improved estimation. Note that the minimum phase variance after estimation is always less than the a priori variance, σ^2, on the phase parameter.

We can also derive an interpretation for the MAP estimator suggested by (9.2.16) for this example. In this case we seek the solution for θ such that we achieve

$$\max_\theta \left\{ \int_0^T x(t) \log[(n_s + n_b) + n_s b \cos(\omega_s t + \theta)] \, dt - \frac{\theta^2}{2\sigma^2} \right\} \qquad (9.2.24)$$

This equation suggests a phase estimator in which we compare the cross correlation of the receiver shot noise with a replica of the expected log intensity at each phase angle θ. The phase angle leading to the maximum value for the braces is the MAP estimate. These comparisons can be made simultaneously in parallel channels, although theoretically we would require a continuum of such channels. When the phase angle θ is discretized to a finite number of values, $(\theta_1, \theta_2, \ldots)$, the estimator takes the form in Figure 9.4. The subtraction of the second term in (9.2.24) accounts for the weighting by the a priori density on θ. Note that the strength of the background (n_b) determines how much emphasis is placed on the correlation operation. Roughly speaking, the correlating system in Figure 9.4 determines the maximum from all possible values, while the tracking system in Figure 9.3 automatically searches out the maximum.

Example 9.3

Consider a problem similar to Example 9.2, except we wish to estimate frequency. Let

$$n(t, \theta) = n_s[1 + b \cos(\theta t)], \qquad 0 \leq t \leq T \qquad (9.2.25)$$

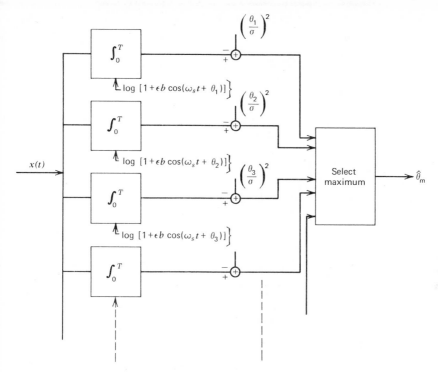

Figure 9.4. The correlator bank MAP phase estimator. ($\epsilon = n_s/n_s + n_b$)

The parameter θ now corresponds to the frequency of the field intensity, and is assumed to have the density

$$p(\theta) = \frac{1}{\sqrt{2\pi}\,\sigma} \exp\left[-\frac{(\theta - \omega_s)^2}{2\sigma^2} \right] \qquad (9.2.26)$$

The MAP estimate is that for which we achieve

$$\max_{\theta}\left\{ \int_0^T x(t)\, \log[(n_s + n_b) + n_s b\, \cos(\theta t)]\, dt - \frac{(\theta - \omega_s)^2}{2\sigma^2} \right\} \qquad (9.2.27)$$

This again corresponds to the maximum output of a bank of correlators. In each element of the bank, the detector output is correlated with the output of a log amplifier fed by a free running oscillator at each particular frequency θ. The MAP estimator selects for $\hat{\theta}_m$ that maximizing the braces in (9.2.27).

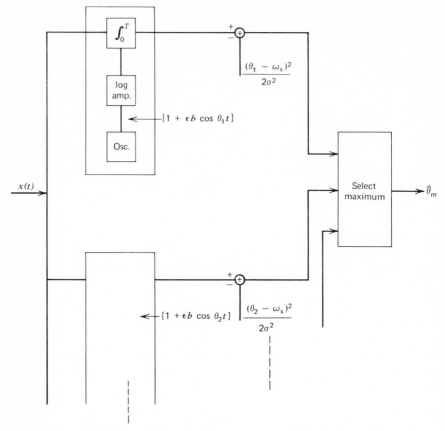

Figure 9.5. A MAP frequency estimator for direct detection optimal systems. ($\epsilon = n_s/n_s + n_b$)

A block diagram for this estimator is shown in Figure 9.5. Note that frequencies farthest from the a priori mean frequency ω_s subtract the most from the correlation value, due to a priori weighting. The system can be realized by a simpler system in which a single oscillator is stepped through all frequencies, T sec at a time, but the resulting estimation time is increased.

The CRB for frequency estimation is obtained identical to (9.2.23). For the case $b = 1$ and $n_b = 0$,

$$\text{CRB} = \left\{ \frac{1}{\sigma^2} + \int_0^T n_s t^2 [1 - E(\cos \theta t)] \, dt \right\}^{-1} \qquad (9.2.28)$$

Now $E(\cos \theta t) = $ Real part $E(e^{j\theta t}) = $ Real part $[\Psi_\theta(jt)]$ where $\Psi_\theta(j\omega)$ is now the characteristic function of a Gaussian random variable. Since $\Psi_\theta(j\omega) =$

$\exp[j\omega\omega_s - \sigma^2\omega^2/2]$, the above reduces to

$$\text{CRB} = \left\{\frac{1}{\sigma^2} + n_s\left[\int_0^T t^2\,dt - \int_0^T t^2\exp\left(-\frac{\sigma^2 t^2}{2}\right)\cos(\omega_s t)\,dt\right]\right\}^{-1}$$

$$\approx \left\{\frac{1}{\sigma^2} + \frac{T^3}{3}n_s\right\}^{-1} \tag{9.2.29}$$

where the approximation holds if $\sigma T \gg 1$. Thus the frequency variance has the approximate bound

$$\text{CRB} \cong \frac{\sigma^2}{1 + \dfrac{K_s(\sigma T)^2}{3}} \tag{9.2.30}$$

where $K_s = n_s T$. The variance bound therefore is inversely proportional to the collected energy K_s and to the square of the σT product. Since σ is the uncertainty "spread" on the unknown frequency, estimation improves as we make $T \gg 1/\sigma$; that is, we should observe over many times the period of the expected frequency offset from the mean frequency. Alternatively, it requires that the correlation bandwidth $(1/T)$ be much smaller than the uncertainty bandwidth to be examined. This means we should frequency estimate by searching the unknown frequency range with as small a bandwidth as possible. Of course, this increases the number of correlators needed in Figure 9.5.

These results also apply to the heterodyning case. Let us consider the receiver in Figure 9.6, in which optical heterodyning over a single spatial mode is performed prior to photodetection. From (6.1.26), the intensity of the detected process is

$$n(t, \theta) = \alpha P_L[1 + \beta\cos(\omega_d t + \theta_d)] \tag{9.2.31}$$

where P_L is the local oscillator power, β^2 is the ratio of received field power of the heterodyning mode to P_L, and ω_d and θ_d are the difference frequency

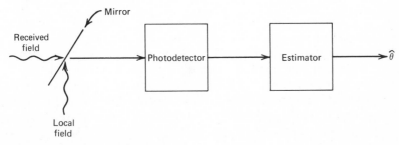

Figure 9.6. The heterodyned estimator.

and phase of the received and local optical carriers. This intensity is similar to that in Examples 9.2 and 9.3, and the ensuing discussion of each is applicable here for estimating optical carrier phase or carrier frequency. We emphasize that these are parameters of the optical carrier itself, and not of its intensity modulation. Since generally $\beta \ll 1$, the MAP estimator of optical carrier phase approaches a phase lock tracking system, in which the local oscillator is itself a voltage controlled optical source. The implementation and analysis of such devices have been discussed by Thompson and Pratt [7].

When the number of modes in each counting interval is sufficiently reduced, the multimode model may be invalid, and single mode (Laguerre) counting should be inserted. In this case, from (8.3.2),

$$P(\mathbf{k}|\theta) = \prod_{i=1}^{M} \text{Lag}(k_i, \mu_i(\theta), \mu_{b0}) \tag{9.2.32}$$

where μ_{b0} is the noise count per mode. The single mode equivalent of (9.2.3) takes the form

$$\max_{\theta} \left\{ \sum_{i=1}^{M} \log L_{k_i} \left[\frac{-\mu_i(\theta)}{\mu_{b0}(1 + \mu_{b0})} \right] - \int_{0}^{T} \frac{n(t, \theta)}{1 + \mu_{b0}} \, dt + \log p(\theta) \right\} \tag{9.2.33}$$

and involves the computation of Laguerre functions for each k_i, rather than a correlation. Similarly, (9.2.4) becomes

$$\frac{p'(\theta)}{p(\theta)} - \sum_{i=1}^{M} \left\{ k_i c(k_i, \theta) \left[\frac{\mu_i'(\theta)}{\mu_i(\theta) + \mu_{b0}} \right] - \frac{\mu_i(\theta)}{1 + \mu_{b0}} \right\} = 0 \tag{9.2.34}$$

where

$$c(k_i, \theta) = 1 - \frac{L_{k_i - 1}\{-\mu_i(\theta)/[\mu_{b0}(1 + \mu_{b0})]\}}{L_{k_i}\{-\mu_i(\theta)/[\mu_{b0}(1 + \mu_{b0})]\}}$$

Thus the loop correlation involving the counts in multimode estimation is modified by the term $c(k_i, \theta)$ when single mode counting is used. This term must be continuously computed since it depends on both the parameter θ and the counts k_i. The modified estimator is shown in Figure 9.7.

It should be emphasized that the estimators discussed in this section were based strictly on photodetected observables. That is, the estimator processing followed optical reception of the received field. Extension to estimator processing over detector arrays is an obvious extension of the results here, similar to that discussed in Chapter 7 (see Problem 9.12). A completely

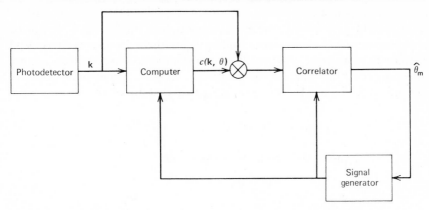

Figure 9.7. The MAP estimator for single mode direct detection optical systems.

separate topic is that of estimating field parameters when the observable is the field itself, rather than the photodetected field. Such analysis is generally confined to Gaussian fields, and the interested reader is referred to the work of Hoversten, Harger, and Halme [8].

9.3 MAP ESTIMATION OF ARRIVAL TIME

In certain communication systems, the receiver has the task of having to determine the time at which a known transmitted signal field arrives at the receiver. The problem has obvious applications to radar, ranging, and synchronization systems. In this section we examine specifically the time of arrival measurement in optical systems using a MAP estimation approach.

Let $n(t)$ be the count intensity of the transmitted optical field and let θ be its arrival time at the receiver, measured with respect to a fixed transmission time. Any receiver operation that requires knowledge of this arrival time must accurately estimate θ. In a typical system this estimation is often continually repeated by retransmitting the optical field. For this reason we consider the intensity $n(t)$ to be a periodic waveform in t with repetition period T and average value n_s. A receiver observation time of T sec therefore corresponds to one period of the intensity waveform. The estimation problem is therefore one of observing over $(0, T)$ the photodetector output due to an optical field with intensity $n(t - \theta)$, and estimating the delay (arrival time) variable θ. The resulting processing may then be repeated over subsequent periods to obtain updated estimates. We concentrate only on the quantum limited case, where the background noise is negligible and the received field is the delayed transmitted field.

The MAP estimate of θ under quantum limited conditions follows directly from (9.2.16) with $n_b = 0$. Since $n(t)$ is periodic with period T, it can be expanded into a Fourier series which will contain an average term and harmonics of frequency $1/T$. Each harmonic integrates to zero in the second integral of (9.2.16), and the contribution from the average value does not depend on θ. The MAP delay estimate is then that for which

$$\max_{\theta}\left\{\int_0^T x(t) \log[n(t - \theta)]\, dt + \log p(\theta)\right\} \tag{9.3.1}$$

If the intensity $n(t)$ is differentiable, we note that

$$\frac{dn(t - \theta)}{d\theta} = -\frac{dn(t - \theta)}{dt} \tag{9.3.2}$$

and the estimate $\hat{\theta}_m$ equivalently satisfies

$$\frac{d \log p(\hat{\theta}_m)}{d\hat{\theta}_m} = \int_0^T x(t) \frac{d}{dt}[\log n(t - \hat{\theta}_m)]\, dt \tag{9.3.3}$$

The estimator in (9.3.1) again has the interpretation as a bank of correlators using the log intensity at each value of θ, followed by a selection of the maximum. Equation 9.3.3 defines a general tracking correlator loop involving the time derivative of the log intensity. The actual form of the loop depends on the periodic intensity involved. If, for example, $n(t)$ is sinusoidal, the loop reduces to the phase tracking loops in Section 9.2.

Sometimes (9.3.3) will have an explicit solution. Consider the case where the delay is a priori Gaussian distributed about a mean delay m_θ, with variance σ^2. Further, let us assume the intensity corresponds to a periodic sequence of intensity pulses, where each pulse is described by the function

$$n(t) = \frac{n_s T}{\sqrt{2\pi}\, W/2} \exp\left[-\frac{t^2}{W^2/2}\right], \qquad -\frac{T}{2} \le t \le \frac{T}{2} \tag{9.3.4}$$

That is, the pulses are Gaussian shaped with "width" W, as shown in Figure 9.8a. We assume T is many times larger than W so that the pulse occupies a relatively small portion of the observation interval and end effects can be neglected. For this case,

$$\frac{d \log n(t)}{dt} = \frac{d}{dt}\left[-\frac{t^2}{W^2/2}\right] = -\frac{t}{W^2/4} \tag{9.3.5}$$

and (9.3.3) becomes

$$\frac{\hat{\theta}_m - m_\theta}{\sigma^2} = \frac{4}{W^2}\int_0^T x(t)[t - \hat{\theta}_m]\, dt \tag{9.3.6}$$

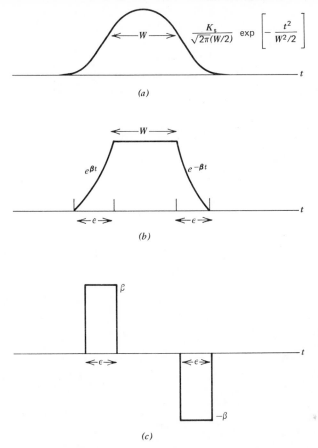

Figure 9.8. Examples of timing pulse waveforms. (a) Gaussian pulse, (b) exponential pulse, (c) log derivative of the exponential pulse.

Hence,

$$\hat{\theta}_m = \frac{\int_0^T tx(t)\,dt + m_\theta(W/2\sigma)^2}{\int_0^T x(t)\,dt + (W/2\sigma)^2} \tag{9.3.7}$$

The integral in the denominator is the observed total number of counts over $(0, T)$. The numerator integral is the "center of gravity" of the observed detector process $x(t)$. The MAP estimator therefore computes the mean or "center of gravity" of the shot noise locations over $(0, T)$ and uses it in (9.3.7). In the typical situation the initial delay uncertainty σ is many times the pulse width so that $\sigma^2 \gg W^2$, and the MAP estimate is approximately the ratio of the integrals above.

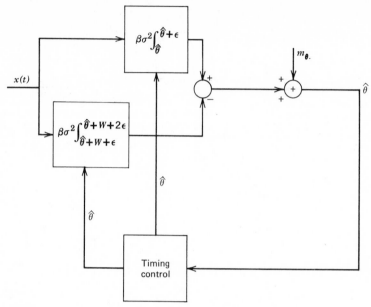

Figure 9.9. The early-late gate loop for estimating pulse arrival time.

The estimator changes form as the intensity pulses become sharper in form. Consider the exponential pulse in Figure 9.8b, with its log derivative shown in Figure 9.8c. Equation 9.3.6 becomes instead

$$\frac{\hat{\theta}_m - m_\theta}{\sigma^2} = \beta \int_{\theta_m}^{\theta_m + \epsilon} x(t)\, dt - \beta \int_{\theta_m + \epsilon + W}^{\theta_m + 2\epsilon + W} x(t)\, dt \qquad (9.3.8)$$

The feedback estimator now corresponds to a short-term integration over the front and back end of the expected optical pulse. The estimate is that value of $\hat{\theta}_m$ that "locks up" equal ϵ sec integrations separated by W sec, as shown in Figure 9.9. The tracking loop that implements (9.3.8) is often called an *early-late gate loop* [2, Chap. 7].

The dependence on intensity waveform can be further pursued by investigating the Cramer–Rao bound for delay estimation. For a given density $p(\theta)$, the CRB decreases as the integral in (9.2.18) increases. Using (9.3.2) and the fact that $n(t)$ is periodic, this integral can be rewritten as

$$\int_0^T \frac{(d[n(t - \theta)]/dt)^2}{n(t - \theta)}\, dt = \int_{-\theta}^{T - \theta} \frac{(dn(u)/du)^2}{n(u)}\, du$$

$$= \int_0^T \frac{(dn(u)/du)^2}{n(u)}\, du \qquad (9.3.9)$$

which does not depend on the delay variable θ. In addition, we can write

$$\int_0^T \frac{(dn(t)/dt)^2}{n(t)}\, dt = \int_0^T \left[\frac{dn(t)}{dt}\right]\left[\frac{d\log n(t)}{dt}\right] dt \tag{9.3.10}$$

By applying the Schwarz inequality to the right integral, we note that it is maximized if

$$\frac{dn(t)}{dt} = \frac{d\log n(t)}{dt}, \qquad n(t) \neq 0 \tag{9.3.11}$$

In this case (9.3.9) becomes

$$\int_0^T \left[\frac{dn(t)}{dt}\right]\left[\frac{d\log n(t)}{dt}\right] dt = \int_0^T \left[\frac{dn(t)}{dt}\right]^2 dt \tag{9.3.12}$$

Thus the integral in the CRB is upper bounded by the energy of the time derivative of the transmitted intensity. By applying Fourier transforms, we can further write

$$\int_0^T \left(\frac{dn(t)}{dt}\right)^2 dt = \frac{1}{2\pi}\int_{-\infty}^{\infty} \omega^2 |F_n(\omega)|^2\, d\omega \tag{9.3.13}$$

where $F_n(\omega)$ is the Fourier transform of $n(t)$ over one period. The integral on the right can be interpreted as the mean squared frequency of the transform of the intensity. Thus the CRB for delay estimation is minimized if a transmitter intensity $n(t)$ is used that satisfies (9.3.11) while having the largest possible mean square bandwidth in (9.3.13). The equality in (9.3.11) occurs only if $n(t) = \log n(t) + (\text{constant})$ when $n(t) \neq 0$. This can be satisfied only if $n(t)$ is constant whenever it is nonzero. Thus, (9.3.11) and (9.3.13) together suggest that best estimation (minimal CRB) corresponds to "flat" intensities, with as wide a frequency bandwidth as possible. The limiting form of such waveforms would be a periodic train of rectangular, narrow pulses in time, although theoretically, (9.3.12) is not valid for such intensities (the derivative of a pulse is not squared integrable). This pulsed intensity corresponds to transmission of narrow bursts of light and we intuitively expect such optical fields to indeed yield best delay estimation.

Since the rectangular pulse intensity is not differentiable, the correlating tracking loop loses its physical interpretation as an early-late gate estimator. However, the estimator in (9.3.1) retains its meaning as a MAP estimator. For a pulsed $n(t)$, the correlator reduces to a short-term integrator over the pulse width, starting at each value of delay θ. This often is called a "sliding window" integrator, and the delay point where the window maximizes (9.3.1) is the MAP estimate. Unfortunately, this theoretically requires a search over all values of θ in $(0, T)$. This search can often be implemented by dividing the total search into a "coarse" search followed by a "fine" search. Mathematically, this can be formulated as follows. Let us write the delay θ in the form

$$\theta = jW + \Delta \qquad (9.3.14)$$

where j is an integer, W is the intensity pulse width, and $0 \le \Delta \le W$. We are in effect dividing the delay into an integer multiple of pulse widths plus an additive excess portion Δ, as shown in Figure 9.10. We can now show that

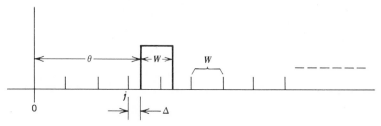

Figure 9.10. Delayed pulse arrival: θ = delay, j = integer pulse width delay, Δ = excess delay.

the MAP estimate of θ can be obtained as $\hat{\theta}_m = \hat{j}W + \hat{\Delta}$, that is, by simultaneously determining MAP estimates of j and Δ and substituting into (9.3.14). This follows since the joint MAP estimate of j and Δ must satisfy the simultaneous equations

$$\frac{\partial}{\partial \hat{j}} p[\hat{j}, \hat{\Delta} | x(t)] = 0$$

$$\frac{\partial}{\partial \hat{\Delta}} p[\hat{j}, \hat{\Delta} | x(t)] = 0 \qquad (9.3.15)$$

where $p[j, \Delta | x(t)]$ is the conditional density of θ given $x(t)$, with $\theta = jW + \Delta$. On the other hand, the MAP estimate of θ satisfies $\partial[(p(\theta)|x(t)]/\partial\theta = 0$. However,

$$\frac{\partial}{\partial \hat{\theta}} p[\hat{\theta} | x(t)] = \frac{\partial}{\partial \hat{j}} p[\hat{\theta} | x(t)] \cdot \frac{d\hat{j}}{d\hat{\theta}} + \frac{\partial}{\partial \hat{\Delta}} p[\hat{\theta} | x(t)] \frac{d\hat{\Delta}}{d\hat{\theta}}$$

$$= \frac{\partial}{\partial \hat{j}} p[\hat{\theta} | x(t)] \cdot \frac{1}{W} + \frac{\partial}{\partial \hat{\Delta}} p[\hat{\theta} | x(t)] = 0 \qquad (9.3.16)$$

If $\hat{j}$ and $\hat{\Delta}$ simultaneously satisfy (9.3.15), then (9.3.16) is also satisfied with $\hat{\theta} = \hat{j}W + \hat{\Delta}$. Thus delay estimates $\hat{\theta}$ can be obtained by estimating individually the number of pulse shifts j and the amount of excess Δ. The estimation of j can be considered a coarse estimation problem (determining which interval the pulse is in), while estimation of Δ can be considered a fine estimation problem (estimating the excess shifts within a pulse interval). The overall estimator has the form of Figure 9.11.

In practice the time delay θ would not be expected to vary more than a pulse width from one observation interval to the next. This suggests an alternative procedure in which we obtain first a pure MAP estimate of j alone in one period, then use $\hat{j}$ to estimate Δ in the subsequent period. The system achieves coarse estimation first, then carries out finer estimation over later observation periods. The system is often easier to implement, at the expense of increased estimation time, but we point out that it generally does not yield the joint MAP estimates required in (9.3.15). To formulate the coarse acquisition problem we model the observable as a vector sequence $\mathbf{k}$ of counts k_i over disjoint pulse widths W in $(0, T)$. If we assume an initial a priori joint density $p(j, \Delta)$, then we can determine the MAP estimate of j alone from

$$\max_{j} P(j|\mathbf{k}) = \max_{j} \int_0^W P(\mathbf{k}|j, \Delta)P(j, \Delta)\, d\Delta \qquad (9.3.17)$$

where $P(\mathbf{k}|j, \Delta)$ is the conditional probability of $\mathbf{k}$, given j and Δ, and j enumerates each of the W sec intervals over the observation period.

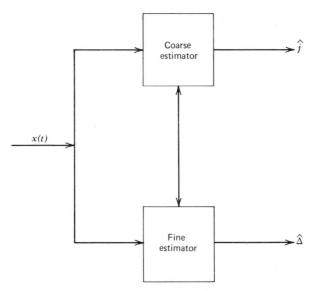

Figure 9.11. Coarse and fine estimator of delay ($\theta = jW + \Delta$).

When conditioned on a particular j and Δ, the received rectangular pulse will influence only the j and $j + 1$ interval counts, all others producing zero counts in the quantum limited case. Thus,

$$P(\mathbf{k}\,|\,j,\Delta) = \frac{[K_s(1 - \Delta/W)]^{k_j}[K_s\,\Delta/W]^{k_{j+1}}}{k_j!\,k_{j+1}!}\,\exp[-K_s] \qquad (9.3.18)$$

where K_s is the received pulse count energy. The MAP estimate of j is that value at which a maximum occurs in (9.3.17). Clearly, if we observe a count sequence of which only two are nonzero, (9.3.17) is always maximum for the nonzero k_i for any Δ (i.e., $\hat{j}$ is the index of the first nonzero k_i). If only one count is nonzero, it can be labeled either by k_j or k_{j+1}, and the MAP estimate is that producing the maximum. Thus if the qth count is nonzero, we must compare $P(j = q\,|\,\mathbf{k})$ with $P(j + 1 = q\,|\,\mathbf{k})$. In (9.3.17) we must therefore compare

$$\int_0^W \left(1 - \frac{\Delta}{W}\right)^{k_q} P(j = q, \Delta)\,d\Delta = P(j = q)\int_0^W \sum_{i=0}^{k_q}\binom{k_q}{i}\left(\frac{-\Delta}{W}\right)^i p(\Delta\,|\,j = q)\,d\Delta$$

$$= P(j = q)\sum_{i=0}^{k_q}\binom{k_q}{i}\left(\frac{-1}{W}\right)^i m_i(q) \qquad (9.3.19)$$

to

$$\int_0^W \left(\frac{\Delta}{W}\right)^{k_q} p(j = q - 1, \Delta)\,d\Delta = P(j = q - 1)\int_0^W \left(\frac{\Delta}{W}\right)^{k_q} p(\Delta\,|\,j = q - 1)\,d\Delta$$

$$= P(j = q - 1)\left(\frac{1}{W}\right)^{k_q} m_{k_q}(q - 1) \qquad (9.3.20)$$

where $P(j = q)$ is the a priori probability that $j = q$ (pulse in interval q) and $m_i(q)$ is the ith moment of the conditional density $p(\Delta\,|\,j = q)$. Thus if only one count is nonzero, the moment sequences of the a priori conditional density on Δ must be computed to determine initial MAP acquisition. If we assume the most practical case where j is uniform over the integers, and $p(\Delta\,|\,j)$ is uniform over $(0, W)$ for any j [a priori delay is uniformly distributed over $(0, T)$], then $m_i(q) = W^{i+1}/i + 1$ for all q, and both (9.3.19) and (9.3.20) have the value $W/k_q + 1$. Thus, in the uniform case, we can equally likely select $\hat{j} = q$ or $\hat{j} = q - 1$. If no counts are nonzero, we can only estimate j from its a priori density.

Once $\hat{j}$ has been determined (coarse estimation achieved) in a particular observation interval, it can be used as the true j in subsequent observation intervals in which the estimation of Δ is accomplished. With $\hat{j}$ given, the estimate $\hat{\Delta}$ is that value for which $d \log p(\mathbf{k}\,|\,\hat{j}, \hat{\Delta})/d\Delta = 0$, or that satisfying

$$\frac{-k_j}{1 - \hat{\Delta}/W} + k_{j+1}\left(\frac{W}{\hat{\Delta}}\right) = 0 \qquad (9.3.21)$$

The solution is then

$$\hat{\Delta} = \left(\frac{k_{j+1}}{k_{j+1} + k_j}\right)W \qquad (9.3.22)$$

Thus estimation of delay with rectangular pulses in quantum limited detection can therefore be achieved by first determining $\hat{j}$ during one observation period, then computing (9.3.22) in the next. The latter uses the observed count ratio as the fraction of the pulse width for the excess shift. As observations are made over subsequent intervals, $\hat{\Delta}$ can be continually recomputed to keep track of changes. We emphasize that we have assumed that j does not change throughout all intervals. If for some reason the delay jumps by several pulse positions, j must be reestimated before $\hat{\Delta}$ is computed.

The variance of the estimate $\hat{\theta}$ in (9.3.22) is difficult to determine explicitly since $\hat{\Delta}$ involves a ratio of random counts. In addition, as stated before, the CRB is hampered by the nondifferentiability of the pulsed intensities. However, a variance upper bound on $\hat{\Delta}$ can be determined by noting that Var $\Delta \leq W^2$. Furthermore, even if all counts are zero, the variance on $\hat{\Delta}$ is at most that of the a priori density on θ (i.e., σ^2), if we use the mean as the delay estimate. Thus,

$$\text{variance } \hat{\theta} = \sigma^2[\text{Prob } \mathbf{k} = 0] + (\text{Var } \Delta)[\text{Prob } \mathbf{k} \neq 0]$$
$$\leq \sigma^2 \exp[-K_s] + W^2[1 - \exp(-K_s)] \qquad (9.3.23)$$

This shows the estimator variance will be bounded by the square of the pulse width W by increasing pulse energy K_s.

When background noise is present, the coarse estimation operation is more complicated since the nonsignal intervals produce noise counts also. In this case (9.3.18) is replaced by

$$P(\mathbf{k}|j, \Delta) = \frac{\exp(-K_s/(1 + \mu_{b0}))}{1 + \mu_{b0}} \left(\frac{\mu_{b0}}{1 + \mu_{b0}}\right)^k L_{k_j}^{D-1}(A) L_{k_{j+1}}^{D-1}(B) \qquad (9.3.24)$$

where $k = k(0, T)$, $A = -K_s(1 - \Delta/W)/\mu_{b0}(1 + \mu_{b0})$, $B = -K_s \Delta/W \mu_{b0}$ $(1 + \mu_{b0})$, D is the number of time-space modes per interval W over the detector area, and μ_{b0} is the noise count per mode. For a given count sequence $\mathbf{k}$ over a T sec interval, we must again determine the j maximizing (9.3.17), which is equivalent to determining

$$\max_{j}\left\{P(j)\int_0^W L_{k_j}^{D-1}(A) L_{k_{j+1}}^{D-1}(B) p(\Delta|j)\, d\Delta\right\} \qquad (9.3.25)$$

Unfortunately, this maximization must be found after integration over Δ. However, we note that in comparing two different pairs of indices, say

(j_1, j_2) and (j_3, j_4), maximization of (9.3.25) is equivalent to the comparison

$$\frac{\int_0^W L_{k_{j_1}}^{D-1}(A) L_{k_{j_2}}^{D-1}(B) p(\Delta | j_1) \, d\Delta}{\int_0^W L_{k_{j_3}}^{D-1}(A) L_{k_{j_4}}^{D-1}(B) p(\Delta | j_3) \, d\Delta} \gtrless 1 \qquad (9.3.26)$$

when each j is equally likely. We now see that for any Δ density, if $k_{j_1} > k_{j_3}$ and $k_{j_2} > k_{j_4}$, then (9.3.26) must exceed unity, due to the positiveness and monotonicity of Laguerre functions with their indices. Thus, if any pair of successive counts is each greater than the corresponding members of any other pair of counts, the optimal estimate of j is always the index of the first of the larger. If no one pair dominates any other pair in this way, then one must integrate first in (9.3.25). When Δ does not depend on j, and is uniformly distributed over W, Mohanty [9] has shown that for the case $D = 1$, the integrations can be performed, using identities (C.5.7) and (C.5.10) of Appendix C, in the form

$$\int_0^y L_m(y - x) L_n(x) \, dx = L_{m+n+1}(y) \qquad (9.3.27)$$

After substituting, and using again the monotonicity of the Laguerre functions, (9.3.25) becomes

$$\max_j \left\{ L_{k_j + k_{j+1}} \left[\frac{-K_s}{\mu_{b0}(1 + \mu_{b0})} \right] \right\} = \max_j \{k_j + k_{j+1}\} \qquad (9.3.28)$$

Thus j is the index of the pair of consecutive counts having the largest sum, and coarse estimation is achieved by determining the maximum consecutive count pair. We point out that the procedure of basing coarse estimation on the largest of the counts (selecting $\hat{j}$ as that j for which k_j is maximum) is equivalent to imposing a basic assumption that $\Delta = 0$. For then in (9.3.25) $B = 0$ and A does not depend on Δ, and subsequent maximization over j is equivalent to maximization over $\{k_j\}$.

An application closely related to the arrival time estimation problem just given is that of estimating excess delay during PPM communications. In this operation an optical pulse W sec wide is sent in one of M possible W sec time intervals, and the excess time shift Δ is added during transmission, independent of which pulse position is used. This added time shift will cause PPM detection errors if not compensated as discussed in Section 8.6. A sync subsystem of the receiver must measure the added shift during each word interval for proper receiver compensation. This measurement must be made, however, without regard to the pulse position modulation. Thus during each word interval the transmitted pulse arrives with a total delay $\theta = jW + \Delta$ as before, where j is the integer position due to the modulation and Δ is the added excess delay during transmission. The estimation problem can be formulated as one of estimating Δ in the presence of the parameter j. Because of the position modulation, however, j must be considered independent from one observation period to the next, and estimates of j in one period cannot be used in subsequent periods. Thus, during each observation of $\mathbf{k}$, Δ must be reestimated in the presence of j. The resulting MAP tracking system for estimating Δ depends on the manner in which the index j is modeled. If j is considered an unknown parameter (no a priori density specified), then the maximization over Δ must take into account all the possible values that j can take on. Thus, $\hat{\Delta}$ is the value for which we achieve

$$\max_{\Delta} p(\Delta|\mathbf{k}) = \max_{j}\left[\max_{\Delta} p(\Delta|\mathbf{k}, j)\right]$$

$$= \max_{j, \Delta}[p(\Delta|\mathbf{k}, j)] \qquad (9.3.29)$$

This is equivalent to determining simultaneous maximizing values of Δ and j, and therefore corresponds to simultaneous estimates of these parameters. In other words, the MAP estimator must estimate both parameters, even though only the estimate of Δ is of interest. Furthermore, both estimates must be obtained during each observation interval, and previous estimates cannot be used in subsequent intervals.

If a delay of one word interval is acceptable, a suboptimal estimator procedure would be one that first estimates j during the original observation, stores the observation (detector output) for one word length, then reuses the stored observables, along with the estimate $\hat{j}$, to determine $\hat{\Delta}$, as shown in Figure 9.12. The estimate of j can be made using the techniques similar to the coarse estimation previously discussed. The estimator is therefore attempting to first detect which interval contains the pulse (i.e., decode the PPM word) then uses the decoded word to estimate Δ. This is referred to as *decision*

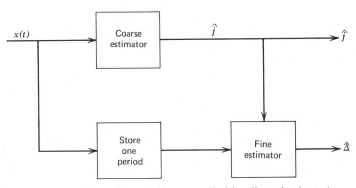

Figure 9.12. Coarse and fine estimator with storage (decision directed estimator).

directed estimation [4, 6], and the resulting estimates are often called data aided estimates. Such estimators are inherently linked to the decoding subsystem.

If the word interval delay in decision directed systems is prohibitive, an alternative scheme is shown in Figure 9.13. Here estimates of Δ are made consecutively from each successive pair of observed counts, using (9.3.22), and each result is stored until the end of the observation period. The estimate of j from the decoder at the end of the period is then used to select the $\hat{\Delta}$ corresponding to the most likely count pair. This operation avoids the word interval delay, but requires a bank of simultaneous estimators.

If, instead of treating j as an unknown parameter, we model it as a random variable taking on the values $1, 2, 3, \ldots, M$ with equal probability, the MAP

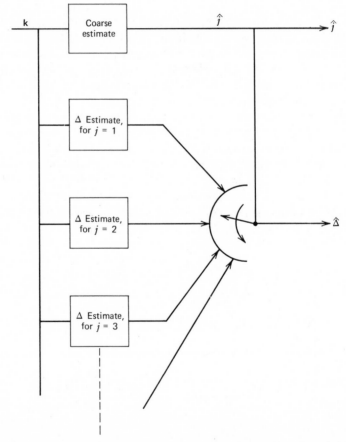

Figure 9.13. Coarse and fine estimator with parallel processing (decision directed estimator).

estimate of Δ can be obtained by averaging over these j values. Hence we write

$$\max_{\Delta} p(\Delta \,|\, \mathbf{k}) = \max_{\Delta} \left[\sum_{j=1}^{M} p(\Delta \,|\, \mathbf{k}, j) \right] \qquad (9.3.30)$$

Since each term $p(\Delta \,|\, \mathbf{k}, j)$ is the conditional density of Δ when the pulse is transmitted in the jth position, only the k_j and k_{j+1} counts are necessary to estimate Δ. (All other counts are either zero in the quantum limited case, or contain only noise counts, when background is present.) Hence, $p(\Delta \,|\, \mathbf{k}, j)$ can theoretically be computed immediately after k_j and k_{j+1} are observed. The summation in (9.3.30) is therefore a superposition of all such a posteriori densities, each delayed until the end of the observation interval. The estimate Δ is then made from this superposition. Note that the delaying of the a posteriori densities can be considered as a form of modulation removal (eliminating the position shift due to PPM) and shifting the excess delay Δ to the end of the interval where the estimate is made. Note that this latter estimate is not simply the average of the individual MAP estimates at each value of j. Instead, we must sum the densities prior to estimating. If it is known that Δ is confined to a narrow region about each pulse position, then (9.3.30) is approximately

$$\max_{\Delta} p(\Delta \,|\, \mathbf{k}) \approx \max_{\Delta} [p(\Delta \,|\, \mathbf{k}, j_{\max})] \qquad (9.3.31)$$

where $j_{\max}$ is the j maximizing $p(\Delta \,|\, j, \mathbf{k})$ over all Δ. The last term is identical to the simultaneous estimate of j and Δ, and therefore corresponds to the optimal MAP tracker defined in (9.3.29).

The estimation of arrival time has its most important application to time synchronization. In synchronizing systems, the initial coarse and fine pulse estimation operations are referred to as *pulse acquisition* (acquiring the interval in which the arrival pulse occurs). In such systems the estimate must be periodically updated, and the continual estimation of the excess shift Δ is called *tracking*. In Chapter 10, the design and performance analysis of time synchronization subsystems utilizing the concepts just mentioned is investigated in detail for optical receivers.

9.4 CONDITIONAL MEAN ESTIMATION IN OPTICAL RECEIVERS

The MAP estimation procedures of the preceding section depend on our ability to derive properties of the maximum point of the a posteriori density. To determine conditional mean estimators, however, an average over this density must be taken. This required integration is often difficult to perform and the associated properties of the optimal estimator structure are generally no longer obvious. In addition, the denominator of the density $p(\theta|y)$, which does not depend on θ and therefore did not influence the MAP estimate, plays the role of a normalizing factor and must be carefully evaluated in conditional mean estimators. Let us consider again the formulation of the optical system of the preceding sections. A multimode field involving a parameter θ is photodetected over M time intervals. We write the intensity as $n(t, \theta)$ and again let $p(\theta)$ be the a priori density on θ. The a posteriori density conditioned on the count sequence $\mathbf{k}$ is given by (9.2.1). The conditional mean estimate of θ is then

$$\hat{\theta}_c = \int_{-\infty}^{\infty} \theta p(\theta|\mathbf{k})\, d\theta$$

$$= \frac{e^{-K_b}}{CP(\mathbf{k})} \int_{-\infty}^{\infty} \theta \prod_{i=1}^{M} (\mu_i(\theta) + K_b)^{k_i} e^{-\mu_T(\theta)} p(\theta)\, d\theta \tag{9.4.1}$$

where $\mu_T(\theta) = \int_0^T n(t, \theta)\, dt$ and $C = \prod k_i!$. Note the integral must be evaluated directly (we do not have the advantage of taking logarithms as before). The factor $P(\mathbf{k})$ must be determined for the observed sequence. This can be most easily computed from

$$P(\mathbf{k}) = E_\theta[P(\mathbf{k}|\theta)p(\theta)]$$

$$= \frac{e^{-K_b}}{C} \int_{-\infty}^{\infty} \prod_{i=1}^{M} (\mu_i(\theta) + K_b)^{k_i} e^{-\mu_T(\theta)} p(\theta)\, d\theta \tag{9.4.2}$$

Thus, $\hat{\theta}_c$ in (9.4.1) is actually the ratio of two integrals, each of which is somewhat difficult to evaluate in the general case.

Example 9.4

Consider again the intensity level estimation problem of Example 1 where $n(t, \theta) = \theta n_s$ and $\mu_i(\theta) = \theta n_s \tau$. Substituting into (9.4.1) yields

$$\hat{\theta}_c = \frac{\int_{-\infty}^{\infty} \theta(\theta n_s \tau + K_b)^k e^{-\theta K_s} p(\theta)\, d\theta}{\int_{-\infty}^{\infty} (\theta n_s \tau + K_b)^k e^{-\theta K_s} p(\theta)\, d\theta} \tag{9.4.3}$$

where $k = k(0, T)$ and $K_s = n_s T$. Note that $\hat{\theta}_c$ depends only on the total count k observed over all time intervals and on the total field count energy K_s. The specific form for $\hat{\theta}_c$ depends on the density $p(\theta)$. When θ has an exponential density as in (9.2.9), (9.4.3) becomes

$$\hat{\theta}_c = \frac{\int_{K_b}^{\infty} u^{k+1} e^{-\beta u}\, du}{n_s \tau \int_{K_b}^{\infty} u^k e^{-\beta u}\, du} - \frac{K_b}{n_s \tau} \tag{9.4.4}$$

where

$$\beta = \frac{K_s}{n_s \tau} + \frac{1}{2\sigma^2 n_s \tau} \qquad (9.4.5)$$

Use of the incomplete gamma function complement:

$$\Gamma_c(a, b) \triangleq \int_b^\infty e^{-t} t^{a-1} \, dt \qquad (9.4.6)$$

then yields

$$\hat{\theta}_c = \left(\frac{1}{n_s \tau \beta}\right) \left[\frac{\Gamma_c(k + 2, \beta K_b)}{\Gamma_c(k + 1, \beta K_b)}\right] - \frac{K_b}{n_s \tau} \qquad (9.4.7)$$

The conditional mean intensity estimate again depends on the total count $k = k(0, T)$, but in a more complicated manner than for the MAP estimator in (9.2.10). For $K_b = 0$ (quantum limited detection), (9.4.7) reduces to a linear function in k (Problem 9.11).

For single mode conditions with Laguerre counting the conditional mean estimator is even more difficult to evaluate than (9.4.1). In this case, we have

$$\hat{\theta}_c = \frac{\int_0^\infty \theta \prod_{i=1}^M L_{k_i}(\hat{\mu}_i(\theta)) \exp[-\mu_i(\theta)/1 + \mu_{bo}] \, p(\theta) \, d\theta}{\int_0^\infty \prod_{i=1}^M L_{k_i}(\hat{\mu}_i(\theta)) \exp[\mu_i(\theta)/1 + \mu_{bo}] \, p(\theta) \, d\theta} \qquad (9.4.8)$$

which must be evaluated for the observed sequence $\mathbf{k} = (k_1, \ldots, k_m)$. For intensity estimation in Example 9.4, (9.4.8) simplifies somewhat to

$$\hat{\theta}_c = \frac{\mu_{bo}(1 + \mu_{bo}) \int_0^\infty t \prod_{i=1}^M L_{k_i}(t) e^{-t\mu_{bo}T} \, dt}{n_s \tau \int_0^\infty \prod_{i=1}^M L_{k_i}(t) e^{-t\mu_{bo}T} \, dt} \qquad (9.4.9)$$

Thus, general forms for CM estimators in this case are quite difficult to derive, even in relatively straightforward problems.

9.5 RECURSIVE PARAMETER ESTIMATION

We have concentrated on determining optimal estimators from observing photodetector outputs over a fixed time interval. In each case, the estimator observes the complete count or shot noise process first, then produces the estimate of the desired parameter. Although this procedure was useful for suggesting optimal or near optimal design, a more practical system would be one that continually generates estimates while the count observation is being made. To do this, the estimator must compute the a posteriori density after each individual interval count, from which MAP or CM estimates can be produced. The practical method for accomplishing this is to update the density from the previous observation. Thus the estimator recursively generates the desired density from observation to observation, starting from the a priori parameter density, and each time producing the associated optimal estimate.

To examine the computation involved, consider again an intensity $n(t, \theta)$ and an observation time discretized into τ sec count intervals. We assume we have observed the sequence of counts up to the nth one, $K_n \triangleq (k_1, k_2, \ldots, k_n)$, under quantum limited conditions. We would like to relate the conditional density $p(\theta | K_n)$ to that at the preceding interval, $p(\theta | K_{n-1})$. This can be done by writing

$$
\begin{aligned}
p(\theta | K_n) &= \frac{P(K_n | \theta)p(\theta)}{P(K_n)} \\
&= \frac{P(k_n | K_{n-1}, \theta)P(K_{n-1} | \theta)p(\theta)}{P(k_n | K_{n-1})P(K_{n-1})} \\
&= \left[\frac{P(k_n | K_{n-1}, \theta)}{P(k_n | K_{n-1})} \right] p(\theta | K_{n-1})
\end{aligned}
\tag{9.5.1}
$$

The bracket indicates the factor by which the preceding density is modified after observing the nth count k_n. Thus the desired density can be produced in discrete real time by recursively carrying out (9.5.1), for $n = 1, 2, 3, \ldots$, with $p(\theta | K_0) \triangleq p(\theta)$. However, the computation of the bracket may not be as simple as it may first appear. The numerator can be easily written since k_n is independent of K_{n-1} when conditioned on θ. Hence,

$$
P(k_n | K_{n-1}, \theta) = \text{Pos}(k_n, \mu_n(\theta))
\tag{9.5.2}
$$

where again $\mu_n(\theta) = \int_{t_n - \tau}^{t_n} n(t, \theta) \, dt$ is the integrated signal intensity over the nth interval. The denominator, however, requires computing

$$
\begin{aligned}
P(k_n | K_{n-1}) &= E_\theta[P(k_n | K_{n-1}, \theta)] \\
&= \int_{-\infty}^{\infty} \text{Pos}(k_n, \mu_n(\theta))p(\theta | K_{n-1}) \, d\theta
\end{aligned}
\tag{9.5.3}
$$

Thus the denominator of the upgrading factor requires an average using the density of the preceding step. It is this integration at each step that often places a computational burden on the recursive estimator. As the observation interval τ is decreased, $t_n \to t_{n-1}$, and the upgrading is theoretically occurring instantaneously with the detector output observation. In this case (9.5.1), with appropriate modification, indicates how the desired density evolves in time from one instant to the next, approaching a continuous functional in time. To derive the computation involved now, we must reconsider (9.5.2) and (9.5.3) as τ approaches a differential element dt. Those terms can be modified by writing $p(\theta|K_{n-1})$ as $p[\theta|x(t)]$ and $p(\theta|K_n)$ as $p[\theta|x(t+dt)]$, where $x(t)$ is the detector shot noise observed up to time t. For a differential interval, each k_n takes on integer values greater than one with probability approaching zero, while $\mu_n(\theta) \to n(t_n, \theta)\tau$. Hence, the Poisson density in (9.5.2) approaches the binary density as $dt \to 0$:

$$P(k_n|x(t), \theta) = \begin{cases} 1 - n(t, \theta)\, dt, & k_n = 0 \\ n(t, \theta)\, dt, & k_n = 1 \end{cases} \tag{9.5.4}$$

Similarly,

$$P(k_n|x(t)) = \begin{cases} 1 - \hat{n}(t)\, dt, & k_n = 0 \\ \hat{n}(t)\, dt, & k_n = 1 \end{cases} \tag{9.5.5}$$

where now

$$\hat{n}(t) = E_\theta[n(t, \theta)|x(t)]$$

$$= \int_{-\infty}^{\infty} n(t, \theta)p[\theta|x(t)]\, d\theta \tag{9.5.6}$$

Thus, as $dt \to 0$, the upgrading factor in (9.5.1) is then a ratio of terms that can take on one of two values at each t. That is,

$$\frac{P(k_n|K_{n-1}, \theta)}{P(k_n|K_{n-1})} \xrightarrow[dt \to 0]{} \begin{cases} \dfrac{1 - n(t, \theta)\, dt}{1 - \hat{n}(t)\, dt}, & k_n = 0 \\[2ex] \dfrac{n(t, \theta)\, dt}{\hat{n}(t)\, dt}, & k_n = 1 \end{cases} \tag{9.5.7}$$

Since k_n only has the value zero or one, we can write (9.5.7) in functional form form as

$$\frac{P(k_n|K_{n-1}, \theta)}{P(k_n|K_{n-1})} = \left[\frac{1 - n(t, \theta)\, dt}{1 - \hat{n}(t)\, dt}\right](1 - k_n) + \frac{n(t, \theta)}{\hat{n}(t)} k_n \tag{9.5.8}$$

Expanding the bracket and dropping all terms of order greater than dt, and substituting into (9.5.1) yields

$$p(\theta|x(t+dt)) = p(\theta|x(t)) + p(\theta|x(t))\left[\frac{n(t, \theta) - \hat{n}(t)}{\hat{n}(t)}\right][k_n - \hat{n}(t)\, dt] \tag{9.5.9}$$

Thus, over a differential time interval dt at time t, $p[\theta|x(t)]$ evolves according to (9.5.9) as long as dt is small enough for (9.5.4) to be valid. Note that the computation required for the recursive procedure is the determination of $\hat{n}(t)$ defined in (9.5.6). We recognize this as the conditional mean value of $n(t, \theta)$, conditioned upon $x(t)$. As such, it is precisely the CM estimate of the intensity

$n(t, \theta)$, given the observable process up to time t. That is, to produce the evolution in time of the a posteriori density of θ, needed to obtain either CM or MAP recursive estimates of θ at each t, a simultaneous CM estimate of the signal intensity must be made at each t. Note that a CM estimate specifically must be used to construct the density, no matter what type estimate for θ is being recursively produced from that density. In the literature it is common practice to consider $p[\theta|x(t)]$ as a function in t, and place (9.5.9) into the form of a differential equation by subtracting $p[\theta|x(t)]$ from both sides and dividing by dt. This then produces

$$\frac{dp[\theta|x(t)]}{dt} = p[\theta|x(t)]\left[\frac{n(t, \theta) - \hat{n}(t)}{\hat{n}(t)}\right][x(t) - \hat{n}(t)] \qquad (9.5.10)$$

where we have used the interpretation

$$\lim_{dt \to 0} \frac{k_n}{dt} = x(t)$$

$$p[\theta|x(0)] = p(\theta)$$

Equation 9.5.10 now defines a dynamical system equation for the time continuous evolution of the desired a posteriori density of θ, from which MAP and CM estimates can be generated. The input, or forcing function, is the delta function shot noise process $x(t)$ produced by the detector operating over receiver area. The equation is nonlinear, and more complicated than it might appear, since $\hat{n}(t)$ is actually itself a function of $p[\theta|x(t)]$. Therefore this equation, though perhaps useful for developing mathematical insight into the dynamics of the recursive estimator, is not easily implementable into an actual system without a considerable number of simplifications. For this reason, (9.5.9) may be more useful for designing a discrete time (sampled data) implementation. Generalizations of (9.5.10) to the recursive estimation of vector parameters and vector waveforms in optical communications have been developed rigorously by Snyder [10] and Clark [11]. Their work has significant application to the derivation of optimal demodulators (estimators) for optical carriers modulated with random waveforms, rather than the basic parameter estimators considered here. Analytical procedures along these lines are restricted to specific modulating waveform models, and again the complexity of the resulting recursive equations often tends to mask the implications of the approach. Nevertheless, by utilizing appropriate and practical assumptions, useful forms of phase and frequency waveform demodulators have been derived in this way, and lead to recursive extensions of the phase and frequency estimators in Section 9.2. The interested reader should pursue further discussion of optical recursive estimation, along with specific applications and solutions, in the works of Clark, Snyder, Hoversten, Harger, and others [12–15].

9.6 ESTIMATION BASED ON VOLTAGE OBSERVABLES

Although estimation based on electron counts formulates into a well-defined estimation problem with meaningful interpretations, there is the prime assumption that the receiver has the capability of observing exactly the number or locations of all electron occurrences. However, as pointed out in Section 7.8, the photoelectron effects are actually observed through a photo-detector voltage output, which often includes the effect of nonideal photo-multiplication and added circuit noise. Thus, in a more practical problem formulation, the true observable is not the count sequence but rather the sequence of voltage values generated from these counts. For ideal photo-multiplication the observable is more accurately given as $\mathbf{y} = \{y_i\}$, where

$$y_i = Gk_i + c_i \qquad (9.6.1)$$

G is the photomultiplier conversion factor from counts to voltage, and $\{c_i\}$ is an additive, independent, Gaussian circuit noise sequence, each component of zero mean. We would like to determine the compensation that must be made on our previously derived count estimators in order to account for this added randomness of the voltage observation. Formally, an estimation of θ, based on $\mathbf{y}$, requires the computation of the density $p(\theta|\mathbf{y})$. To relate this to count observations we write

$$p(\theta|\mathbf{y}) = \int p(\theta|\mathbf{y}, \mathbf{k})p(\mathbf{k}|\mathbf{y})\, d\mathbf{k} \qquad (9.6.2)$$

where the above is a vector integration over the components of $\mathbf{k}$. The density $p(\theta|\mathbf{y}, \mathbf{k})$ is the a posteriori density of θ given both the voltage observable $\mathbf{y}$ and count observable $\mathbf{k}$. However, since θ influences only $\mathbf{k}$ (the added noise c_i does not depend on θ) we have

$$p(\theta|\mathbf{y}, \mathbf{k}) = \frac{p(\mathbf{y}|\theta, \mathbf{k})P(\mathbf{k}|\theta)p(\theta)}{p(\mathbf{y}, \mathbf{k})}$$

$$= \prod_i \frac{p_{c_i}(y_i - Gk_i)P(k_i|\theta)p(\theta)}{p_{c_i}(y_i - Gk_i)P(k_i)}$$

$$= \prod_i \frac{P(k_i|\theta)p(\theta)}{P(k_i)} = p(\theta|\mathbf{k}) \qquad (9.6.3)$$

where $p_{c_i}(x)$ is the probability density of c_i in (9.6.1). In other words, the density of θ conditioned on both $\mathbf{y}$ and $\mathbf{k}$ is the same as that conditioned on $\mathbf{k}$ alone. This means in (9.6.2) we instead write

$$p(\theta|\mathbf{y}) = \int p(\theta|\mathbf{k})P(\mathbf{k}|\mathbf{y})\, d\mathbf{k} \qquad (9.6.4)$$

It then follows that the CM estimate of θ is

$$\hat{\theta}_c = \int_{-\infty}^{\infty} \theta \left[\int p(\theta|\mathbf{k}) P(\mathbf{k}|\mathbf{y}) \, d\mathbf{k} \right] d\theta$$

$$= \int (\hat{\theta}_c|\mathbf{k}) P(\mathbf{k}|\mathbf{y}) \, d\mathbf{k} \qquad (9.6.5)$$

where $\hat{\theta}_c|\mathbf{k}$ is the CM estimate conditioned on $\mathbf{k}$. Thus the optimal CM estimate based on voltage observables $\mathbf{y}$ is the CM estimate conditioned on $\mathbf{k}$ averaged over the density on $\mathbf{k}$ induced from $\mathbf{y}$. The integration can in fact be rewritten as summations since the density on each k_i is only over the integers. Thus,

$$P(\mathbf{k}|\mathbf{y}) = \prod_i \sum_{q_i=0}^{\infty} \frac{p_{c_i}(y_i - Gq_i)}{p(y_i)} P(q_i) \delta(q_i - k_i) \qquad (9.6.6)$$

where

$$P(q_i) = E_\theta[\text{Prob}(k_i = q_i|\theta)]$$

$$p(y_i) = \sum_{q_i=0}^{\infty} p_{c_i}(y_i - Gq_i) P(q_i)$$

The delta functions in (9.6.6) reduce (9.6.5) to summations over all possible M-fold integer vectors and their corresponding probabilities. It is usually difficult to obtain a closed form expression for the resulting sum (note it is a multidimensional sum and requires explicit solution of $\hat{\theta}_c|\mathbf{k}$ in terms of $\mathbf{k}$).

If the CM estimate is linear in the observed counts that is, $\hat{\theta}_c|\mathbf{k} = \sum a_i k_i + d_i$, then using (9.6.6) in (9.6.5) yields

$$\hat{\theta}_c = \sum_i a_i [E(k_i|y_i)] + d_i$$

$$= \sum_i a_i [(\hat{k}_i)_c|y_i] + d_i \qquad (9.6.7)$$

In this case, the optimal CM estimator uses the same linear form as that using the count observables, but substitutes for the observed count the CM estimate of the count based on the voltage observable. For y_i in (9.6.1) $E(k_i|y_i) = y_i/G$ and the normalized observed voltage value is used directly as the count value for each interval. It is interesting that (9.6.7) is valid for any density on the additive circuit noise terms $\{c_i\}$.

Another simplification occurs if the added noise is weak, so that $p_c[k_i - (y_i/G)]$ takes on significant values only in the neighborhood of the nearest integer to y_i/G, as shown for example in Figure 9.14. (It can be seen

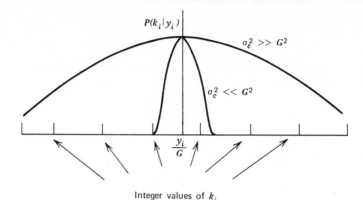

Figure 9.14. Possible forms for the conditional density $p(k_i/y_i) = p_c(k_i - y_i/G)$.

that this will approximately be true if the variance of each c_i is much less than G^2.) In this case (9.6.6) reduces to

$$P(\mathbf{k}|\mathbf{y}) \cong \prod_i \delta(k_i - [y_i/G])$$

where $[y_i/G]$ is the nearest integer to y_i/G. Thus,

$$\hat{\theta}_c \cong (\hat{\theta}_c | \mathbf{k} = [\mathbf{y}/G]) \tag{9.6.8}$$

Again the estimator is the same as for the count observables and treats the nearest integer to y_i/G as if it were the true count k_i; that is, essentially it neglects the added noise.

For nonideal photomultipliers, where each electron is susceptible to a random gain, the model described in Section 7.8 must be used for the voltage value. That is, we write instead of (9.6.1),

$$y_i = \sum_{j=1}^{k_i} g_j + c_i \tag{9.6.9}$$

where $\{g_j\}$ are independent random electron gain variables with mean G and variance σ_g^2. Equation 9.6.6 can now be used with $p_c(y_i - Gq_i)$ replaced by the k_i-fold convolution of the densities $\{g_j\}$ and $p_c(x)$. If the gains and the noise variables are assumed to be Gaussian, then this latter convolution produces a Gaussian density with mean Gk_i and variance $(k_i\sigma_g^2 + \sigma_c^2)$.

While CM estimates lend themselves to this interpretation, no such equivalent statements can be made for MAP estimates. The latter requires determining

$$\max_\theta \left\{ \int p(\theta|\mathbf{k})P(\mathbf{k}|\mathbf{y}) \, d\mathbf{k} \right\} \tag{9.6.10}$$

where the integration must be performed before the maximization. Hence, there is no obvious relation between the MAP θ estimates conditioned on $\mathbf{k}$ in Section 9.2 and those condition on $\mathbf{v}$. If y_i is related to k_i by (9.6.1), then we obtain simplification only for the high gain-to-circuit noise ratio case. For then $p(\mathbf{k}|\mathbf{v}) \to \prod \delta(k_i - [y_i/G])$ and (9.6.10) requires

$$\max_{\theta} \{p(\theta|[\mathbf{y}/G])\} \tag{9.6.11}$$

This again leads to the same MAP estimator as that based on counts, only with $[y_i/G]$ substituted for k_i.

REFERENCES

[1] Van Trees, H., *Detection, Estimation, and Modulation Theory*, Part I, John Wiley and Sons, New York, 1968.

[2] Stiffler, J., *Theory of Synchronous Communications*, Prentice-Hall, Englewood Cliffs, New Jersey, 1971.

[3] Viterbi, A., *Principles of Coherent Communication*, McGraw-Hill Book Co., New York, 1966, Chap. 5.

[4] Hancock, J. and Wintz, P., *Signal Detection Theory*, McGraw-Hill Book Co., New York, 1966, Chap. 5.

[5] Middleton, D., *Introduction to Statistical Communications*, McGraw-Hill Book Co., New York, 1960, Chap. 21.

[6] Lindsey, W. and Simon, M., *Telecommunication Systems Engineering*, Prentice-Hall, Englewood Cliffs, New Jersey, 1973.

[7] Thompson, G. and Pratt, W., "Optical Heterodyne Receiver Design Using Nonlinear Recursive Estimation," *Proc. IEEE*, **58-10**, 1727 (October 1970).

[8] Hoversten, E., Harger, R. O., and Halme, S. J., "Communication Theory For The Turbulent Atmosphere," *Proc. IEEE*, **58-10**, 1626 (October 1970).

[9] Mohanty, N., "Estimation of Delay of M PPM Signals in Laguerre Communications," *IEEE Trans. Commun. Tech.*, **COM-22**, 713 (May 1974).

[10] Snyder, D. L., "Filtering and Detection for Doubly Stochastic Poisson Processes," *IEEE Trans. Inf. Theory*, **IT-18**, 91 (January 1972).

[11] Clark, J. R., "Estimation for Poisson Processes with Applications in Optical Communications," Ph.D. Dissertation, Dept. of Electrical Engineering, Massachusetts Institute of Technology, Cambridge, September 1971.

[12] Clark, J. R., "Direct Detection Optical Receivers for Angle Modulated Signals," Proc. Int. Conf. Commun., Philadelphia (June 1972).

[13] Clark, J. R. and Hoversten, E. V., "Performance of Demodulators for Gaussian Messages and Doubly Stochastic Poisson Processes," Proc. 1972 Inf. Theory Symposium, Pacific Grove, California, February 1972.

[14] Hoversten, E., Snyder, D., Harger, R., and Kurimota, K., "Direct Detection Optical Communication Receivers," *IEEE Trans. Communication*, **COM-22** (January 1974).

[15] Forrester, R., Jr., and Snyder, D., "Phase Tracking Performance of Direct Detection Optical Receivers," *IEEE Trans. Communications*, **COM-21**, 1037 (September 1973).

PROBLEMS

1. Show that the CM estimate of a parameter θ, based on an observable y, $\hat{\theta}_c$, is that value of θ that minimizes $E(\theta - \hat{\theta}_c)^2$, when the average is taken over all observables y and θ. [*Hint*: Expand first with a conditional average, the minimize by differentiation.]

2. Consider the problem of estimating the intensity level θ of a monochromatic field in background noise of count level n_b (Example 9.1). Determine the equation that the MAP estimate of θ must satisfy for the following a priori densities

(a) Rayleigh:

$$p(\theta) = \frac{\theta}{\sigma^2} e^{-\theta^2/2\sigma^2}$$

(b) chi-squared:

$$p(\theta) = \frac{(2\sigma^2)^{-D}}{(D-1)!} \theta^{D-1} e^{-\theta/2\sigma^2}, \qquad D \geq 2$$

(c) uniform:

$$p(\theta) = \frac{1}{\theta_0}, \qquad 0 \leq \theta \leq \theta_0$$

3. Determine the MAP estimate and estimate variance for the three cases in Problem 9.2 under the quantum limited condition $n_b = 0$.

4. Show that the estimator in (9.2.10) is asymptotically efficient as the observation time is increased. That is, (9.2.13) approaches the CRB for large values of K_s.

5. Determine a simple rule for determining how many correlators should be used in Figure 9.5 if the frequency is to be estimated to within a frequency range given by $\pm (CRB)^{1/2}$, where CRB is given in (9.2.30). Assume high K_s and neglect frequency uncertainty beyond $\pm 2\sigma$.

6. (a) Rederive the MAP estimator equations in (9.3.1) and (9.3.3) for arrival time estimation when background noise is present with count rate n_b.

(b) Apply the result to the case when the signal intensity is given in (9.3.4) and solve. Explain the difference from the quantum limited result in (9.3.6).

7. Show that under a bandwidth constraint $(F_n(\omega) = 0, |\omega| > 2\pi B)$, (9.3.13) is maximized with $n(t)$ a sine wave at frequency B.

8. Let θ be a binary random variable $(0, 1)$ with equal probability and define the signal intensity

$$n(t) = \theta n_1(t) + (1 - \theta) n_2(t)$$

where $n_1(t), n_2(t)$ are deterministic intensities. For an observed count vector $\mathbf{k}$, show that, for any count statistics on $\mathbf{k}$ with intensity $n(t)$,

(a) $$\hat{\theta}_c = \text{Prob}[\theta = 1|\mathbf{k}]$$

(b) $$\hat{\theta}_c = \frac{\Lambda}{1+\Lambda}, \qquad \Lambda = \frac{P(\mathbf{k}|n=n_1)}{P(\mathbf{k}|n=n_2)}$$

(c) The likelihood test in deciding whether $n = n_1$ or $n = n_2$ is equivalent to a threshold test on $\hat{\theta}_c$.

9. An optical field with intensity $n(t, \theta)$ is received in background noise under condition H_1. Under another condition H_0, only background noise is received. An optical receiver observes the field and attempts to estimate θ without knowing which condition is true. (If the receiver knows H_0 is true, it can only rely on a priori information about θ.) Show that the CM estimate of θ, after observing the count vector $\mathbf{k}$, is

$$\hat{\theta}_c = (\hat{\theta}_c|H_1)P(H_1|\mathbf{k}) + (\hat{\theta}_c|H_0)P(H_0|\mathbf{k})$$

where $\hat{\theta}_c|H_i$ is the CM estimate of θ under H_i, and $P(H_i|\mathbf{k})$ is the a posteriori probability that H_i is true.

10. Consider again Problem 9.9. Assume the receiver operates by first performing a test to determine which H_i is true. It then selects $\hat{\theta}_c|H_i$ if $p(H_i|\mathbf{k})$ is larger. Is the estimate the same as the $\hat{\theta}_c$ in Problem 9.9? Explain.

11. Show that if $K_b = 0$ in (9.4.7), $\hat{\theta}_c$ reduces to

$$\hat{\theta}_c = \frac{k+1}{K_s + 1/2\sigma^2}$$

12. Let $n(t, \mathbf{r}, \theta)$ be a time-space field intensity defined over an area A, containing the parameter θ. Subdivide A into $\{A_i\}$ and consider a photodetector placed at each A_i. (That is, each A_i is the area of the detector.) Let $\mathbf{k}_i$ be the output count sequence of detector i over $(0, T)$. Derive the MAP time-space estimator processing of the set of $\mathbf{k}_i$ for estimating θ, after observing the received field over A and $(0, T)$ with the detector array. Assume constant noise count rate over A and $(0, T)$.

10   Time Synchronization

In the transmission of digital data, time synchronization between the received signal and the receiver decoder is needed to perform the decoding operation. This receiver timing is provided by a synchronization (sync) subsystem operating at the receiver. In our preceding discussions we have, for the most part, ignored the analysis, design, and performance of this subsystem. We have shown, however, that imperfect timing can cause a deleterious effect on overall digital transmission. In this chapter we examine several procedures and design equations associated with typical timing subsystems of an optical communication link. We should also point out that accurate timing is also needed in two-way optical ranging systems in which a receiver ranging signal must be retransmitted exactly when a transmitted range signal is received.

The time synchronization problem has of course received considerable attention in the past for the additive Gaussian noise channel, and the interested reader is referred to the presentations in the books by Viterbi [1], Stiffler [2], Lindsey [3], and Lindsey and Simon [4]. Although the approach here parallels these earlier studies, the optical system presents a completely different model for the detector output than the conventional signal noise models. This is due to the quantum (shot noise) nature of the photodetector output signal. As a result, classical analysis procedures often lead to design equations that differ markedly from their well-known Gaussian counterparts This requires reinvestigation for the optical case in order to obtain proper design directions. This is the objective of this chapter.

10.1   THE TIMING SUBSYSTEM

Receiver timing is accomplished in a sync subsystem following the photodetector, and generally operating in parallel with the information channels, as shown in Figure 10.1. Timing information generated is then used to clock the data decoders and/or provide timing markers for ranging. In some receivers the decoded data may in fact be used to further improve the timing

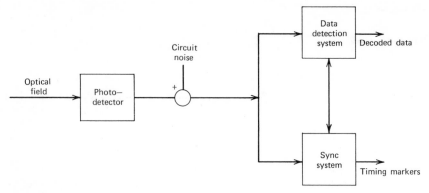

Figure 10.1. The data synchronization subsystem.

operation. Timing information is generally provided to the receiver in the form of a timing signal sent from the transmitter. This timing signal is used by the receiver to generate timing markers indicating the beginning of bits, words, or data frames. These timing waveforms are often superimposed upon the information waveforms prior to transmission, and then are separated out at the receiver. This separation can be achieved in either time or frequency. In the former, the timing information is sent during one time interval and the data sent in others. In frequency separation, the data and timing information are placed in nonoverlapping parts of the intensity spectrum. In both cases, the timing subsystem can extract the timing waveform (either time gating or frequency filtering) and use the resulting signal for time synchronization. If the waveform separation is successful, timing is achieved with a clean waveform and synchronization is obtained virtually independent of the transmitted data. Time synchronization under these conditions is generally described as "pure" synchronization. An alternate procedure is to not transmit a separate timing signal with the data, but instead rely on the sync subsystem to generate its timing markers directly from the data waveform itself. The system no longer benefits from a clean timing waveform, but will have advantages in power and spectral savings. Systems that derive timing directly from the data must have provisions for stripping off the modulation, or must be aided by the decoder decisioning. In the case of impure synchronization, timing and data detection are directly interconnected.

In general synchronization studies, the timing waveform used to indicate the timing markers can take on a variety of forms. For the optical system operating in a pure sync format, it was shown in Section 9.3 that the most efficient waveform corresponds to a periodic train of narrow intensity pulses.

The timing markers therefore become the pulses themselves, and timing is obtained by measuring the arrival time of the light pulses. An analytical study of the optical time synchronization procedure for intensity pulse trains was formulated in Section 9.3 as a problem in estimating the time of arrival of the synchronizing signal. It was shown that the receiver operation of estimating the arrival time of a periodic pulse waveform can be divided into a coarse acquisition operation (discretizing the period of the timing signal and determining the pulse location interval) and fine acquisition (measurement of the excess fractional shift that occurs).

After time acquisition has been achieved, and compensation is made for excess shift, the arrival time (pulse location) of the received optical pulse is determined for each period. In an ideal system with no transmission effects, this timing marker occurs in exactly the same location during every subsequent frame. Theoretically, once time acquisition is achieved time synchronization should remain perfect and no more time markers need be transmitted. Unfortunately transmission effects introduce varying transmission delays, signal generators and oscillators tend to drift, and relative motion may exist between transmitter and receiver. Each of these will affect the time of marker arrival, causing it to shift slightly from one frame to another. As a result the transmitter must continually send its markers and the receiver must continually readjust for its new location. This receiver operation is called tracking. Tracking is, in general, much simpler to perform than acquisition since the extraneous shifts are generally small compared to the location interval itself, and the receiver essentially knows approximately where the marker is. Thus only a slight adjustment is usually needed to maintain the desired timing from frame to frame.

Sync subsystems generally contain separate circuitry for the acquisition operation and for performing the tracking, as shown in Figure 10.2. The timing procedure begins with an initial acquisition operation in which the timing pulse intervals are searched until both coarse and fine acquisition are completed. An indication of acquisition is then used to initiate the tracking operation. Tracking is then continually maintained and updated throughout the communication time period. This tracking is generally accomplished by having the receiver also generate a periodic timing signal from a signal generator, and using a feedback tracking loop to keep the signal generator in sync with the received pulses. As long as the tracking loop is synchronized, the receiver will generate accurate timing markers for the decoding and ranging operations. If the tracking system "loses" the received timing waveform, the sync subsystem must return to the acquisition operation until the received waveform is "found" and tracking is reinstated. Loss of lock in the tracking loop is generally indicated by lock detector circuitry monitoring the tracking error. Continual or updated timing is necessary to overcome the

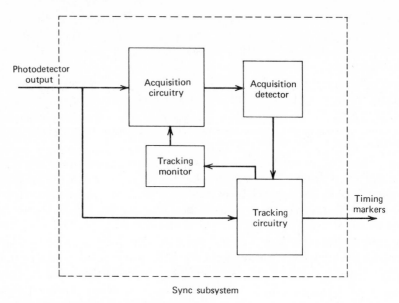

Sync subsystem

Figure 10.2. The synchronization subsystem.

unintentional variations in transmission delay. If the basic assumption is made that these delay variations are slow relative to the width of the timing pulses, then their only effect is to vary their time location without distorting their shape. Thus, if $n(t)$ represents the periodic count intensity of the timing signal at the receiver with no delay variations, and if τ_d is the time varying delay occurring during transmission, then the recovered count intensity is given by $n(t - \tau_d)$. Here it is tacitly implied that τ_d is a function of t that changes slowly with respect to the pulse widths of $n(t)$. Note that this latter condition is equivalent to the assumption that the bandwidth of the delay variations is much smaller than the bandwidth of the timing waveform. This is a prime assumption in our ensuing analysis.

10.2 TIMING ACQUISITION

When a periodic sequence of narrow intensity pulses is transmitted as the timing waveform in a pure sync system, the acquisition operation requires a search over all possible pulse locations. The pulse train may be sent prior to data transmission, so that timing can be accomplished first, or they can be sent continuously with the data, either over a separate channel or as a subcarrier pulse. Pulse location intervals are determined by electron counting over each consecutive pair of intervals. Intervals must be searched in pairs since the pulse may be offset from the defined intervals, due to the fractional excess shift, as shown in Figure 10.3. This search may physically be accomplished by stepping a two-interval counter over each pair of intervals one at a time. A threshold test can be used at each interval pair position, with the test continuing until threshold is crossed, that is, coarse acquisition is indicated. An alternative procedure is to store the counts from each position over a fixed search period, and select the interval pair having the largest count as the acquisition marker location. If the decided pair is correct, time acquisition is successfully achieved. Adjustment for excess shift can then be made by counting over each interval of the pair and linear extrapolating [see (9.3.22)]. Tracking can then commence. If the acquisition is incorrect, the ensuing tracking operation is lost, and acquisition is eventually restarted. It is evident that acquisition circuit design involves a tradeoff of length of time to acquire versus the probability of a successful acquisition.

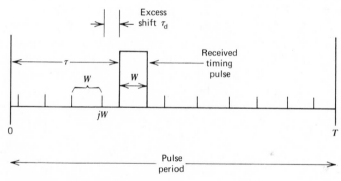

Figure 10.3. The received timing pulse.

Consider an acquisition system using W sec pulses producing a signal count of K_s and a noise count of K_b at the receiver during each W sec interval. We assume first that a continual search type of acquisition is used with a threshold test for each interval pair. The probability of successful acquisition of the correct-interval pair, PAC, is then the probability that the threshold k_T is crossed when the acquisition pulse is present. The threshold is set by the requirement that the probability that k_T is crossed when no pulse is present (false alarm probability, PFA) is at some suitably small value. If k_i is the detected count over the ith interval, and if the pulse overlaps the i and $i + 1$ intervals (see Figure 10.3), then

$$\text{PAC} = \text{Prob}[(k_i + k_{i+1}) \geq k_T] \tag{10.2.1}$$

For a Poisson counting, this becomes

$$\text{PAC} = \sum_{k=k_T}^{\infty} \text{Pos}(k, K_s + 2K_b)$$

$$= \frac{\Gamma(k_T, K_s + 2K_b)}{\Gamma(k_T, \infty)} \tag{10.2.2}$$

where $\Gamma(a, b)$ is the incomplete gamma function in (7.4.6). The threshold k_T is determined from the condition

$$\text{PFA} = \sum_{k=k_T}^{\infty} \text{Pos}(k, 2K_b) \tag{10.2.3}$$

for the desired PFA. A set of curves generated in this way is shown in Figure 10.4. The acquisition test continues until threshold is crossed, with the performance governed by these probabilities. The aquisition probabilities are reminiscent of OOK error probabilities, and the system suffers from the same basic disadvantage—the noise level must be known in order to set the test threshold properly. Another serious disadvantage of a threshold acquisition search is that the acquisition time may become quite lengthy. Since PAC in (10.2.2) is the probability of successful acquisition in the correct interval, the probability of acquiring the pulse in a single period search, PAC_1, is then obtained by averaging over all possible pulse positions. The probability of acquiring when the pulse is in the jth position is (PAC) $(1 - \text{PFA})^{j-1}$, and the average probability over all positions is then

$$\text{PAC}_1 = \frac{1}{M} \sum_{j=1}^{M} (\text{PAC})(1 - \text{PFA})^{j-1}$$

$$= \frac{\text{PAC}}{M} \left[\frac{1 - (1 - \text{PFA})^M}{\text{PFA}} \right] \tag{10.2.4}$$

where M is the number of pulse pair positions in one period. If the acquisition is not successful in a single period, it must be repeated in the next period. The probability it will take i periods to acquire is then $(\text{PAC}_1)(1 - \text{PAC}_1)^{i-1}$, and the average number of periods that will be searched is then

$$N_s = \sum_{i=1}^{\infty} i(\text{PAC}_1)(1 - \text{PAC}_1)^{i-1}$$

$$= \frac{1}{\text{PAC}_1} \tag{10.2.5}$$

Thus the average acquisition time is $TN_s = T/\text{PAC}_1$. If PFA in (10.2.4) is close to unity, the average test length may become quite long.

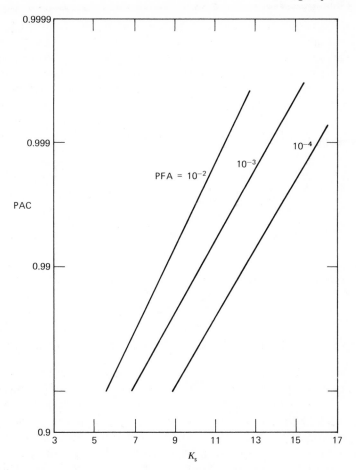

Figure 10.4. Single search acquisition probabilities: PFA = false alarm probability, $K_b = 1$.

When a comparison test is used, no threshold is required and an acquisition decision is always made after a fixed time period. The probability of acquisition is given by the probability that the correct interval pair exceeds the remaining pairs. A question may arise concerning the handling of count equalities for two separate pairs. A random choice can be made in this case, just as in binary word detection, in which case the acquisition probabilities are identical to the word detection probabilities in an M-level PPM block coded test. Here M is now the number of interval pairs in T sec (i.e., $M = T/2W$). Thus,

$$PAC_1 = 1 - PWE \qquad (10.2.6)$$

where PWE is obtained from (8.2.1) by replacing K_b by $2K_b$, with K_b interpreted as the noise count over a W sec counting interval. However, it may also be argued that a false acquisition is more serious than a word error in digital systems, and a more appropriate acquisition model is to consider count equalities as an acquisition error. The system therefore chooses to rescan in the next interval rather than guess. Hence, correct acquisition occurs only if the correct interval count truly exceeds all other counts. The acquisition probability in (10.2.6) now simplifies when all count equality possibilities are removed. Hence,

$$
\begin{aligned}
\text{PAC}_1 &= \sum_{k_1=1}^{\infty} \text{Pos}(k_1, K_s + 2K_b)\left[\sum_{k_2=0}^{k_1-1} \text{Pos}(k_2, 2K_b)\right]^{M-1} \\
&= \sum_{k=1}^{\infty} \text{Pos}(k, K_s + 2K_b)\left[\frac{\Gamma(k-1, 2K_b)}{(k-1)!}\right]^{M-1}
\end{aligned}
\tag{10.2.7}
$$

These acquisition probabilities are easily computed for specific values of M, K_s, and K_b.

10.3 PULSE EDGE TRACKING

In Section 9.3 the tracking problem is formulated as arrival time estimation, and optimal estimators often have the form of feedback tracking loops. When tracking a periodic sequence of pulses, the tracking loop takes the form of a pulse edge tracker, in which receiver integrators are timed to operate over the received pulse edges. The time difference obtained from these integrations is then used to readjust the receiver timing. When this timing difference is zero the receiver is in sync with the received pulses. This edge tracking can be accomplished by an early-late gate pulse tracker, as shown in Figure 10.5. (It may be remembered that such a device was in fact a MAP

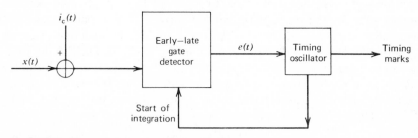

Figure 10.5. The early-late gate pulse tracking loop.

estimator of excess delay). The output of the photodetector provides the input to the tracker. We concentrate here on the pure tracking operation and again assume the timing waveform is a T sec periodic sequence of intensity pulses, each W sec wide. Thus, over a single period, we write the detected signal count intensity as

$$
\begin{aligned}
n(t) &= n_s, & 0 \leq t \leq W \\
&= 0, & W \leq t \leq T
\end{aligned}
\tag{10.3.1}
$$

where n_s is the average count rate over the detector area. The intensity (10.3.1) is assumed to be received with the delay variation τ_d, which represents the excess shift in pulse location relative to its known acquisition interval, as shown in Figure 10.3. The delay τ_d, which may change with time, is to be tracked by the loop. The output of the photodetector is taken as the shot noise current process

$$
x(t) = Ge \sum_{j=1}^{k(0,t)} \delta(t - t_j)
\tag{10.3.2}
$$

where $\delta(t)$ is the detector impulse function, e is the electron charge, G is the photomultiplication gain, $\{t_j\}$ are the random event times, and k(0, t) is the electron counting process. The photodetector output $x(t)$ will have added to it a white circuit noise current $i_c(t)$, and the resulting signal, $x(t) + i_c(t)$, provides the input to the tracking loops. Within the loop, the pulse integrators are time controlled by the receiver timing oscillator, generating the error voltage used to readjust the oscillator. The latter, in addition, provides the timing markers for the receiver. As long as the received pulse train and oscillator signals are in alignment, the receiver markers are being produced exactly in time synchronism with those of the received signal. The pulse integrators consist simply of two $W/2$ sec integrations offset by $W/2$ sec, as shown in Figure 10.6. The first integrator operates over the "early" part of the pulse, the second over the trailing or "late" part of the pulse. If these integrations have been timed to begin exactly with the leading edge of the received pulse, the resulting integration subtraction would be zero, no oscillator correction is necessary, and the system is in time sync. If a time difference occurred between the start of the early integration and the received pulse, a proportional error signal is generated whose polarity depends on the direction of the time difference. This error signal can be used to adjust the loop timing oscillator in the proper direction. Unfortunately, in the optical system the input to the loop is not a clean pulse train, but rather the shot noise process of (10.3.2), containing the optically pulsed intensity of (10.3.1) and any background effect. In addition, this shot noise has added to

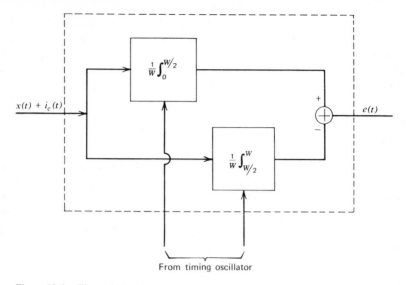

From timing oscillator

Figure 10.6. The early-late gate detector.

it the additive circuit white noise current $i_c(t)$. Hence, the error signal generated after subtraction of the loop integrations is

$$e(t) = \frac{1}{W} \int_{\tau_1}^{\tau_1 + W/2} [x(t) + i_c(t)] \, dt - \frac{1}{W} \int_{\tau_1 + W/2}^{\tau_1 + W} [x(t) + i_c(t)] \, dt \quad (10.3.3)$$

where τ_1 represents the start of the loop integration, that is, the timing of the loop. The dependence of the right-hand side on t is implicit in the parameter τ_1, which varies as the loop attempts to track out the varations τ_d. Although τ_1 is actually a function of t it is treated as a constant when integrating over the period T sec long. This latter fact is simply a restatement of the fact that the bandwidth of τ_1, which is roughly the same as that of τ_d, is much less than the pulse repetition frequency $1/T$. After substituting the input processes, $e(t)$ can be rewritten as

$$e(t) = \frac{Ge}{W} \left[k\left(\tau_1, \tau_1 + \frac{W}{2}\right) - k\left(\tau_1 + \frac{W}{2}, \tau_1 + W\right) \right] - c(\tau_1) \quad (10.3.4)$$

where $c(\tau_1)$ is the Gaussian circuit noise random process, obtained by integrating $i_c(t)$ over $(\tau_1, \tau_1 + W)$, and has zero mean and variance $N_{0c}/2T$, with N_{0c} the one-sided circuit noise spectral level.

The performance of the tracker can be directly related to the instantaneous timing error between the received and the oscillator signal. This timing error is defined by

$$\tau \triangleq \tau_d - \tau_1 \tag{10.3.5}$$

where all parameters are actually functions of t. Using straightforward analog loop analysis, and recalling that the oscillator timing depends on the integral of the voltage controlling it, the timing error τ satisfies the integral-differential equation

$$\frac{d\tau}{dt} = \frac{d\tau_d}{dt} - G_L e(t) \tag{10.3.6}$$

where G_L is the oscillator timing gain. Since the error signal $e(t)$ depends on both τ_d and τ_1, the equation is in general nonlinear in τ. Clearly the solution for $\tau(t)$ necessarily evolves as a stochastic process due to the randomness of the error process $e(t)$. This is true even if the additive circuit noise $i_c(t)$ is set equal to zero (i.e., only background interference), due to the randomness of the shot noise process. Although the complete operating statistics of $\tau(t)$ are of ultimate interest, the behavior of the instantaneous mean value of $\tau(t)$ can be derived from (10.3.6). If we statistically average both sides, interchanging differentiation and averaging on the left, we obtain the equation

$$\frac{dE[\tau]}{dt} = \frac{dE[\tau_d]}{dt} - (G_L)E[e(t)] \tag{10.3.7}$$

where the E denotes statistical averaging. Since the additive circuit noise has zero mean, the mean error voltage is given by the mean shot noise count. The latter is the integrated count intensity over the integration intervals. Hence,

$$E[e(t)] = \frac{Ge}{W} \left[E\left[k\left(\tau_1, \tau_1 + \frac{W}{2}\right)\right] - E\left[k\left(\tau_1 + \frac{W}{2}, \tau_1 + W\right)\right]\right]$$

$$= \left(\frac{Ge}{W}\right) E_\tau \left\{ \int_{\tau_1}^{\tau_1 + W/2} [n(t - \tau_d) + n_b]\, dt - \int_{\tau_1 + W/2}^{\tau_1 + W} [n(t - \tau_d) + n_b]\, dt \right\} \tag{10.3.8}$$

where E_τ is the expectation operator over the random variable τ. The integrals above can be rewritten in terms of a receiver correlation. Define the function $y(t)$ by

$$y(t) \triangleq \begin{cases} 1, & 0 \le t \le \dfrac{W}{2} \\[2ex] -1, & \dfrac{W}{2} < t \le W \end{cases} \tag{10.3.9}$$

Then (10.3.8) becomes

$$E[e(t)] = \left(\frac{TGe}{W}\right) E_\tau[R_{yn}(\tau)] \qquad (10.3.10)$$

where

$$R_{yn}(\tau) \triangleq \frac{1}{T} \int_0^T y(t)n(t + \tau)\, dt \qquad (10.3.11)$$

Hence, the mean of the error process can be related to the correlation of the periodic intensity modulation $n(t)$ with the time function $y(t)$. The latter function can therefore be considered as the receiver "timing" signal produced by the loop. The correlation function $R_{yn}(\tau)$ for the functions $n(t)$ in (10.3.1) and $y(t)$ in (10.3.9) is plotted in Figure 10.7. This correlation function is the

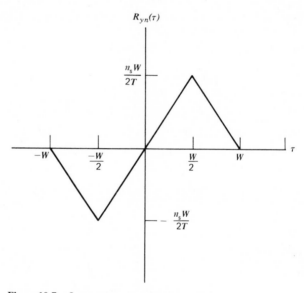

Figure 10.7. Loop mean error response curve.

mean error function of the tracking loop, and is often referred to as the loop error curve, relating voltage error to timing error.

Equation 10.3.7 therefore becomes

$$\frac{dE[\tau]}{dt} = \frac{dE[\tau_d]}{dt} - \frac{(TGeG_L)}{W} E_\tau[R_{yn}(\tau)] \qquad (10.3.12)$$

This is the differential equation of the mean timing error variable in the tracking loop. If τ is confined to the linear range of $R_{yn}(\tau)$ (i.e., if $\tau \approx 0$ and the loop is tracking well), then we can approximate $R_{yn}(\tau) \approx [n_s \tau/T]$ and $E_\tau[R_{yn}(\tau)] \approx E_\tau[n_s \tau/T] = (n_s/T)E[(\tau)]$. Equation 10.3.12 then becomes a linear differential equation in terms of the mean error process $E[\tau]$. Furthermore, this linear equation corresponds to that of the linear feedback system in Figure 10.8. The latter can be considered the linear mean equivalent loop to Figure 10.5, and is useful for analyzing or synthesizing based upon the mean timing error process. Note that in this equivalent system, the mean input variation $E(\tau_d)$ appears as the loop input, and the loop timing oscillator becomes a feedback loop integrator whose output is the timing process

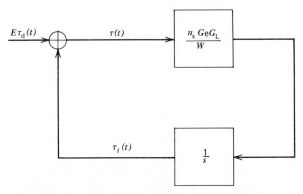

Figure 10.8. The linear model for the early-late gate pulse tracking loop.

$\tau_1(t)$. The linear equivalent loop has a loop gain of $G_L\, Gen_s/W$ and a one-sided loop bandwidth† of

$$B_L = \frac{G_L\, Gen_s}{4W} \qquad (10.3.13)$$

Note that the loop bandwidth depends directly on the received optical power n_s, which therefore appears as a parameter of the equivalent system. The loop bandwidth must be sufficiently wide (i.e., there must be sufficient loop gain) to track the expected time variations in τ_d. The use of equivalent loop models to analyze and design signal tracking capabilities has received considerable attention in the literature [3, Chap. 4], and is not pursued here.

---

† In a linear feedback system, if $H(s)$ is the transfer function from loop input to feedback signal, then the loop bandwidth is defined by $B_L = \int_0^\infty |H(\omega)|^2\, d\omega/2\pi$. It essentially represents the bandwidth that the loop exhibits to the input.

Although mean error performance in tracking the received delay variations can be determined from the linear mean system, the adverse effects arising from the random nature of the optical field and circuit noise cannot be derived (note that the linear system is noiseless). In this case, the dynamical equation of the true system, (10.3.6), must be examined in detail for a complete statistical analysis. Unfortunately, the discrete nature of the timing error equation indicates that the statistics of the solution $\tau(t)$ are highly nonstationary as the process evolves in time. An indication of the statistical properties of $\tau(t)$ can be obtained by examining the steady state probability density of τ. This latter density, $p(\tau)$, is known to satisfy the Kolmogorov–Smoluchowski steady state equation [3, Chap. 7; 5, Chap. 10]

$$\sum_{j=1}^{\infty} \left(\frac{(-1)^j}{j!} \right) \frac{\partial^{j-1}}{\partial \tau^{j-1}} [C_j(\tau)p(\tau)] = 0 \qquad (10.3.14)$$

where

$$C_j(\tau) \triangleq \lim_{\Delta t \to 0} \frac{E[\Delta \tau^j]}{\Delta t} \qquad (10.3.15)$$

$$\Delta \tau = \tau(t + \Delta t) - \tau(t)$$

with suitable initial conditions and with the condition that $p(\tau)$ integrate to unity. When the coefficients $C_j(\tau)$ exist, this equation provides a relation that must be satisfied by the steady state density of the process $\tau(t)$. The equation is a partial differential equation with variable coefficients and, in general, involves all orders of derivatives. The principal usefulness of (10.3.14), however, occurs when only the first few coefficients are nonzero. In particular, if $C_j(\tau) = 0, j \geq 3$, the resulting equation is the steady state Fokker–Planck equation, and has been extensively studied [1, Chap. 4; 3, Chap. 7].

The calculation of the sequence of coefficients $C_j(\tau)$ requires determination of the moments of the error increment $\Delta \tau$ in (10.3.15). Consider the system of (10.3.6) when tracking an intensity pulse having a constant time shift $\tau_d(t) \triangleq \tau_0$. The timing error $\tau(t)$ therefore satisfies (10.3.6) with $d\tau_d/dt = 0$. The timing error variation $\Delta \tau$ is then

$$\Delta \tau = \int_t^{t+\Delta t} \left(\frac{d\tau}{dz} \right) dz$$

$$= -G_L \int_t^{t+\Delta t} e(z)\, dz \qquad (10.3.16)$$

The coefficients in (10.3.15) can now be determined by using (10.3.4) in (10.3.16). Unfortunately, an exact calculation of the coefficients is hampered by the sampled data (short-term integrated) nature of the loop. (The integrator smooths the error signal and produces, in effect, a second-order loop.) We consider instead a continuous first-order loop in which the integrator is neglected, and the resulting time averaged (over a pulse interval W) coefficients are used as an approximation to the desired steady state coefficients. This is equivalent to assuming that the integration over the pulse width is extremely short relative to the time variations of the input intensity process, and can be neglected. The subsequent time averaging of the coefficients is similar to the smoothing produced by the loop. The error signal $e(t)$ is therefore taken as the signal prior to the early-late integrations, or

$$e(t) \approx [x(t) + i_c(t)]y(t - \tau_1) \tag{10.3.17}$$

where $y(t)$ is defined in (10.3.9). Substituting for $x(t)$ from (10.3.2), and evaluating (10.3.16), yields

$$\Delta\tau = -(G_eG_L)\sum_{j=1}^{k(\Delta t)} y(t_j - \tau_1) - G_L \int_t^{t+\Delta t} i_c(z)y(z - \tau_1)\, dz \tag{10.3.18}$$

where $k(\Delta t)$ is the count occurring over $(t, t + \Delta t)$. To determine the C_j coefficients in (10.3.15) the time averaged moments of $\Delta\tau$ must be calculated. The first moment becomes

$$E[\Delta\tau] = -(G_eG_L)E\sum_{j=1}^{k(\Delta t)} y(t_j - \tau_1) - G_L \int_t^{t+\Delta t} E[i_c(z)]y(z - \tau_1)\, dz \tag{10.3.19}$$

Now $E[i_c(z)] = 0$ and we recall that the shot noise term has mean [see (4.3.4)]

$$E\sum_{j=1}^{k(\Delta t)} y(t_j - \tau_1) = \int_t^{t+\Delta t} y(t_j - \tau_1)[n(t_j - \tau_d) + n_b]\, dt_j \tag{10.3.20}$$

In the limit as $\Delta t \to 0$,

$$\lim_{\Delta t \to 0} \int_t^{t+\Delta t} y(a - \tau_1)[n(a - \tau_d) + n_b]\, da = \Delta t[y(t - \tau_1)[n(t - \tau_d) + n_b]] \tag{10.3.21}$$

The first coefficient is derived as the time averaged value of this moment over an integration period of T sec, assuming $(\tau_1 - \tau_d)$ does not change during this interval. Hence

$$C_1 = \frac{1}{T}\int_{\tau_1}^{\tau_1 + W}\left[\lim_{\Delta t \to 0}\frac{E[\Delta\tau]}{\Delta t}\right] dt$$

$$= -(G_eG_L)\left[\frac{1}{T}\int_0^T y(t)n(t + \tau)\, dt\right] \tag{10.3.22}$$

The calculation of the mean squared value of $\Delta\tau$ requires computation of the cross products involved. However, noting that the eventual computation of the C_j requires a division by Δt, followed by a limit as $\Delta t \to 0$, only terms of order Δt need be retained. In particular, we see from (10.3.21) that any product of averages of the shot noise summation is always at least of order $(\Delta t)^2$. Hence,

$$E[\Delta\tau^2] = (GeG_L)^2 E \sum_{j=1}^{k(\Delta t)} y^2(t_j - \tau_1)$$

$$+ G_L^2 \int\int_t^{t+\Delta t} E[i_c(z_1)i_c(z_2)]y(z_1 - \tau_1)y(z_2 - \tau_1)\, dz_1\, dz_2 + O(\Delta\tau^2)$$

$$\text{(10.3.23)}$$

The average of the first summation was shown in (4.3.12) to yield

$$E \sum_{j=1}^{k(\Delta t)} y^2(t_j - \tau_1) = \int_t^{t+\Delta t} y^2(t_j - \tau_1)[n(t_j - \tau_d) + n_b]\, dt_j + O(\Delta t^2)$$

$$= [y^2(t - \tau_1)n[(t - \tau_d) + n_b]]\,\Delta t + O(\Delta t^2) \quad \text{(10.3.24)}$$

for $\Delta t \to 0$. Since the circuit noise $i_c(t)$ has a flat spectrum of one-sided level N_{0c}, the double integral in (10.3.23) can be shown to be $[G_L^2 W N_{0c}\,\Delta t/2T]$ (Problem 10.4). Therefore,

$$C_2 = \frac{1}{T} \int_0^T \left[\lim_{\Delta t \to 0} \frac{E[(\Delta\tau)^2]}{\Delta t} \right] dt$$

$$= (GeG_L)^2 \left[\frac{1}{T} \int_0^T y^2(t)n(t + \tau)\, dt + \frac{W n_b}{T} + \frac{W N_{0c}}{2T(Ge)^2} \right] \quad \text{(10.3.25)}$$

Similarly, for higher moments we use the fact that

$$E[\Delta\tau^j] = -(GeG_L)^j \sum_{j=1}^{k(\Delta t)} y^j(t_j - \tau_1) + O(\Delta t^2), \quad j \geq 3 \quad \text{(10.3.26)}$$

to derive

$$C_j = (-GeG_L)^j \left[\frac{1}{T} \int_0^T y^j(t)n(t + \tau)\, dt + \frac{n_b}{T} \int_0^T y^j(t)\, dt \right] \quad \text{(10.3.27)}$$

The required coefficients computed in this way can therefore be collected from (10.3.22), (10.3.25), and (10.3.27) and summarized as

$$
C_j(\tau) = \begin{cases}
-(GeG_L)R_{yn}(\tau), & j = 1 \\[2ex]
(GeG_L)^2\left[R_{y^2n}(\tau) + \dfrac{Wn_b}{T} + \dfrac{N_{0c}W}{2T(Ge)^2}\right], & j = 2 \\[2ex]
-(GeG_L)^j R_{y^jn}(\tau), & j \geq 3, \text{ odd} \\[2ex]
(GeG_L)^j\left[R_{y^jn}(\tau) + \dfrac{Wn_b}{T}\right], & j \geq 4, \text{ even}
\end{cases}
\tag{10.3.28}
$$

where now

$$
R_{y^jn}(\tau) = \frac{1}{T}\int_{-\infty}^{\infty} y^j(t)n(t+\tau)\,dt \tag{10.3.29}
$$

For the loop signal $y(t)$ in (10.3.9), (10.3.29) can be further simplified by noting that

$$
R_{y^jn}(\tau) = \frac{1}{T}\int_0^W n(t+\tau)\,dt \triangleq R_0(\tau), \qquad j \text{ even}
$$

$$
= R_{yn}(\tau), \qquad j \text{ odd} \tag{10.3.30}
$$

Thus the steady state Kolmogorov equation becomes

$$
0 = (GeG_L)[R_{yn}(\tau)p(\tau)] + \frac{(GeG_L)^2}{2}\frac{d}{d\tau}\left\{\left[R_0(\tau) + n_b + \frac{N_{0c}}{2(Ge)^2}\right]p(\tau)\right\}
$$

$$
+ \sum_{\substack{j=3 \\ \text{odd}}}^{\infty} \frac{(GeG_L)^j}{j!}\frac{d^{j-1}}{d\tau^{j-1}}\{R_{yn}(\tau)p(\tau)\} + \sum_{\substack{j=4 \\ \text{even}}}^{\infty} \frac{(GeG_L)^j}{j!}\frac{d^{j-1}}{d\tau^{j-1}}\{[R_0(\tau) + n_b]p(\tau)\}
$$

$$
\tag{10.3.31}
$$

The infinite number of derivatives manifests the discontinuities of the error process caused by the quantum nature of the detected optical process. It is the form of this equation vis-à-vis the Fokker–Planck equation that theoretically separates optical tracking from tracking in additive Gaussian noise. This complication was previously noted by Ohlson [6] when dealing with additive shot noise. (However, the optical model here involves an input shot process with signal embedded within. Both models lead to infinite-order equations but with slightly different coefficients.) Although an exact solution for $p(\tau)$ is somewhat ambitious, some meaningful information and approximating solutions can be derived. In particular, consider the case where the system operates in near lock operation, so that it may be assumed that $\tau \approx 0$. The instantaneous tracking error can therefore be considered to be confined to the linear range of $R_{yn}(\tau)$ and

$$R_{yn}(\tau) \approx \frac{n_s \tau}{T}$$

$$R_0(\tau) \approx \frac{n_s W}{T}$$

$$(10.3.32)$$

Substituting into (10.3.31) and dividing by the coefficient of the second term yields the modified equation:

$$0 = \rho \tau p(\tau) + \frac{d}{d\tau} p(\tau) + \sum_{\substack{j=3 \\ (\text{odd})}}^{\infty} \tilde{C}_j \frac{d^{j-1}}{d\tau^{j-1}} [\tau p(\tau)] + \sum_{\substack{j=4 \\ (\text{even})}}^{\infty} \tilde{C}_j \frac{d^{j-1}}{d\tau^{j-1}} [p(\tau)] \quad (10.3.33)$$

where

$$\rho = \frac{n_s}{(W G e G_L / 2)[n_s + n_b + [N_{0c}/2(Ge)^2]]} \quad (10.3.34)$$

$$\tilde{C}_j = \left(\frac{2}{j!}\right) \left(\frac{2}{W\rho}\right)^{j-2} \left[\frac{n_s^{j-1}}{[n_s + n_b + [N_{0c}/2(Ge)^2]]^{j-1}} \right] \quad (10.3.35)$$

Note that the coefficients $\tilde{C}_j$ vary as $1/\rho^{j-2}$, exhibiting a decreasing importance of the higher derivatives as the parameter ρ in (10.3.34) is increased. The bracketed term in $\tilde{C}_j$ is bounded by one and approaches one as the system approaches quantum limited operation, that is, when $n_s \gg [n_b + N_{0c}/(Ge)^2 2]$. A physical interpretation to ρ can be introduced by noting the linear mean equivalent loop of Figure 10.8. Since it is often desirous to operate the tracking loop with a given loop bandwidth B_L, the loop gain G_L is generally adjusted so as to obtain this value in (10.3.13). Thus, if B_L is the desired bandwidth, then we set the loop gain $G_L = 4B_L W/Gen_s$, and (10.3.34) becomes

$$\rho = \frac{1}{W^2} \left[\frac{n_s^2}{2B_L[n_s + n_b + (N_{0c}/2(Ge)^2)]} \right] \quad (10.3.36)$$

When written in this way, the numerator in the brackets is the square of the signal intensity rate of the received signal, while the denominator is effectively the total noise power occurring in the bandwidth B_L (since the level $n_s + n_b$ is the two-sided shot noise spectral level). Thus, ρ is proportional to a signal-to-noise power ratio, and the coefficients in $\tilde{C}_j$ in (10.3.35) vary as an inverse power of this ratio. We therefore expect that solutions for $p(\tau)$ can be suitably approximated by solving truncated versions of (10.3.33) with fewer and fewer terms, as we increase the optical signal-to-noise ratio in (10.3.36). As a limiting case we see that as $\rho \gg 1$, $\tilde{C}_j \to 0$ for all $j \geq 3$ and the differential equation is approximated by its first two terms. This becomes

$$0 \cong \rho \tau p(\tau) + \frac{dp(\tau)}{d\tau} \tag{10.3.37}$$

The equation is easily solved as

$$p(\tau) = \frac{1}{(2\pi/\rho)^{1/2}} \exp\left[-\frac{\tau^2}{2/\rho}\right] \tag{10.3.38}$$

This corresponds to a zero mean Gaussian density for τ, having variance $1/\rho$. Thus, for high values of signal-to-noise ratio, the tracking error approaches a Gaussian variable. For ρ values not particularly high, the higher terms in (10.3.33) can no longer be neglected, and higher order truncations must be considered. Of course, as ρ is decreased in value, the increasing variance will cause the loop error to exceed the linear range of $R_{yn}(\tau)$, violating our assumption that the loop is in fact completely linear. Assuming shot noise limited operation $[(n_s + n_b) \gg N_{0c}/(Ge)^2]$, the loop error variance for the normalized delay variable τ/W is therefore

$$\sigma_{\tau/W}^2 = \frac{1/\rho}{W^2} = \frac{2B_L}{n_s}\left[1 + \frac{n_b}{n_s}\right]$$

$$= \frac{2B_L}{n_s}\left[1 + \left(\frac{B_L}{n_s}\right)\left(\frac{n_b}{B_L}\right)\right] \tag{10.3.39}$$

This is plotted in Figure 10.9 as a function of the parameter n_s/B_L for several values of normalized noise energy n_b/B_L. The curves, in essence, summarize the performance of a pure sync system operating with signal count rate n_s in a tracking bandwidth of B_L Hz. The rapid increase in the normalized error variance as the parameter n_s/B_L is decreased represents the deterioration of the timing performance. The presence of background noise (n_b) causes the increase to occur at higher values of n_s/B_L. Since n_s represents the rate of occurrence of signal photoelectrons during pulse transmission, the parameter n_s/B_L can be considered as an indication of the "denseness" of signal counts, indicating the accumulation of electrons over a time period equal to the reciprocal of the loop bandwidth. Alternatively, by substituting for n_s in terms of the received optical pulse power P_s we can write $n_s/B_L = \eta P_s/hf B_L$, which has now the familiar interpretation as the ratio of detected pulse power to the quantum noise in the loop B_L bandwidth. The ratio

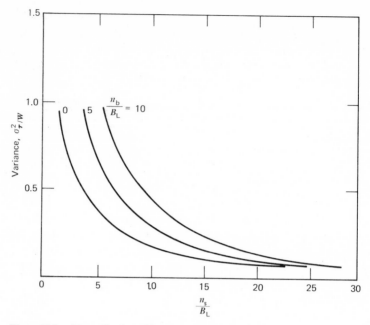

Figure 10.9. Normalized tracking error variances.

n_b/B_L has a similar interpretation in terms of received background noise power.

It should be pointed out that the variances plotted in Figure 10.9 correspond to the relatively simply tracking loop in Figure 10.5. Some improvement in performance can often be attained by designing more complicated tracking systems. For example, by simultaneously processing with a parallel integrator over the leading edge of the subsequent sync pulse [i.e., over the time interval $(3W/2, 5W/2)$] one can delay and negatively combine with $e(t)$ in Figure 10.5 to strengthen (double) the error amplitude. This effectively doubles the signal power and theoretically produces a 3 dB power savings. The interested reader is referred to the book by Lindsey and Simon [4] for further discussion of these possible modifications.

10.4 DECISION DIRECTED SYNCHRONIZATION (PULSE EDGE TRACKING WITH PPM)

In the preceding section we assumed that the periodic sync signal was transmitted continuously as a pulse train to allow the synchronization subsystem to maintain tracking. Any data waveforms transmitted along with these sync pulses had to be separated out, either by frequency filtering or by time separation (i.e., sending the data during the time intervals separating the pulse positions). Since a portion of the transmitted power must be allocated to the sync pulses, the data signals therefore use only a fraction of the total available power. That is, the power must be divided among the sync and data signals. An alternative procedure is not to use a sync pulse, but devote the total transmitter to the data waveform, and to attempt to derive the tracking operation directly from the data waveform itself. The latter, of course, exhibits the same random delays during transmission as the sync pulses themselves and can be used for receiver tracking. Unfortunately, however, the data waveform also contains the information modulation that tends to obscure the synchronization operation. In the following we examine the pulse tracking operation when the pulse sequence contains digital PPM modulation. Recall the latter format is a favorable method for digital data transmission, and yet still utilizes pulse-type waveforms that can be tracked by early-late gate loops.

In a PPM systems the received optical intensity is still pulse like, but no longer periodic, and varies in position according to the data bit sequence. For example, in binary PPM, if the optical intensity is written as in (10.3.1), then its modulation during a bit period is given by $n(t)$ in (10.3.1), with $T = 2W$, if a binary one is sent, but is given by $n(t - W)$ if a binary zero is sent, as is shown in Figure 10.10. A receiver tracking loop can maintain timing if it can generate a correction voltage proportional to the timing offset between the loop timing signal and the received pulse sequence. However,

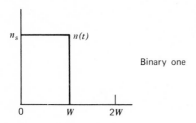

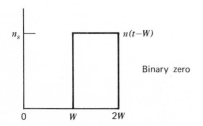

Figure 10.10. The PPM signal set.

the tracking problem is hampered by the fact that modulation is present and adjacent symbols may effect the error signal if loop integration is too long. This latter point can be avoided by using timing signals that integrate only over a narrow region about the midpoint edge. (Note that a pulse edge is always present at the midpoint of the bit interval no matter which bit is sent in Figure 10.10.) Consider the modified timing signal

$$y(t) = \begin{cases} 0, & 0 \le t \le \dfrac{W}{3}, \quad \dfrac{5W}{3} < t \le 2W \\[2ex] -1, & \dfrac{W}{3} < t \le \dfrac{2W}{3} \\[2ex] +1, & \dfrac{2W}{3} < t \le \dfrac{4W}{3} \\[2ex] -1, & \dfrac{4W}{3} < t \le \dfrac{5W}{3} \end{cases}$$

(10.4.1)

which is shown in Figure 10.11. Such a timing signal can be generated by a proper combination of early and late gate integrations. The "dead zones" at the edges will prevent adjacent bit interference if the timing error is large. The $y(t)$ above is timed by the loop oscillator and correlates with the input signal as discussed previously. If a data one is sent with a time offset of τ sec, the error voltage generated from the correlation $R_{yn}(\tau)$ is shown in Figure 10.12. Note the curve is identical over the linear range to that shown in Figure 10.7, except the linear range extends over a narrower region of τ. However, if a zero is sent, an error voltage of $-R_y(\tau)$ is generated instead.

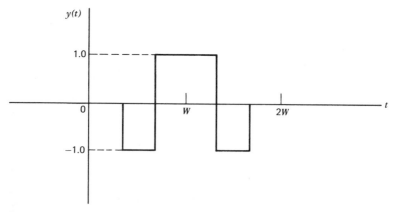

Figure 10.11. Modified timing signal for PPM pulse tracking.

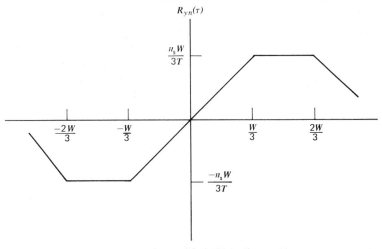

Figure 10.12. Loop error curve for modified PPM pulse tracking loop.

Hence, for equally likely data bits, the average error voltage within the loop is then $\frac{1}{2}(R_{yn}(\tau)) + \frac{1}{2}(-R_{yn}(\tau)) = 0$. That is, on the average, no loop error is generated for controlling the receiver timing oscillator during modulation reception.

To compensate for this modulation, an augmented decision directed edge tracking system can be used. Such a system employs the auxiliary decisioning operation shown in Figure 10.13. The decision loop attempts to determine

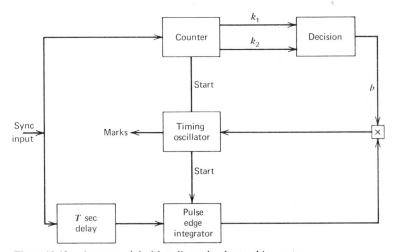

Figure 10.13. Augmented decision directed pulse tracking system.

the true data bit, using this result to properly modify the sign of the loop error voltage. This can be implemented by multiplying the generated error in a delayed (by one bit period) tracking loop by either a plus or minus one, depending on the data bit. This latter decision is made from a count comparison over each possible bit subinterval as they arrive. Thus the error in the delayed edge tracking loop becomes $be(t)$, where

$$b = \begin{cases} +1 & \text{if} \quad \text{one is decided} \\ -1 & \text{if} \quad \text{zero is decided} \end{cases} \tag{10.4.2}$$

or equivalently,

$$b = \begin{cases} +1 & \text{if} \quad k_1 \geq k_2 \\ -1 & \text{if} \quad k_1 < k_2 \end{cases} \tag{10.4.3}$$

where k_1, k_2 are the counts over the first and second subintervals of each bit period. The loop differential equation for the tracking loop error now becomes

$$\frac{d\tau}{dt} = \frac{d\tau_j}{dt} - G_L[be(t)] \tag{10.4.4}$$

Since the counts in (10.4.3) are random counts, the parameter b is a random variable. Thus the coefficients $C_j(\tau)$ in (10.3.15) for the steady state density are a function of this variable, and therefore require a subsequent average over its statistics. When a one is sent the probability that $k_1 \geq k_2$ is equivalent to the probability that the one is correctly detected, whereas the probability that $k_1 < k_2$ corresponds to the probability that an error is made. Hence, when a one is sent,

$$b = \begin{cases} +1 & \text{with probability } 1 - PE_1 \\ -1 & \text{with probability } PE_1 \end{cases} \tag{10.4.5}$$

where PE_1 is the bit error probability when a one is sent. When a zero is sent, the signs are reversed and PE_1 is replaced by PE_0. It should be remembered, however, that the timing for this subinterval counting is in turn controlled by the receiver loop timing signal, which has the loop timing errors incorporated within it. Thus the bit error probabilities in (10.4.5) must include these timing error effects. Recall that a timing error between the true bit arrival time and the start of subinterval counting will cause the counting to occur over an offset interval; the effect of these timing errors on bit decisioning was derived in Section 7.7. Thus the tracking and decisioning are not independent operations and in fact are intercoupled through the timing error.

The steady state coefficients of the Kolmogorov equation describing the tracking performance can be formally evaluated from (10.3.15) and (10.4.4) by first conditioning on b, then averaging over the probabilities in (10.4.5). Denote by $C_j(\tau)$ the Kolmogorov coefficients when a data one is sent and no decisioning is used (i.e., pure tracking), and denote by $D_j(\tau)$ the coefficients in the modulated case with decisioning. Noting that $b^j = 1$ for all j even, we see that

$$D_j(\tau) = \begin{cases} C_j(\tau), & j \text{ even} \\ \frac{1}{2}C_j(\tau)\{+1[1 - PE_1] + (-1)PE_1 - (-1)[1 - PE_0] \\ \quad - (+1)[PE_0]\}, & j \text{ odd} \end{cases}$$

$$= \begin{cases} C_j(\tau), & j \text{ even} \\ C_j(\tau)[1 - 2PE(\tau)], & j \text{ odd} \end{cases} \tag{10.4.6}$$

where $PE(\tau) = \frac{1}{2}[PE_1 + PE_0]$. Note that the dependence of PE on τ has been emphasized. The resulting steady state density equation is again given by (10.3.14) with $C_j(\tau)$ replaced by the $D_j(\tau)$ above. Note that the coefficients are now more complicated functions of τ due to the auxiliary decisioning, and approach the pure tracking as $PE(\tau) \to 0$. In this latter case, the system is correctly identifying the true bit during each period, and essentially "removing" the binary modulation. The first coefficient,

$$D_1(\tau) = GeG_L R_{yn}(\tau)[1 - 2PE(\tau)],$$

is the average loop error function, and represents the modified nonlinearity of the mean equivalent loop. This coefficient is plotted in Figure 10.14 as a function of τ and the received pulse count K_s, obtained from Figure 10.7 and the PPM decisioning curves of Figure 7.16. Note that the effect of the decision process is to reduce the width and amplitude of the tracking error function. As $PE(\tau) \to \frac{1}{2}$, there is no average error being generated for loop tracking, and the system cannot track.

For the strong signal-near lock assumption [use of (10.3.32) and $\rho \gg 1$ in (10.3.34)], the steady state equation is approximated by

$$0 = \rho\tau[1 - 2PE(\tau)]p(\tau) + \frac{dp(\tau)}{d\tau} \tag{10.4.7}$$

Even with this simplification, the solution density cannot be generated as easily as in Section 10.3, since the first coefficient is now more complicated. However, the fractional variance for this solution can be estimated by approximating the coefficient $D_1(\tau)$ in Figure 10.14 by a single cycle sine wave of proper amplitude. This amplitude depends on the count energy K_s

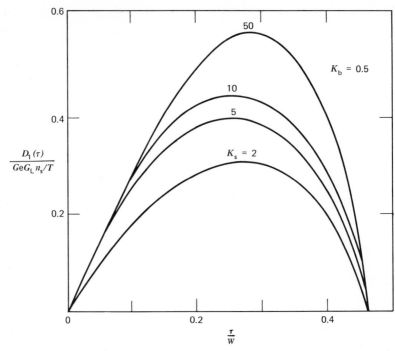

Figure 10.14. Normalized first coefficient for decision directed tracking. ($K_b = n_b W$)

per data bit used for decisioning, which in turn is related to the n_s/B_L parameter by

$$K_s = n_s W = (B_L W)\frac{n_s}{B_L} \tag{10.4.8}$$

where $B_L W = B_L/2R_b = \frac{1}{2}(B_L/R_b)$. The parameter (B_L/R_b) is the ratio of tracking loop bandwidth to the data bit rate R_b, and is typically less than unity. When written as in (10.4.8), $B_L W$ can also be interpreted as the fraction of the sync energy n_s/B_L appearing in the data pulse, and therefore used in the auxiliary decisioning. For a fixed value of $B_L W$, each value of n_s/B_L generates a corresponding value of K_s, to which an effective sine wave can be fitted to the corresponding curves of $D_1(\tau)$. The variance can then be determined at each (n_s/B_L) value by numerically solving (10.4.7). The resulting normalized variance computed in this way is plotted in Figure 10.15 as a function of n_s/B_L for several values of $B_L W$. The curve for the noiseless, pure

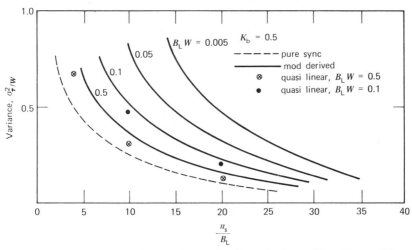

Figure 10.15. Tracking error variances for decision directed pulse tracking. ($K_b = n_b W$)

sync operation from Figure 10.9 is superimposed. The results show that a deterioration of performance occurs over pure sync, because of the decisioning process, which can therefore be considered as the price to be paid for impure synchronization. Note that the decisioning causes system degradation similar to an effective loss in signal-to-noise ratio (reduced n_s) and can therefore be interpreted as a power loss in the sync subsystem.

Although the use of the curves in Figure 10.15 is convenient for assessing performance, their derivation requires a somewhat lengthy computation. Furthermore, this computation must be repeated at each desired value of background noise K_b. However, a simple method can be used, at the expense of analytical accuracy, to derive similar curves. This method makes use of a form of truncated quasi-linear solution, which basically amounts to replacing the first coefficient in (10.4.7) by a modified linear coefficient, as in (10.3.37), but retaining its dependence on the decision error probability PE. To accomplish this linearization, we first recognize that PE depends on both pulse energy K_s and timing error τ, and we write this as $PE(K_s, \tau)$. To linearize we replace the functional dependence on τ by the root mean square value of τ in (10.3.38) that is, $\tau_{rms} = (1/\rho)^{1/2}$. Thus we consider instead $PE(K_s, (1/\rho)^{1/2})$. The truncated quasi-linear differential equation for the timing error density $\rho(\tau)$ is then

$$\frac{dp(\tau)}{d\tau} + \rho[1 - 2PE(K_s, (1/\rho)^{1/2})]\tau p(\tau) = 0 \qquad (10.4.9)$$

Note the equation is linear in $p(\tau)$, but the coefficients are nonlinear in ρ. The solution for $p(\tau)$ again yields a Gaussian density with normalized variance

$$\sigma_{\tau/W}^2 = \frac{1}{W^2 \rho[1 - PE(K_s, (1/\rho)^{1/2})]} \qquad (10.4.10)$$

For the shot noise limited case, $1/W^2\rho$ is given in (10.3.39), K_s is related to n_s/B_L by (10.4.8), and the variance depends only on the parameters n_s/B_L, $B_L W$, and n_b/B_L. The values of PE at any value of $K_s = B_L W(n_s/B_L)$ and τ_{rms} can be obtained from curves similar to Figure 7.16. Several points of this variance for the shot noise limited case are superimposed in Figure 10.15. The results tend to display the same behavior for tracking performance, although the variance values are slightly lower than the more accurate results determined earlier.

10.5 TRACKING SINUSOIDAL SYNCHRONIZATION INTENSITIES

The use of sync signals in the form of optical pulses was considered primarily because of certain estimation advantages of such waveforms, as discussed in Section 9.4. However, the important information in the tracking operation is not the sync signal waveform itself but rather its location in time. It may therefore be advantageous to use other types of synchronization waveforms that provide satisfactory tracking capability. For example, it is evident that with time acquisition achieved, we can replace the sync pulses with a sinusoidal intensity having a positive zero crossing at the point where the pulse leading edge would occur, as shown in Figure 10.16. Synchronization

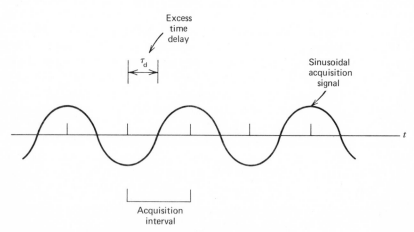

Figure 10.16. Sinusoidal acquisition signal with excess time delay.

is then maintained by tracking this zero crossing within the acquisition interval, rather than tracking pulse locations. This has the advantage of using narrow band synchronization intensities that are relatively easy to generate, rather than pulsed, wide band intensities. Note that the sinusoidal intensity must have a period that is at least twice the pulse width in order to guarantee only one positive zero crossing in an acquisition interval (recall the acquisition pulse is located to within a pair of pulse location intervals).

Determining the zero crossing location of a sine wave is equivalent to determining its phase angle relative to a point in time at the beginning of the known acquisition interval (see Figure 10.16). In Section 9.2 it was shown that, under certain conditions, an optimal measurement of phase angle of a sinusoidal intensity is obtained by phase locking with a generalized tracking loop. The latter is a modified form of a phase tracking loop, used extensively in phase modulated communications [1, Chaps. 1–5]. In practical implementations of tracking subsystems, it would be expected, however, that standard phase tracking loops would be preferred because of their ease of implementation and their wide usage. For this reason, we concentrate in this section on examining phase lock loop tracking of a sinusoidal intensity modulated optical field [7].

The overall system is shown in Figure 10.17. Again we disregard the presence of data and consider only pure tracking behavior. The loop is completely analogous to the early-late gate loop in its operation. The loop correlating signal is the output of a voltage controlled oscillator (VCO), and the previous early-late gate integration is replaced by a phase comparator. The latter measures the phase difference between incoming and VCO signals. This error voltage is then used to control the phase of the VCO. Theoretically, when the loop is in lock with an incoming sinusoid the VCO

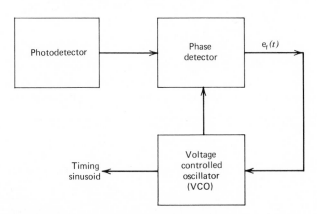

Figure 10.17. The phase locked tracking loop.

produces an identical sinusoid whose zero crossings are identical to those of the input. As the input sinusoid undergoes phase changes (or equivalently, time shifts) the loop attempts to track out these changes, keeping the VCO sinusoid in phase (time) agreement. In the optical system of Figure 10.17 this operation is again complicated by the fact that the input is not a pure sinusoid, but rather the output of a photodetector, so that the phase comparator does not measure the exact phase difference between received intensity signal and VCO.

Analytically, the system can be investigated similarly to the procedure in Section 10.3. The input to the phase tracking loop is given by (10.3.3) except the intensity $n(t)$ in (10.3.1) is replaced by

$$n(t) = n_s[1 + \sin(\omega_s t + \theta_1)] \tag{10.5.1}$$

where ω_s is the frequency of the synchronizing sinusoid, θ_1 is the phase (time) delay to be tracked by the loop, and n_s is again the average count rate over the detector area. The VCO output is represented as

$$\text{VCO output} = \cos(\omega_s t + \theta_2) \tag{10.5.2}$$

where θ_2 is its instantaneous phase variation. The loop phase detector generates the product term

$$\begin{aligned} e_m(t) &= [x(t) + i_c(t)][\text{VCO output}] \\ &= x(t)\cos(\omega_s t + \theta_2) + i_c(t)\cos(\omega_s t + \theta_2) \end{aligned} \tag{10.5.3}$$

and uses as its output, $e_f(t)$, the filtered version of $e_m(t)$. [We assume here that the phase detector contains internal filtering that responds to only the slowly varying portions of $e_m(t)$, which we label as $e_f(t)$.] The VCO output phase responds to its input control voltage $e_f(t)$ through the relation

$$\frac{d\theta_2}{dt} = G_L e_f(t) \tag{10.5.4}$$

where G_L is the VCO proportionality constant. We define the loop phase error ϕ as the phase difference between the synchronizing signal phase and the VCO phase:

$$\phi = \theta_1 - \theta_2 \tag{10.5.5}$$

Then, from (10.5.1) to (10.5.5), the phase error satisfies the differential equation

$$\frac{d\phi}{dt} = \frac{d\theta_1}{dt} - G_L e_f(t) \tag{10.5.6}$$

By making the substitution $e_f(t) = \overline{e_m(t)}$, where the overbar denotes low pass filtering, we can rewrite (10.5.6) as

$$\frac{d\phi}{dt} = \frac{d\theta_1}{dt} - G_L \overline{e_m(t)} \tag{10.5.7}$$

10.5.7 is the loop phase error differential equation, corresponding to (10.3.6) of the pulse tracking loop. Note that it is an identical type of dynamical equation, and can be analyzed in a way similar to the methods in Section 10.3. In particular we note that by interchanging statistical averaging and filtering (integration) we can write

$$
\begin{aligned}
E[\overline{e_m(t)}] &= \overline{\{E[e_m(t)]\}} \\
&= \overline{E\{\cos(\omega_s t + \theta_2)[((x(t) + i_c(t))]\}} \\
&= \overline{E\{\cos(\omega_s t + \theta_2)[n(t) + n_b]\}} \\
&= \left(\frac{Gen_s}{2}\right) E[\sin \phi] \tag{10.5.8}
\end{aligned}
$$

where we retained only lower frequency terms because of the filtering. Note that this result compares directly with (10.3.12) with $R_{yn}(\tau)$ replaced by $\sin \phi$. Thus the phase tracking loop replaces the triangular error curve by a sinusoidal error curve. The loop differential equation for the mean phase error therefore satisfies

$$\frac{dE[\phi]}{dt} = \frac{dE[\theta_1]}{dt} - (GeG_L n_s)E[\sin \phi] \tag{10.5.9}$$

which can be compared with (10.3.12). The loop can be effectively linearized by the assumption $\sin \phi \approx \phi$, which now defines the equivalent loop in Figure 10.18. This linear loop has slightly different loop gain than the pulse tracking loop in Figure 10.8, since the loop variable is now phase instead of time. The linear loop can be used to determine mean tracking performance (Problem 10.3). Note that the loop has an effective bandwidth of $B_L = GeG_L n_s/4$.

The Kolmogorov equation for the steady state phase error density can be obtained from (10.3.14). If we again assume a constant phase offset so that $d\theta_1/dt = 0$, the phase increment in (10.5.7) becomes

$$
\begin{aligned}
\Delta\phi &= \int_t^{t+\Delta t} \left(\frac{d\phi}{dz}\right) dz \\
&= -GeG_L \sum_{j=1}^{k(\Delta t)} \cos[\omega_s t_j + \theta_1 - \phi(t_j)] + \int_t^{t+\Delta t} i_c(z) \cos[\omega_s z + \theta_2(z)] \, dz \tag{10.5.10}
\end{aligned}
$$

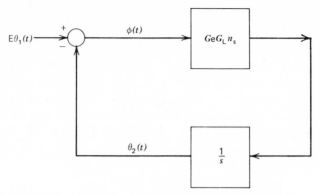

Figure 10.18. The phase locked loop linear equivalent system.

where $k(\Delta t)$ is again the number of electron occurrences in the interval $(t, t + \Delta t)$. Note that (10.5.10) has the same form as (10.3.18), except the receiver feedback signal $y(t)$ in pulse tracking is replaced by $\cos(\omega_s t)$. Under the assumption $\omega_s T \gg 1$, the coefficients for $\Delta\phi$ can therefore be derived exactly as in (10.3.18) to (10.3.28), yielding (Problem 10.5)

$$
C_j = \begin{cases}
-c_1(GeG_L n_s)\sin\phi, & j = 1 \\[2mm]
c_2(GeG_L)^2\left[n_s + n_b + \dfrac{N_{0c}}{2(Ge)^2}\right], & j = 2 \\[2mm]
-c_j(GeG_L)^j n_s \sin\phi, & j \geq 3,\ \text{odd} \\[2mm]
c_j(GeG_L)^j[n_s + n_b], & j \geq 4,\ \text{even}
\end{cases}
\tag{10.5.11}
$$

where $c_j = [(\tfrac{1}{2})(\tfrac{3}{4})(\tfrac{5}{6}) \cdots (j/j + 1)]$, j odd, and $c_j = c_{j-1}$, j even. The infinite-order steady state probability density equation now becomes

$$
0 = \sum_{\substack{j=1 \\ \text{odd}}}^{\infty} \frac{c_j(GeG_L)^j n_s}{j!}\frac{d^j}{d\phi^j}[\sin\phi p(\phi)] + \sum_{\substack{j=4 \\ \text{even}}}^{\infty} \frac{c_{j-1}(GeG_L)^j(n_s + n_b)}{j!}
$$

$$
+ \frac{1}{2}(GeG_L)^2\left[n_s + n_b + \frac{N_{0c}}{2(Ge)^2}\right]\frac{d^2 p(\phi)}{d\phi^2}
\tag{10.5.12}
$$

as in (10.3.31). The steady state error density $p(\phi)$ can be determined (theoretically) by solving Equation (10.5.12). The equation is again of infinite order and approximating solutions must be found from suitable truncations. If we divide through by the second coefficient, the first term will have a coefficient ρ, and the remaining terms for $j \geq 3$ will have coefficients that behave as $(1/\rho)^{j-2}$, where now

$$
\rho = \frac{n_s}{(GeG_L/2)[n_s + n_b + (N_{0c}/2(Ge)^2)]}
$$

$$
= \frac{n_s^2}{2B_L[n_s + n_b + (N_{0c}/2(Ge)^2)]} \tag{10.5.13}
$$

and $B_L = GeG_L n_s/4$ is the loop noise bandwidth of the equivalent system of Figure 10.18. Note that ρ can again be interpreted as a count signal-to-noise ratio. For the quantum limited case, $n_s \gg [n_b + (N_{0c}/2G^2e^2)]$, $\rho = n_s/2B_L$, which is again the average signal count in $1/2B_L$ sec. When $\rho \gg 1$, (10.5.12) reduces to

$$
0 = \rho \frac{d}{d\phi}[\sin \phi p(\phi)] + \frac{d^2 p(\phi)}{d\phi^2} \tag{10.5.14}
$$

This equation can be easily solved over $(-\pi, \pi)$, with the conditions that $p(\pi) = p(-\pi)$ and $p(\phi)$ integrates to unity, yielding

$$
p(\phi) = \frac{\exp[\rho \cos \phi]}{2\pi I_0(\rho)} \tag{10.5.15}
$$

where $I_0(\rho)$ is the zero order imaginary Bessel function of argument ρ. This solution is shown plotted in Figure 10.19 for several values of ρ. Note that the density approaches a Gaussian shape with variance $1/\rho$ for large ρ, and tends to spread as ρ is decreased. Haney [8] has investigated computer solutions of truncated versions of (10.5.12). His results have indicated that ρ need not be particularly large ($\rho \geq 5$) before (10.5.15) is a fairly accurate solution. The phase error variance computed from (10.5.15) is shown plotted in Figure 10.20, along with the variance $1/\rho$ and that computed from a third order truncation solution.

It is of interest to point out that the solution (10.5.15) is the same solution obtained for first-order phase locked loop when driven by a sinusoidal signal in additive white Gaussian noise [1, Chap. 4]. Thus the phase error differential equation due to shot noise becomes identical to that due to additive input Gaussian noise as the higher coefficients are eliminated. In essence, this serves as an apparent justification for the truncation of (10.5.12) at large ρ values, since it is shown in Chapter 4 that a Poisson shot noise process approaches a continuous Gaussian process as the arrival rate increases. Note that the solution in (10.5.15) depends only on ρ, which in turn depends only on the received sync power level, noise levels, and loop bandwidth. In particular, the synchronizing carrier frequency ω_s in (10.5.1)

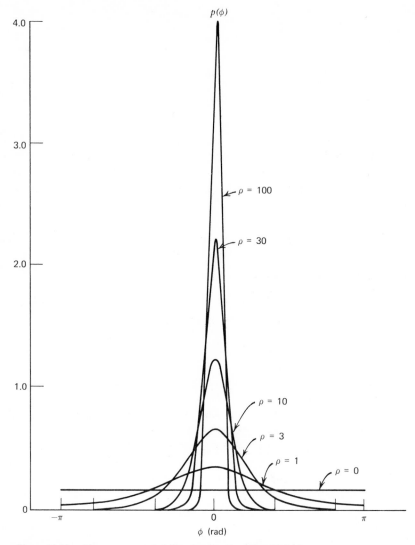

Figure 10.19. Phase error probability density, Equation 10.5.15.

does not appear in the solution. The sync frequency is important, however, in converting phase errors in radians to timing errors in seconds. That is, if ϕ_0 is a particular loop phase error, the corresponding timing error in seconds is then

$$\tau_0 = \frac{\phi_0}{\omega_s} \tag{10.5.16}$$

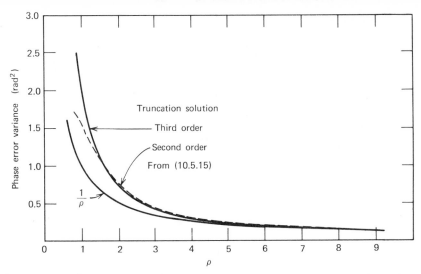

Figure 10.20. Phase error variances in phase lock tracking. (Order of truncation solution is the order of the highest derivative used.)

This indicates that phase tracking should be accomplished at as high an intensity frequency as possible. Unfortunately, Doppler offsets also increase with frequency, and tend to increase the loop lockup time (i.e., the time before linear loop analysis is valid).

We have concentrated here on only relatively simple phase tracking systems, confining analysis to only first-order loops and constant phase intensity inputs. Extension to higher order loop analysis has been explored in the work of Gagliardi and Haney [9]. Higher order loops require vector Kolmogorov equations to be solved. Linearizing assumptions produce steady state error densities similar to (10.5.15) with modified versions of the loop bandwidth B_L inserted. As a good approximation when $\rho \geq 5$, the density in (10.5.15) can be used for higher order loops with the B_L used to define ρ in (10.5.13) taken to be the actual loop noise bandwidth (see Problem 10.6 and the footnote associated with (10.3.13)).

When input intensities with frequency offsets are to be tracked the preceding results must be modified. Let the detected intensity be a sine wave as in (10.5.1) with a frequency $(\omega_s + \omega_d)$ and zero phase offset, and again let the VCO signal be given as in (10.5.2). The input intensity is therefore of the same form as (10.5.1) with $\theta_1 = \omega_d t$. Thus, in solving the system equation in (10.5.6), $d\theta_1/dt = \omega_d$, and the corresponding differential variation in (10.5.10) will have the term $\omega_d \, \Delta t$ added to it. Since the additional term is of order Δt, it will affect only the first coefficient in (10.5.11), which now becomes

$$C_1 = \omega_d - \tfrac{1}{2} G e G_L n_s \sin \phi \qquad (10.5.17)$$

The effect is to modify the steady state density equation in (10.5.12). In particular, the third-order truncation in (10.5.10) is now

$$0 \approx \frac{d}{d\phi} \{[\rho \sin \phi - \beta] p(\phi)\} + \frac{d^2}{d\phi^2} p(\phi) + \tilde{C}_3 \frac{d^3}{d\phi^3} [\sin \phi p(\phi)] \qquad (10.5.18)$$

where

$$\beta = \frac{\omega_d}{\tfrac{1}{2}(GeG_L)^2 [n_s + n_b + (N_{0c}/2(Ge)^2)]} = \frac{2\omega_d \rho}{B_L}$$

$$\tilde{C}_3 = \left(\frac{3}{2}\right)\left(\frac{1}{\rho}\right) \frac{n_s}{[n_s + n_b + (N_{0c}/2(Ge)^2)]}$$

Equation 10.5.18 was solved on a computer under quantum limited conditions $[\tilde{C}_3 = 3/2\rho, \rho = n_s/2B_L]$ and assuming $\beta = 0.7\rho$. The solutions are shown in Figure 10.21 for $\rho = 1.5$ and 3. Note that the solution densities are no longer symmetric about $\phi = 0$, but instead are offset due to the fre-

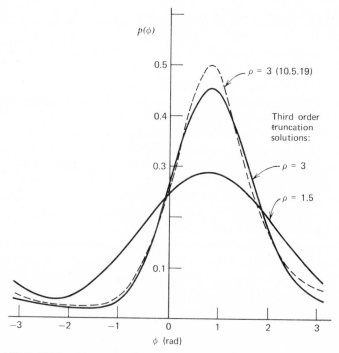

Figure 10.21. Phase error density due to frequency offset.

quency difference ω_d. This means that an average offset phase error is needed to obtain frequency locked conditions between the incoming frequency and the receiver VCO. For $\rho \geq 3$ the density $p(\phi)$ is well approximated by the solution with only the first two terms in (10.5.18). This can be solved exactly as

$$
p(\phi) = D_1 \exp(\rho \cos \phi + \beta\phi)\left[1 + D_2 \int_{-\pi}^{\phi} \exp(-\rho \cos x - \beta x)\, dx \right]
$$

$$(10.5.19)$$

for $|\phi| \leq \pi$, and D_1 and D_2 are evaluated from initial conditions. When the same initial conditions used in (10.5.15) are inserted, the resulting solution is included in Figure 10.21. For $\rho \gg 1$, the density is approximately Gaussian with variance $1/\rho$, and centered at the phase offset β/ρ (Problem 10.8). Note that this phase increases the mean squared tracking phase error, and therefore a larger value of ρ may be required in frequency locking in order to maintain a specified error requirement.

REFERENCES

[1] Viterbi, A., *Principles of Coherent Communications*, McGraw-Hill Book Co., New York, 1960.

[2] Stiffler, J., *Theory of Synchronous Communication*, Part 2, Prentice-Hall, Englewood Cliffs, New Jersey, 1971.

[3] Lindsey, W. C., *Synchronization Systems in Communication and Control*, Prentice-Hall, Englewood Cliffs, New Jersey, 1972.

[4] Lindsey, W. C. and Simon, M. K., *Telecommunication Systems Engineering*, Prentice-Hall, Englewood Cliffs, New Jersey, 1973.

[5] Middleton, D., *Introduction to Statistical Communication Theory*, McGraw-Hill Book Co., New York, Chap. 10.

[6] Ohlson, J. E., "Phase-Locked Loop Operation in the Presence of Impulsive and Gaussian Noise," *IEEE Trans. Commun. Tech.*, **COM-21**, 991–996 (September 1973).

[7] Snyder, D. L. and Rhodes, I. B., "Phase and Frequency Tracking Accuracy in Direct-Detection Optical Communication Systems," *IEEE Trans. Commun. Tech.*, **COM-20**, 1139–1142 (December 1972).

[8] Haney, G., "Phase Lock Tracking of Shot Noise Processes," Dissertation presented to the University of Southern California, January 1971.

[9] Gagliardi, R. and Haney, M., "Optical Synchronization—Phase Locking with Shot Noise Processes," USCEE Report 396, Electronic Sciences Dept., University of Southern California, August 1970.

PROBLEMS

1. (a) Determine the probability that a threshold acquisition test will terminate in n repetition periods of the marker pulse, given the parameters $K_s = 9$, $K_b = 1$, and PFA $= 10^{-3}$.

(b) Determine how large n should be in (a) for the test to terminate with probability 0.999 or better, given the same parameter values.

2. Using the algebraic identity

$$\sum_{j=1}^{M} x^j = x \frac{(x^M - 1)}{(x - 1)}$$

prove (10.2.4) and (10.2.5).

3. Use the equivalent linear loop model in Figure 10.18.

(a) Show that the transform of the phase error is related to that of the phase input by $F_{\phi}(s) = F_{\theta_1}(s)H(s)$, where $H(s)$ is given by

$$H(s) = \frac{G/s}{1 + (G/s)}$$

(b) Use (a) to determine the error response when $\theta_1(t) = \theta_0$.

(c) Repeat (b) when $\theta_1(t) = \omega_0 t$.

(d) Repeat (c) when $\theta_1(t) = a \sin \omega_0 t$.

4. Let $i(t)$ be a zero mean Gaussian white noise process [correlation given by $R_i(\tau) = (N_0/2)\delta(\tau)$], and let $y(t)$ be an independent process with correlation $R_y(\tau)$. Show that as $\Delta t \to 0$

(a) $E\left[\int_{t}^{t+\Delta t} i(z)y(z) \, dz \right]^2 = \Delta t \left(\frac{N_0}{2} \right) R_y(0)$

(b) $E\left[\int_{t}^{t+\Delta t} i(z)y(z) \, dz \right]^j = O(\Delta t^2), \quad j \geq 3$

Hint: Use the fact that for the zero mean white Gaussian process,

$$E \prod_{j=1}^{M} i(t_j) = 0, \quad M \text{ odd}$$

$$= \left(\frac{N_0}{2} \right)^2 \sum_{j} \sum_{q} \sum_{m} \sum_{n} \delta(t_j - t_q) \delta(t_m - t_n), \quad M \text{ even}$$

5. Starting from (10.5.10), derive the coefficients in (10.5.11), retracing the steps as in (10.3.18) to (10.3.27). Use the fact that $\omega_s T \gg 1$ to eliminate integrals involving frequency ω_s and its harmonics.

6. The linear loop in Figure 10.18 has a loop transfer function from loop input to VCO output given by

$$H(s) = \frac{G/s}{1 + (G/s)}$$

where G is the total loop gain. Show that

$$B_L = \frac{1}{2\pi} \int_0^\infty |H(\omega)|^2 \, d\omega = \frac{G}{4}$$

7. Consider the differential equation

$$0 = Q \sin(2\pi\tau)p(\tau) + \frac{dp(\tau)}{d\tau} + \sum_{i=2}^{N} a_i \frac{d^i p}{d\tau^i}$$

which is a truncated version of (10.5.12). Assume an even periodic solution of the form

$$p(\tau) = \sum_{i=0}^{\infty} b_i \cos(2\pi i\tau)$$

$$b_i = \int_0^T p(\tau) \cos(2\pi i\tau) \, d\tau$$

(a) Derive a recursive equation for the $\{b_i\}$ coefficients by substitution into the above and collecting harmonic terms.
(b) How many coefficients are needed to determine all the remaining ones? Relate to initial conditions.
(c) Assume the $\{b_i\}$ have all been found. Show that the mean squared value of τ can be obtained from these coefficients as

$$E[\tau^2] = \frac{b_0}{12} + \sum_{i=1}^{\infty} b_i \left[\frac{(-1)^i}{2\pi^2 i^2} \right]$$

8. (a) Show that the density in (10.5.15) approaches a Gaussian density as $\rho \to \infty$. Use the fact that

$$I_0(\rho) \xrightarrow[\rho \to \infty]{} \frac{1}{\sqrt{2\pi \rho}} e^\rho$$

and expand the cosine term, noting that $\phi \to 0$ as $\rho \to \infty$.
(b) Show that the same density approaches a uniform density as $\rho \to 0$.
(c) Show that under the conditions that $p(\pi) = -p(\pi)$ and $p(\phi)$ integrates to unity, (10.5.19) is approximately Gaussian with variance $1/\rho$ and mean β/ρ.

11 Pointing, Spatial Acquisition, and Tracking

Before any synchronization or data transmission can occur in a communication system it is of course necessary that the transmitter field power actually reach the receiver detector. This means that the transmitted field, in addition to having to overcome the attenuation effects of the propagation path, must also be properly aimed toward the receiver. Likewise, the receiver detector must determine the direction of arrival of the transmitted field. The operation of aiming a transmitter in the proper direction is referred to as pointing. The receiver operation of determining the direction of arrival of an impinging beam is called spatial acquisition. The subsequent operation of maintaining the pointing and acquisition throughout the communication time period is called spatial tracking. The problems associated with pointing, acquiring, and tracking a transmitted field become particularly acute when dealing with fields having extremely narrow beamwidths and long propagation distances. Since both these properties characterize long-range optical systems, these operations repesent an important aspect of the overall design considerations. In this chapter we discuss pointing, acquisition, and tracking as it applies to a long-range optical communication link, and present several procedures for system implementation. In a sense, it is paradoxical that these key operations, which must be performed successfully before any type of communication is to take place, comprise the last chapter of our study. However, much of the early discussion of system processing techniques is applicable to our development.

11.1 THE OPTICAL POINTING PROBLEM

Recall that in our discussion of optical antennas and beamwidths in Chapter 1, it was shown that a typical optical beam could be confined to an angular beamwidth of a fraction of an arcsecond. If this optical beam is to be detected at a receiver, then this beam must be pointed to within approximately one-half of this beamwidth. Alternatively, if the beam can be aimed toward a desired receiver (considered as point) with an accuracy of only, say, $\pm \psi_e$ rad, then the beamwidth must be at least $2\psi_e$ in order to ensure receiver reception, as shown in Figure 11.1. To emphasize this result numerically, suppose this beam is associated with a satellite-based transmitter aimed at the Earth from a 22,000 mile altitude. A 50 μrad pointing accuracy at this distance corresponds to a distance on Earth of $(50 \times 10^{-6})(22,000) \approx 1$ mile. That is, the center of the satellite beam must intersect the Earth within 1 mile of the receiver. Contrast this to an RF antenna having a typical beamwidth of about $10°$, and an Earth coverage of about 4000 miles. The RF pointing need only be to within 2000 miles, a sizable reduction in required accuracy.

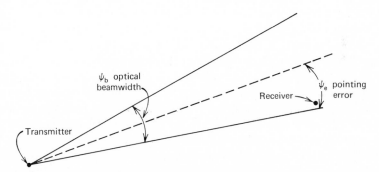

Figure 11.1. Beamwidths and pointing errors in transmitter-receiver systems.

Increasing the optical beamwidth to compensate for pointing errors causes a dilution of transmitter power for point receivers. If P_{rb} is the power collected at the receiver when the beamwidth is ψ_b rad, then compensating for a pointing error of $\pm \psi_e$ rad by increasing the beamwidth to $\psi_b + 2\psi_e$ rad results in a reduced receiver power of

$$P_r = P_{rb}\left(\frac{\psi_b}{\psi_b + 2\psi_e}\right)^2$$

$$= P_{rb}\left(\frac{1}{1 + (2\psi_e/\psi_b)}\right)^2 \tag{11.1.1}$$

Note that a significant power loss will occur when the pointing accuracy approaches or exceeds the optical beamwidth.

Inaccuracies in pointing an antenna beam in a specified direction over long distances are produced by several basic causes. The first major cause is the inability to determine exactly the desired direction. Uncertainty in true line of sight direction is caused by reference frame errors, and pointing can only be established to within the accuracy that a fundamental coordinate system can be established. Usually coordinate systems are oriented relative to a known star or celestial body, in which case it is important that compensation be made for reference motion. Generally this motion is not known exactly. In addition to real reference axis movement, often there is an apparent motion (e.g., the parallactic motion of a star, due to displacement of the Earth from one side of its orbit to the other). Errors in frame reference translate directly to line of sight errors in pointing.

A second major cause of pointing errors is the errors in the actual pointing apparatus. Often the antennas are pointed by means of electronic or mechanical interconnections operated from a remote sensor. Errors in these mechanisms, due to stress, noise, structure fabrication, and so on, will cause the beam to be pointed inaccurately. Such errors are called boresight errors.

A third error source is the inability to compensate exactly for transmitter and/or receiver motion. This motion may occur if either is in orbit, or may be due simply to the rotation of the Earth or hovering motion of a stationary satellite. When we attempt to predict this motion by use of system dynamical equations, coefficient errors lead directly to pointing errors. In addition, actual measurements of position and velocity, made through phase and Doppler frequency tracking operations, inherently contain errors due to system noise, as discussed in Chapter 10.

The Earth's atmosphere is a fourth cause of pointing errors, when dealing with Earth-based transmitters or receivers. Besides the attenuation (absorption), scattering, and cloud cover effects, pointing is drastically affected by beam wander and beam spreading due to turbulence and thermal gradients. (Recall our channel discussion in Chapter 1.) Beam wander is the bending of the light beam from its intended path during propagation, and obviously will directly affect pointing. Beam spreading causes increased beam divergence leading to dilution of available beam power, as in (11.1.1), and therefore has the effect of a pointing error. Since the atmosphere extends only a few hundred miles from the Earth, it appears as a near field effect to an Earth-based transmitter, while appearing as a far field effect to a deep space satellite transmitting to the Earth. Therefore beam wander and divergence are more severe for Earth transmission than Earth reception. Slight angular shifts induced by the atmosphere on Earth-transmitted optical beams project to large error distances after propagating over a long path to a deep space satellite. Uplink beam wander angles may typically vary from ± 1 to ± 15 μrad, but may be as large as 50 μrad for severe temperature gradients. Downlink beams passing through the atmosphere exhibit relatively little beam angular variations, and the predominate atmospheric effect is power absorption. The absence of atmosphere at both the transmitter and receiver (e.g., a satellite to satellite link) greatly reduces the pointing inaccuracies.

Transmitter pointing may be further hindered if the transmission distances are such that propagation transit time is significant and relative motion is involved. In this case the transmitter must actually point the optical beam ahead of the receiver in order to allow reception. That is, the transmitter must allow for the additional motion that occurs during the beam transit time, and point to the projected location. The problem of pointing under these conditions is called the point-ahead problem. Point-ahead is particularly important with Earth-based transmitters communicating with deep space vehicles in orbit. The point-ahead operation is accomplished by directing the transmitted beam toward the point in space where the receiver will be when the transmitted beam arrives. This requires transmitting at a lead angle with respect to the present line of sight position vector. Consider the geometry in Figure 11.2 showing relative motion between an Earth-based

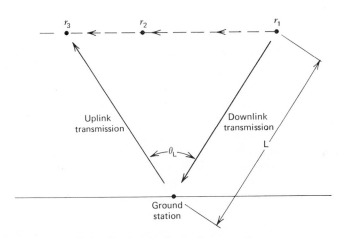

Figure 11.2. Point-ahead angles for moving transmitter.

and satellite link. The satellite transmits at point r_1 and, when received on Earth, defines a vector to the satellite position when it is transmitted. However, at time of reception the satellite has moved to point r_2. The Earth station in retransmitting, must compensate for the motion r_1 to r_2 and for the additional motion to point r_3, where the satellite will be when it receives. This angle from the Earth reception vector to the transmission vector defines the lead angle. Atmospheric effects, if present, cause the transmitter and satellite beams to be distorted and bent. Unfortunately, these effects are not mutually compensating, since one represents a near field and one a far field effect. Both must be properly accounted for if the point-ahead angle is to be exactly determined. Furthermore, since the bending phenomena are generally time dependent, their effects may have to be continually updated. A simplified expression for the point-ahead angle can be derived if it is assumed that the angle will be small. This is generally true as long as the satellite motion is slow with respect to the speed of light. Let L be the distance from transmitter to satellite and let τ_t be the round-trip transit time. Then $\tau_t = 2L/c$, where c is the speed of light. The distance the satellite moves along its orbit during τ_t is $\tau_t v$, where v is satellite tangential velocity. For the small angle assumption, the necessary lead angle θ_L is approximately

$$\theta_L \cong \frac{\tau_t v}{L} = \frac{2v}{c} \tag{11.1.2}$$

This angle must be adjusted for up- and downlink beam bending, if possible. Uncertainty in pointing effects due to other causes must be overcome by increasing beamwidth.

11.2 SPATIAL ACQUISITION

Spatial acquisition requires aiming the receiving antenna in the direction of the arriving optical field. That is, it must align the normal vector to the antenna surface area with the arrival angle of the optical beam. Often this alignment is acceptable to within some degree of accuracy. That is, the arrival angle can be within a specified solid angle from the normal vector. This acceptable angle is called the resolution angle (or resolution beamwidth) of the acquisition procedure, and is denoted Ω_r in subsequent discussion. The minimal resolution angle is obviously the diffraction limited field of view, but in practical design desired resolution is generally larger. This allows for the possibility of the source blurring into many modes, and for compensating for pointing errors and ambiguities. Although resolution angle must be considered a design specification, its value plays an important role in subsequent analysis.

Acquisition can be divided into one-way and two-way procedures. One-way acquisition is shown in Figure 11.3a. A single transmitter, located at one point, is to transmit to a single receiver located at another point. If satisfactory pointing has been achieved (or equivalently, if the transmitter beamwidth covers the pointing errors), the optical beam will illuminate the receiver point. The receiver knows the transmitter direction to within some uncertainty solid angle Ω_u, defined from the receiver location. The receiver would like to aim its antenna normal to the direction of the arriving field to within some preassigned resolution solid angle Ω_r, that is, it wants its antenna normal

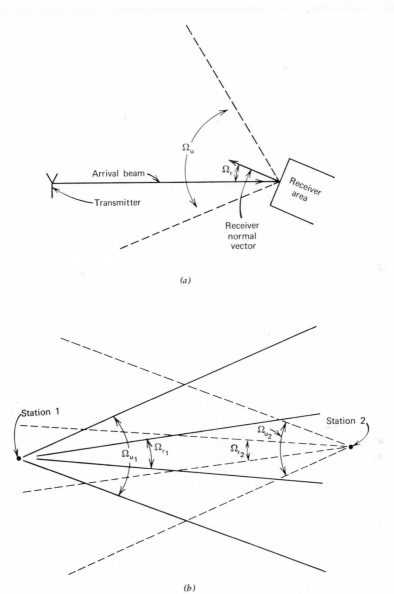

Figure 11.3. (a) One-way acquisition geometry. (b) Two-way acquisition geometry.

vector pointed to within Ω_r sr of the transmitter line of sight vector. In general, $\Omega_r \ll \Omega_u$, so that the receiver must perform an acquisition search over the uncertainty angle in order to acquire the transmitter with the desired resolution.

In two-way acquisition both communicating stations contain both a transmitter and a receiver, as shown in Figure 11.3b. Both must spatially acquire to form a two-way communication link. In typical situations one of the stations has somewhat accurate knowledge of the location of the other, and can therefore transmit a beam wide enough to cover its pointing errors. It uses a receiving antenna with a similar field of view aimed along the line of sight of the transmitted beam. The second station may not have the a priori knowledge for pointing, and must therefore search its uncertainty field of view Ω_{u_2} to acquire. After successful spatial acquisition with resolution Ω_{r_2} the second station transmits with beamwidth Ω_{r_2} to the first station, using the arrival direction obtained from the acquisition. The second station has now acquired and is pointed properly. The first station can now acquire the return beam with its desired resolution Ω_{r_1}. The link is now complete with the desired resolutions, and communication can begin. The operation can be repeated with narrower beams for further refinement, if desired. The first station would now narrow its transmit and receive beam, and the second station would reduce its resolution requirement Ω_{r_2}, reacquire, and retransmit with a narrower beam.

Its is evident that the key operation in either one- or two-way spatial acquisition is the search over the uncertainty angle Ω_u to determine arrival direction. Just as in time synchronization, spatial acquisition is desired in as short a time period as possible, while maintaining a suitable fidelity, that is, probability of successful acquisition. In the following we examine four common acquisition search procedures:

1. antenna scanning, in which the receiving system (antenna lens plus photodetector) is slewed over the uncertainty field of view to find the transmitted beam,

2. focal plane scanning, in which the antenna lens and receiver are rigid, with a wide field of view, and the focal plane is scanned to locate the beam,

3. focal plane arrays, in which an array of fixed detectors is used to cover the focal plane, and

4. sequential searching, in which a fixed detector array is used, and the field of view is readjusted in sequential steps to "zoom in" on the transmitter.

Antenna Scanning. Consider a system using a receiver lens and photodetector with a fixed field of view of Ω_r. The receiving system is swept over the uncertainty region Ω_u, as shown in Figure 11.4. For simplicity we consider only an azimuth (horizontal) search, but the results can be easily

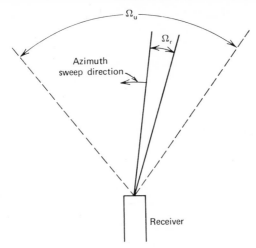

Figure 11.4. Antenna scanning: Ω_u = uncertainty field of view, Ω_r = resolution field of view.

extended to elevation searches as well. As the scanning is made the photo-detector output is continuously monitored until the beam is believed to have been observed. This decisioning is accomplished by a threshold test on the detector output signal. The operation can be modeled as follows. Consider the transmitted field to be a monochromatic point source beacon, transmitted continuously and producing an average count rate at the receiver of $n_s = \alpha P_r$, where P_r is the receiver optical signal power. We assume a multimode counting model with the background adding a noise count rate of n_b. If the transmitted field is in the receiver field of view for T sec, an average signal count of $K_s = n_s T$ is generated. The probability of correctly acquiring the beam is then

$$\text{PAC} = \frac{\Gamma(k_T, K_s + K_b)}{\Gamma(k_T, \infty)} \tag{11.2.1}$$

where $K_b = n_b T$, k_T is the threshold, and $\Gamma(a, b)$ is the gamma function, defined in (7.4.6) as

$$\Gamma(a, b) = \int_0^b e^{-t} t^{a-1} \, dt \tag{11.2.2}$$

The threshold is selected so that the probability of false acquisition PFA is a desired value, where

$$\text{PFA} = \frac{\Gamma(k_T, K_b)}{\Gamma(k_T, \infty)} \tag{11.2.3}$$

Note that the probabilities are identical to those associated with time acquisition in Section 10.2. The signal and noise counts are given, respectively, by

$$K_s = n_s T$$

$$K_b = n_b T = n_{b0} D_{sr} T$$

$$(11.2.4)$$

where n_{b0} is the noise count rate per spatial mode, and D_{sr} is the number of spatial modes in the resolution field of view Ω_r. We can relate T, the time the transmitter is in the field of view, to the azimuth angular slewing rate, S_L rad/sec, by the approximation

$$T \approx \frac{\sqrt{\Omega_r}}{S_L} \sec \qquad (11.2.5)$$

Equations 11.2.1 to 11.2.5 define the relation among the key system parameters. Performance probabilities determine design values of K_s and K_b, from which detector field of view and slewing rate can be determined. Note that both noise and signal counts increase with T but only noise count increases with Ω_r. This implies Ω_r should be as small as possible. Theoretically, Ω_r need be only wide enough to cover a single spatial mode (diffraction limited operation) but propagation turbulence and diffraction blur the source, and larger resolution angles are generally needed to cover the transmitter uncertainty. We also see that reducing Ω_r requires a slower scan rate to generate the desired K_s, which lengthens the acquisition time.

The average acquisition time can be estimated by using a discrete version of the search model in Figure 11.4. We divide the uncertainty region Ω_u into Q disjoint subregions of angle Ω_r, where

$$Q = \frac{\Omega_u}{\Omega_r} \qquad (11.2.6)$$

We assume the receiver spends T sec in each subregion position, and searches over Ω_u until the field is detected. The probabilities at each position are given by (11.2.1) or (11.2.3), depending on whether or not the transmitted beam is present in that position. When formulated in this way the problem is identical to the time synchronization formation in Section 10.2 with spatial solid angles taking the place of time pulse intervals. Our earlier equations (10.2.4) and (10.2.5) can now be used for determining average search time, with the proper change of parameter values. We omit repeating these equations here.

Focal Plane Scanning. In focal plane scanning the search mechanism is different, but the overall effect is identical to antenna scanning. A fixed optical lens transfers the uncertainty field of view to the focal plane, and the searching is accomplished by a single detector scanned over the focal plane. Such an operation can be readily performed by an image dissector system. Focal plane scanning is generally preferable over antenna scanning, since the receiving structure can be made rigid, and no mechanically movable parts are needed. System analysis is identical to that of an antenna scanner, and therefore its performance is again described by (11.2.1) to (11.2.5).

Both focal plane and antenna scanning allow for raster storage in which comparison can be made after the complete focal field has been scanned. This would reduce the acquisition search procedure to a count comparison rather than a threshold test. That is, the scanner would scan the complete field of view while storing the collected signal. Decisioning is postponed until the end of the scan, at which time the most likely position is selected as the beam arrival angle. If the discretized spatial model is used, the test appears as a comparison test over Q spatial cells to determine the maximum count. The acquisition probability is given by a Q-ary orthogonal count test identical to the PPM test in (8.2.1) or the pulse synchronization test in (10.2.7). The parameter Q in (11.2.6) is directly related to the desired resolution angle used for scanning. The probability of a successful acquisition in a single focal plane scan is therefore[†]

$$\text{PAC}_1 = \sum_{k_1=0}^{\infty} \text{Pos}(k_1, K_s + K_b) \left[\sum_{k_2=0}^{k_1-1} \text{Pos}(k_2, K_b) \right]^{Q-1} \qquad (11.2.7)$$

where K_s and K_b are given in (11.2.4). It is convenient to replace D_{sr} by D_{su}/Q and define n_{bu} as the noise count rate over the total uncertainty solid angle Ω_u. (For focal plane scanning, n_{bu} is the total noise over the receiver field of view.) Thus we write

$$K_s = n_s T$$

$$K_b = \frac{n_{bu} T}{Q} \qquad (11.2.8)$$

in (11.2.7). In addition, we denote the total focal plane (uncertainty region) search time by T_t, which is related to T, the time spent observing each resolution position, by

$$T_t = QT \qquad (11.2.9)$$

For a specified Q (i.e., a specified uncertainty region and desired resolution) and fixed power levels of the transmitter and background, the key system parameters, acquisition probability PAC_1, and total search time T_t are directly related to the observation time T. Figure 11.5 is a plot of PAC_1 in

[†] We omit likelihood equalities here in our probabilities, contending that an acquisition operation would rescan rather than guess among possible directions. Thus equalities among maximum counts are accepted as an incorrect acquisition.

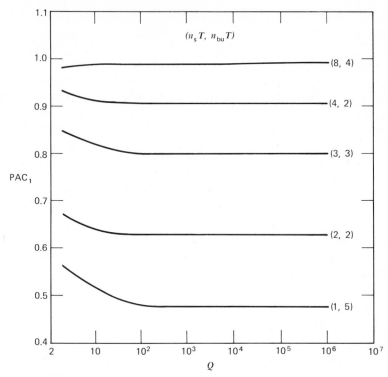

Figure 11.5. Scanning acquisition probabilities: Q = number of uncertainty cells. n_s = signal count rate n_{bu} = background noise count rate.

(11.2.7) for several values of $(n_s T, n_{bu} T)$ as a function of Q. At a particular value of Q, we see that PAC_1 depends quite strongly on the signal and noise energies. For fixed power levels, this means that acquisition probability is critically related to the observation time T. This dependence can be further exhibited as shown in Figure 11.6. where PAC_1 is plotted versus T for several values of Q at a fixed power level. The results indicate a sudden drop in acquisition probability if T is not long enough. This result again stresses the importance of collected signal energy in optical detection problems. Surprisingly, the resultant is somewhat independent of Q when Q is large (typically $Q \geq 10^2$). Thus large Q systems require abnormally long search times to maintain a desired acquisition probability. This is illustrated in Figure 11.7, in which Figure 11.6 is replotted, using (11.2.9). The curves manifest the time needed to perform a complete uncertainty region search for several values of Q. We immediately see a direct tradeoff in search time (T_t) and system resolution (Q) in attaining a prescribed PAC_1.

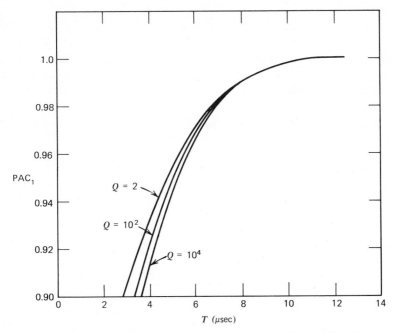

Figure 11.6. Scanning acquisition probabilities: T = time per search position, Q = number of search positions, $n_s = 2 \times 10^6$, $n_{bu} = 2 \times 10^6$.

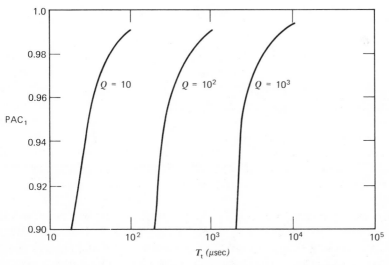

Figure 11.7. Total search time to acquire Q search positions, $n_{bu} = 2 \times 10^6$, $n_s = 2 \times 10^6$.

Focal Plane Arrays. The use of detector arrays in the focal plane allows for parallel processing to be achieved, and therefore reduces the acquisition times. Each detector of the array would examine a certain portion of the uncertainty region, as shown in Figure 11.8. Collection of the individual detector outputs after a fixed observation time permits a count comparison test for transmitter location. The detector with the largest count is considered to be viewing the received beam. Each detector must operate independently and its count must be properly cataloged, thereby increasing the receiver

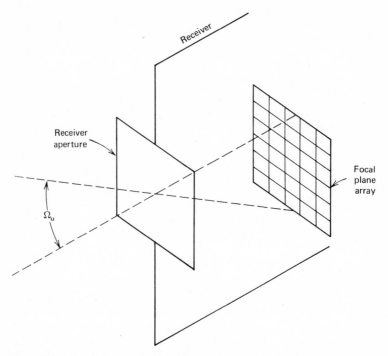

Figure 11.8. Focal plane array for focal scanning.

complexity. However, the acquisition time is reduced to a single observation time T, as opposed to (11.2.9), since the counting is done in parallel. We see therefore that detector arrays can be a powerful tool in fast acquisition applications.

Consider the case of an array having S detectors, so that the uncertainty angle can be subdivided into Ω_u/S resolution areas. It is obvious that S must be extremely large in order to obtain resolution cells on the order of those in the scanning methods. The focal plane is therefore divided into S areas, and the acquisition probability is identical to (11.2.7) with K_s and K_b defined in (11.2.8). However, we now have

$$\Omega_r = \frac{\Omega_u}{S}$$

$$T_t = T \tag{11.2.10}$$

The resulting single scan acquisition probability, PAC_1, can therefore be derived directly from curves similar to Figure 11.7 with Q replaced by S. Since search time is now essentially independent of array size S, it is desirable to use as large an array as possible for best resolution. The tradeoff is directly in terms of receiver complexity, since the array elements must be processed in parallel. Thus, while acquisition by scanning trades off acquisition resolution for search time, acquisition with focal plane arrays trades off resolution for receiver complexity.

Sequential Search with Arrays (Parallel Processing). The difficulty with a single scan array search is that the achieved resolution (11.2.10) is generally larger than that desired, unless S is quite large. This can be improved, however, by repeating the array search until the desired resolution is achieved. Carrying out a sequence of searches with the same array is called *sequential acquisition*. A sequential spatial search is obtained by using a fixed detector array, and continually adjusting the field of view so as to home in on the beam with the desired resolution. At each step of the search, the field of view is divided into sectors by the array, and a parallel decision is made as to which sector is observing the beam. The field of view is then reduced (magnification increased) to the decided sector, and the parallel array decisioning is repeated. The final resolution achieved will therefore depend on the number of times the test is repeated, but the overall operation is restricted by the allowable range of magnification that the receiving lens system will allow, that is, the "zoom" range of the receiver.

Consider a sequential search in which S detectors are used in the array, and the initial uncertainty is Ω_u. At the first step the field of view is divided into S cells of resolution angle Ω_u/S and a S-ary decision is made after T sec. with signal count energy K_s and noise count energy $K_b = n_{bu}T/S$. After the decision is made, the receiver field of view is readjusted to encompass the decided sector only. The test is repeated, with the field of view again divided into S sectors, this time with resolution angle Ω_u/S^2. At the ith step, the resolution angle is Ω_u/S^i, and the number of steps, r, needed to reduce the original uncertainty Ω_u to the desired resolution Ω_r is that for which $\Omega_r = \Omega_u/S^r$ or,

$$r = \frac{\log(\Omega_u/\Omega_r)}{\log S} \tag{11.2.11}$$

The probability of detecting the correct sector at the ith step is then

$$\text{PAC}_i = \sum_{k_1=0}^{\infty} \text{Pos}\left(k_1, n_s T_i + \frac{n_{\text{bu}} T_i}{S^i}\right)\left[\sum_{k_2=0}^{k_1-1} \text{Pos}\left(k_2, \frac{n_{\text{bu}} T_i}{S^i}\right)\right]^{S-1} \quad (11.2.12)$$

where T_i is the observation time of the ith decision. The acquisition probability is then the probability of correct detection at each step, so that

$$\text{PAC} = \prod_{i=1}^{r} \text{PAC}_i \quad (11.2.13)$$

The total acquisition time T_t is the time to perform all r tests. Thus, neglecting the time to adjust the magnification, the parallel processing requires a total time

$$T_t = \sum_{i=1}^{r} T_i \quad (11.2.14)$$

Equations 11.2.11 to 11.2.14 describe the performance parameters of sequential acquisition with parallel processing. For convenience, the operation is generally carried out as either a constant detection probability test, in which PAC_i is adjusted to be equal at each i, or as a constant observation time test, in which T_i is the same at each i. Since the field of view, and therefore the noise count, is reduced at each step, a constant probability test uses less time at each i. Conversely, a constant observation time will operate with improved acquisition probability at each step.

Figure 11.9 summarizes the results for a fixed time sequential test. For a fixed S, $T_i = T$ is adjusted for a desired observation time, and the resulting overall PAC in (11.2.13) is computed for a sequential test. Figure 11.9 shows a typical plot of PAC versus the required time to perform all r tests for several array sizes, S. The result is cross plotted in Figure 11.10, relating array size to search time for a fixed PAC. Note that search time is uniformly decreased as the array size is increased, while achieving the given PAC. However, the improvement attained may not be significant for the larger array sizes. That is, in Figure 11.10, it may not be necessary to consider arrays larger than, say, 100 elements. Note also the significant decrease in total search time T_t for a sequential test, when contrasted with the results of focal plane scanning in Figure 11.8. This can be directly related to the reduction in the number of tests that must be performed. Scanning requires a search time directly proportional to the number of resolution cells that must be examined (Equation 11.2.9) while sequential searching is proportional to the log of this number (Equation 11.2.11).

In a fixed probability sequential search, the acquisition probability at each step is adjusted to a desired value. Thus, if PAC is the desired overall

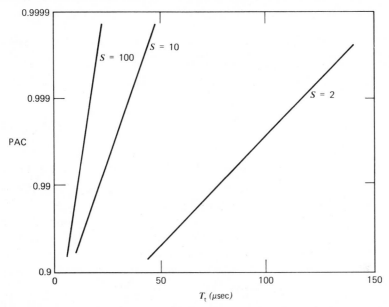

Figure 11.9. Total search time to acquire with a S-detector array and sequential acquisition (fixed time per test, $n_s = 2 \times 10^6$, $n_{bu} = 10^6$).

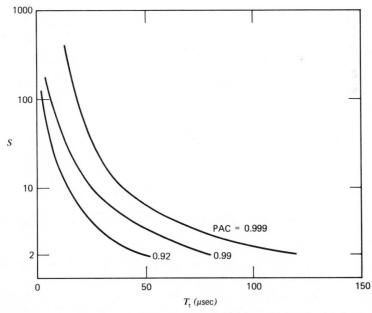

Figure 11.10. Array size versus total search time, sequential acquisition (fixed time per test).

acquisition probability, and if r steps will be performed in the test, then we select

$$PAC_i = (PAC)^{1/r} \qquad \text{for each} \quad i \qquad (11.2.15)$$

and adjust each T_i to achieve this in (11.2.12). The total search time is then found from (11.2.14). As an example, consider the case where

$$n_s = 2 \times 10^6$$
$$n_{bu} = 10^7$$
$$S = 100$$
$$r = 3$$
$$PAC = 0.97 \approx (0.99)^3 \qquad (11.2.16)$$

The curves in Figure 11.11 show plots of PAC_i versus observation time, T_i, in microseconds for these operating parameters. At the first test, $PAC_1 = 0.99$ and $n_{bu} = 10^7$, requiring an observation of about 5.3 μsec. At the second iteration the field of view is reduced and the noise count rate per uncertainty angle is now $n_{bu}/S = 10^5$. We now require an observation time of 3 μsec, as seen from Figure 11.11. The next iteration reduces the noise rate to 10^3, and 2.8 μsec is required. The total search time is then

$$T_t = (5.3 + 3.0 + 2.8)\,\mu\text{sec} = 11.1\,\mu\text{sec} \qquad (11.2.17)$$

The result is about comparable to the corresponding search time in Figure 11.10 for a fixed time sequential test with the same number of detectors, and is several orders of magnitude shorter than a single detector search (Figure 11.7).

Sequential Search with Fixed Arrays (Serial Processing). In the foregoing fixed array test, parallel processing was assumed. That is, all the detectors of the array are examined simultaneously for their outputs, from which a maximum is selected. The field of view corresponding to this maximum is then selected as containing the transmitter. The test is then repeated over this latter field of view. However, parallel processing over large numbers of detectors may become physically difficult. An alternative is to serially scan the array for the maximal output at each iteration rather than parallel examination. The receiver complexity is now eased at the expense of a slightly longer search time. For a fixed observation time of T sec, an array with S detectors now requires ST sec to perform a single iteration, with r separate iterations needed to complete the search. The total search time is now, from (11.2.11),

$$T_t = rST = \left(\frac{S}{\log S}\right)T \log\!\left(\frac{\Omega_u}{\Omega_r}\right) \qquad (11.2.18)$$

For a fixed T (specified PAC) and fixed field of view Ω_u and resolution Ω_r, it is interesting to note that T_t is minimized for the value S such that $dT_t/dS = 0$,

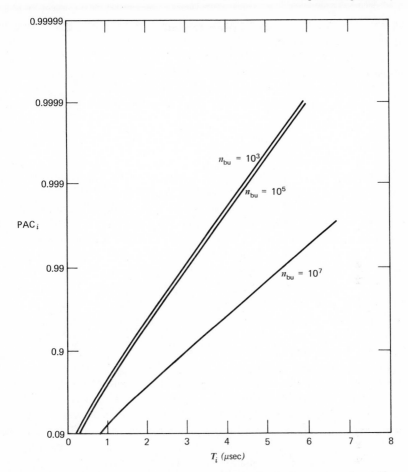

Figure 11.11. Acquisition probability versus search time for single scan. $n_s = 2 \times 10^6$.

which occurs for $\log_e S = 1$, or $S = e$. For square arrays, S must be a squared integer. Hence search time is minimized for $S = 4$ (the nearest square to e) corresponding to a two by two array. The sequential test is then one that divides the field of view in quadrature, serially examines each quadrant for that most likely to contain the transmitter, and then repeats for the decided quadrant. Note that with serial processing the design objective has completely reversed to the use of small detector arrays.

In discussing sequential acquisition we have implicitly required a change of magnification (field of view) at each step. It is important to recognize that time is required to change lens power. In addition, there is a maximum magnification range, $\mathcal{M}$, over which we can expect a telescope system to operate. At the minimum power (first step of the acquisition search) we would cover a field of $D_s = \Omega_u/\Omega_{dL}$ modes, or a solid angle of Ω_u sr. At the maximum power (last step) we cover a field of S modes. Therefore the power change must occur in $\log_S (D_s) - 1$ steps, and we require that at each step the power change by the amount S. We then see that the magnification $\mathcal{M}$ is given by†

$$\mathcal{M} = \left[\frac{D_s}{S}\right]^{1/2} \tag{11.2.19}$$

For example, if $D_s = 10^5$, then a magnification of $\mathcal{M} = 31.6$ is required with an array of 100 elements. However, decreasing the array to 10 elements will require an increased $\mathcal{M}$ of about 100. If we try to change the magnification mechanically, we can almost guarantee that the required time for adjustment will be much greater than the scan times involved. Thus devices or techniques that can significantly reduce magnification adjustment time would be extremely beneficial to sequential acquisition operations in optical search systems. A possible technique might be the use of electronic magnification using larger arrays. In such systems, we use a large array to cover the desired field of view, but we electronically regroup the detectors to effectively change the field of view. In this way we avoid the difficulty associated with the vast amount of components needed to parallel process a large detector array. For the sequential application here, we would only use S processing components at each step, but the processors could be sequentially reconnected to different array elements with proper switching networks, to accomplish the effective change in field of view. It would, of course, still be necessary for each of the array elements to have sufficient gain if quantum limited performance were desired. In general, several orders of magnitude in search time would be lost if the individual array elements had insufficient internal gain.

In conclusion, then, we see that the operation of spatial acquisition involves a tradeoff of system design objectives. Once we specify a receiver aperture (diffraction limited field of view Ω_{dL}) and an uncertainty angle (Ω_u), the total spatial dimension, $D_s = \Omega_u/\Omega_{dL}$, of the search is specified. From our knowledge of orthogonal mode detection, we know that the theoretically best performance, in terms of shortest acquisition time and minimum resolution for a fixed acquisition probability, is obtained when each of the D_s modes is searched in parallel. Any effort to decrease the total number of modes decreases the system resolution. Any effort to decrease the number of modes observed at one time increases acquisition time. From an engineering point of view it may in fact be much more convenient to implement lower

---

† Magnification is usually defined as a ratio of beamwidths. When solid angles are used, a square root is needed in its definition.

resolution or longer search time systems. It must be remembered that the definition of a long acquisition time depends on the system mission. For example, short acquisition time for a communication system that will operate for several hours is not as critical as rapid acquisition of a reentering vehicle or missile. We would of course always desire D_s to be as small as possible by reducing a priori uncertainty and building better components. Reduction of background noise energy during the observation time of any mode requires directly a reduction of optical bandwidth. Since the modulation frequencies will generally be low, we are confined to the achievable minimal optical bandwidth in determining noise per mode. To observe D_s modes simultaneously requires a very large array. The alternative sequential procedures would use as large an array as economically possible, and implement the sequential search algorithms. The resulting performance, while suboptimal, is nevertheless quite good and would probably satisfy most applications.

11.3 SPATIAL TRACKING

After pointing and spatial acquisition have been achieved, there remains the task of maintaining the transmitted beam on the detector area in spite of beam wander or relative transmitter-receiver motion. The operation of keeping the receiving antenna properly oriented relative to the arriving optical signal field is called spatial tracking. Spatial tracking is governed by the same principles as those in pulse tracking described in time synchronization. An optical sensor is used to generate an error voltage relative to any offset alignment error between receiver and beam arrival that may occur. This error voltage is then used to realign the axis of the receiver lens (and the transmitting lens if two-way communication is being used). The system is complicated by the fact that errors are needed to control both azimuth and elevation alignment and two separate tracking operations are needed for each. System considerations and several design implementations are discussed in References 1 to 4.

A typical spatial tracking system is shown in Figure 11.12. The tracking error is determined instantaneously for both azimuth and elevation coordinates by means of the position error sensor. The error voltage of the space coordinates is then used to control the alignment axis of the receiver lens. This is accomplished by some type of control loop dynamics, generally with separate servo loops for individual control of azimuth and elevation axis. The loop control functions are typically of the form of some type of low pass filter that smooths the error signal for proper control. The loop bandwidths must be wide enough to allow the tracker to follow the expected beam motion, yet allowing minimal noise effects within the loop. Note that the optical error sensor, and therefore its characteristics, is an integral part of the spatial tracking loop.

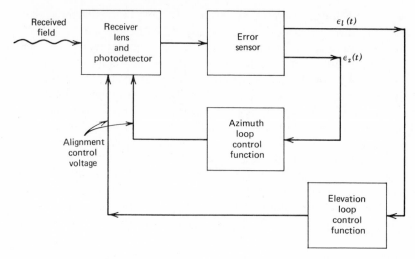

Figure 11.12. Spatial tracking subsystem.

Let (θ_z, θ_l) be the azimuth and elevation angle of the line of sight vector to the transmitter point, with respect to a selected coordinate system, as shown in Figure 11.13. Let (ϕ_z, ϕ_l) be the corresponding angles of the normal vector to the receiver area. Then the instantaneous angular errors in pointing the receiver to the transmitter are given by

$$\psi_z(t) = \theta_z - \phi_z \qquad (11.3.1a)$$

$$\psi_l(t) = \theta_l - \phi_l \qquad (11.3.1b)$$

where the dependence on t emphasizes the changing of the errors in time. Let $\varepsilon_z(t)$, $\varepsilon_l(t)$ be the error voltage generated from the optical sensor for control of the azimuth and elevation angles, respectively. These error signals are used to correct the pointing of the (ϕ_z, ϕ_l) angles. Hence,

$$\phi_z = \overline{\varepsilon_z(t)} \qquad (11.3.2a)$$

$$\phi_l = \overline{\varepsilon_l(t)} \qquad (11.3.2b)$$

where the overbar denotes the effect of the loop filtering. Combining (11.3.1) and (11.3.2) yields the pair of differential equations

$$\frac{d\psi_z}{dt} = \frac{d\theta_z}{dt} - \frac{d}{dt}\left[\overline{\varepsilon_z(t)}\right] \qquad (11.3.3a)$$

$$\frac{d\psi_l}{dt} = \frac{d\theta_l}{dt} - \frac{d}{dt}\left[\overline{\varepsilon_l(t)}\right] \qquad (11.3.3b)$$

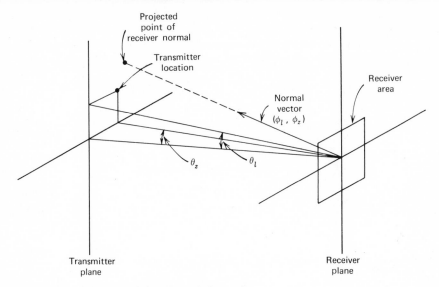

Figure 11.13. Azimuth and elevation pointing angles.

The equations represent a pair of coupled differential equations for the angular pointing errors. The terms $(d\theta_z/dt, d\theta_l/dt)$ represent the movement of the line of sight vector to the transmitter and therefore appear as forcing functions in (11.3.3). The specific form of the equations depends on the relation of the error voltage to the position error (which is specified by the properties of the optical sensor) and the type of filtering within the loop. Note that the form of the equations suggests the equivalent system in Figure 11.14. Here the position angles are the variables of the equivalent system, which involves coupled feedback loops driven by the movement of the transmitter vector. The loop filtering in the equivalent system is identical to that of the actual tracking loop. Note the optical sensor is an integral part of the equivalent loop as well as the actual loop.

The optical sensor is used to generate the error control signal from the received optical beam. The most common type of optical sensor is the quadrant detector, in which four separate photodetectors are used to determine position error. The received optical field is focused onto the center of the detectors placed in a quadrature arrangement in the focal plane, as shown in Figure 11.15a. Error voltages are generated by properly comparing the detector outputs. When the received plane wave field arrives normal to receiver lens it is focused exactly to the center of the quadrant, and theoretically all detectors will receive equal field energy. Offsets in arrival angle cause

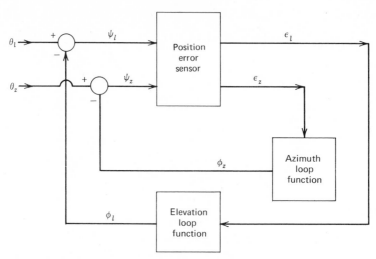

Figure 11.14. Equivalent azimuth-elevation tracking loops.

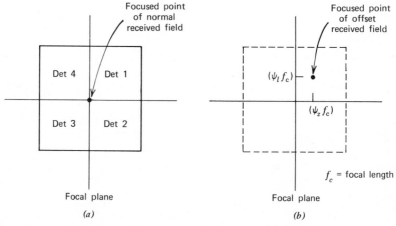

Figure 11.15. The quadrant error sensor. (*a*) Normal arriving field. (*b*) Off angle arriving field.

imbalance in detector energies, which can then be used to generate correcting voltages. An error signal in azimuth and elevation is derived by combining two detector outputs and comparing. Let us represent the individual detector shot noise current outputs as $x_i(t)$, so that the error signals are

$$\varepsilon_z(t) = [x_1(t) + x_2(t)] - [x_3(t) + x_4(t)] \qquad (11.3.4a)$$

$$\varepsilon_l(t) = [x_1(t) + x_4(t)] - [x_2(t) + x_3(t)] \qquad (11.3.4b)$$

In multimode operation, the detector outputs are shot noise processes governed by Poisson counting, whose average values are related to the average intensity collected over their surfaces. Let the received transmitter field be a monochromatic field arriving normal to the receiver lens. Let $I_s(t, \mathbf{r})$ be the intensity of the diffracted signal field in the focal plane. Let the background add a constant intensity I_b over the focal plane. The mean error (11.3.4a) due to the reception of the transmitted and background fields is then

$$E[\varepsilon_z(t)] = E[x_1(t) + x_2(t)] - E[x_3(t) + x_4(t)]$$

$$= \alpha G_d e \int_{\mathscr{A}_1 + \mathscr{A}_2} [I_s(t, \mathbf{r}) + I_b]\, d\mathbf{r} - \alpha G_d e \int_{\mathscr{A}_3 + \mathscr{A}_4} [I_s(t, \mathbf{r}) + I_b]\, d\mathbf{r}$$

$$(11.3.5)$$

where $\mathscr{A}_i$ is the area of the ith detector and G_d is detector gain, assumed equal for each detector. Note that if the detectors are perfectly balanced (equal gain and areas), the background effect cancels out, and the mean error function depends only on the integrated signal intensity. For a circular lens of diameter d, we can approximate the Airy disc in the receiver focal plane by the intensity function

$$I_s(t, \mathbf{r}) = I_s, \qquad |\mathbf{r}| \leq \frac{2\lambda f_c}{\pi d} \triangleq r_0$$

$$= 0 \qquad \text{elsewhere} \qquad (11.3.6)$$

where λ is the field wavelength and f_c is the receiver focal length. If the field arrives exactly normal, the diffraction pattern is centered equally on all four detectors and (11.3.5) produces a zero error. If the signal field arrives at an offset angle (ψ_z, ψ_l) in azimuth and elevation, the intensity pattern in (11.3.6) is centered at the points $(f_c\psi_z, f_c\psi_l)$ in the azimuth and elevation coordinates of the focal plane, as shown in Figure 11.15b. The integrals in (11.3.5) therefore correspond to areas under sectors of an offset circle in the focal plane. For small displacements (ψ_z, ψ_l) such that the intensity pattern is still encompassed by the quadrant detector area, (11.3.5) integrates to (Problem 11.4)

$$E[\varepsilon_z(t)] = \alpha G_d e I_s (\pi r_0^2) \left[1 - \frac{2}{\pi} \cos^{-1}\left(\frac{\psi_z f_c}{r_0}\right) + \frac{2\psi_z f_c}{\pi r_0}\left(1 - \left(\frac{\psi_z f_c}{r_0}\right)^2\right)^{1/2} \right]$$

$$(11.3.7)$$

Note that (11.3.7) does not depend on ψ_l, and the mean azimuth error signal is generated only from the azimuth angle error. Similarly, the mean elevation signal depends only on elevation angle errors, as in (11.3.7), with ψ_z replaced by ψ_l. Thus the quadrant detector "uncouples" the tracking operation by generating independent error signals. This means each tracking loop in

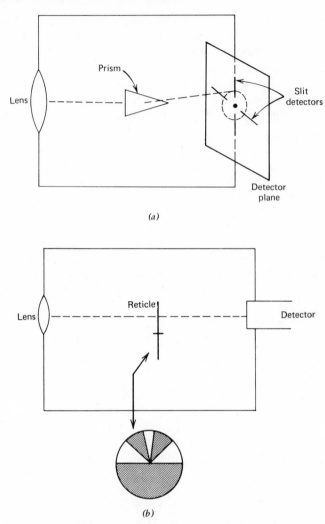

(a)

(b)

Figure 11.16. Optical position error sensors. (a) Rotating (lobing) prism. (b) Rotating reticle.

Figure 11.14 can be considered an independent loop in its tracking behavior. Furthermore, we see that if $\psi_z \ll 1$, (11.3.7) reduces to

$$E[\varepsilon_z(t)] \cong (\alpha G_d e P_r)\left(\frac{\psi_z}{\lambda/d}\right) \qquad (11.3.8)$$

where P_r is the total beam power received. We see therefore that the mean error is linearly proportional to the azimuth angle error. We emphasize that the actual error being generated is random in nature because of the inherent randomness of the photodetector counts, and the tracking errors in azimuth and elevation actually evolve as random processes. Note that if the pointing errors (ψ_z, ψ_l) are zero, (11.3.7) produces a zero error voltage, provided that the quadrant detectors are identical. That is, they must have identical gains and produce identical background and dark current electron average rates. The ability to exactly balance four separate photodetectors is a major problem in quadrant detector design (Problem 11.6).

Other devices are available for generating error signals. A *lobing* detector (Figure 11.16a) rotates a prism in the focal plane so as to cause the diffraction pattern in a plane behind it to mutate, or enscribe a lobe, in the detector surface. Slit detectors symmetrically spaced in the plane will produce pulsed outputs as the pattern rotates past it. If the center of the lobe is directly centered at the center point of the slits, the pulses between detector outputs are equally spaced in time. As the lobe center is offset, the time between pulses becomes asymmetrical. By monitoring the pulse locations (i.e., pulse position demodulating) error voltages can be generated (Problem 11.7). Another device is the rotating reticle (Figure 11.16b) in which a shaped surface with various transparent and opaque areas is rotated in front of the detector in the focal plane. The recticle "chops up" in time the diffraction pattern, as seen by the detector. As the field moves in the focal plane due to offset angles, the intensity modulation imposed changes in form (Problem 11.8). A demodulator following the detector can be used to generate error signals from the detected intensity modulation.

11.4 SPATIAL TRACKING ERRORS

The dynamical behavior of the spatial error during spatial tracking can be examined by the same analysis procedures as in pulse tracking in Chapter 10. In the general approach, the spatial angular errors in azimuth and elevation (ψ_z, ψ_l) must be considered as an error vector, with the set of coupled equations in (11.3.3) describing their dynamical behavior. The resulting spatial error vector then has a joint probability density that changes in time as the error processes evolve. The steady state error vector density can be derived as a solution to a partial differential equation, the vector counterpart to the scalar Kolmogorov–Smoluchoski equation in (10.3.14). The vector equation requires the generation of equation coefficients using joint averages of the differential variation of the error vector [5, Chap. 7]. We omit pursuing the general case here. Instead we concentrate on a more practical situation in which some degree of mathematical simplification occurs. In particular, we assume that an ideal, perfectly matched quadrant detector is used for error generation, so that azimuth and elevation angular errors uncouple, and each pointing error can be determined separately. Furthermore, the individual error equations become scalar equations, and each may be examined independently of the other.

Let us consider the azimuth angle ψ (we drop the subscript for convenience) and model the loop filter of the azimuth tracking loop as a pure integrator (essentially a low pass filter with bandwidth much less than the detector output bandwidth). We wish to determine the statistical behavior of the azimuth error when tracking a stationary (or slowly moving) transmitter beam in the presence of background interference. The background is modeled as contributing a fixed electron rate n_b to each of the individual quadrant detectors. This background rate is independent of the orientation of the receiver. Since the error evolves as a random process in time we again seek its steady state density to describe its statistical behavior. This density can be determined, theoretically, by application of the Kolmogorov theory of Section 10.3. The system differential equation in (11.3.3) under the given assumptions, becomes

$$\frac{d\psi}{dt} = -G\varepsilon(t, \psi) \tag{11.4.1}$$

where G is the integrator gain. The differential angular change in a time interval Δt sec is then

$$\Delta\psi = -G \int_t^{t+\Delta t} \varepsilon(\rho, \psi)\, d\rho$$

$$= -G_d eG[k_{12}(t, t + \Delta t) - k_{34}(t, t + \Delta t)] \tag{11.4.2}$$

where $k_{ij}(t, t + \Delta t)$ is the combined count of detectors i and j over the interval $(t, t + \Delta t)$. The steady state density is obtained by solving the differential equation

$$0 = \sum_{j=1}^{\infty} \frac{(-1)^j}{j!} \frac{\partial^{j-1}}{\partial\psi^{j-1}} [C_j(\psi)p(\psi)] \tag{11.4.3}$$

where

$$C_j(\psi) = \lim_{\Delta t \to 0} \frac{E[(\Delta\psi)^j]}{\Delta t} \tag{11.4.4}$$

with suitable initial conditions. The Kolmogorov coefficients in (11.4.4) are obtained using (11.4.2). The first coefficient C_1 becomes

$$C_1 = \lim_{\Delta t \to 0} \frac{E[\Delta\psi]}{\Delta t} \tag{11.4.5}$$

However, we note that

$$E[\Delta\psi] = -G \int_t^{t+\Delta t} E[\varepsilon(\rho, \psi)] \, d\rho \qquad (11.4.6)$$

For small spatial tracking errors (linear loop operation) we use (11.3.8) to write

$$E[\Delta\psi] = -G \int_t^{t+\Delta t} (G_d e n_s)\psi(\rho) \, d\rho$$

$$= -(GG_d e n_s d/\lambda)\psi(t) \, \Delta t \qquad (11.4.7)$$

for $\Delta t \to 0$. Thus in (11.4.5)

$$C_1 = -(GG_d e n_s d/\lambda)\psi \qquad (11.4.8)$$

where ψ is understood to be $\psi(t)$. To determine higher coefficients we use the fact that detector counts from different spatial areas are statistically independent. Hence,

$$E[\Delta\psi^2] = (-GG_d e)^2[Ek_{12}{}^2 + Ek_{34}{}^2 - 2(Ek_{12})(Ek_{34})]$$
$$= (GG_d e)^2[(n_s + 4n_b) \, \Delta t + O(\Delta t)^2] \qquad (11.4.9)$$

where $O(\Delta t^2)$ represents terms of order Δt^2 or greater and $(n_s + 4n_b)$ is the average count rate collected over all four detectors. We therefore have

$$C_2(\psi) = (GG_d e)^2[n_s + 4n_b] \qquad (11.4.10)$$

In a similar manner, we derive

$$C_j(\psi) = -(GG_d e)^j(n_s d/\lambda), \qquad j \text{ odd}$$

$$= (GG_d e)^j[n_s + 4n_b], \qquad j \text{ even} \qquad (11.4.11)$$

It is again convenient to introduce the tracking loop bandwidth B_L as

$$B_L = \frac{GG_d e n_s}{4} \qquad (11.4.12)$$

Here B_L is the effective loop bandwidth that the azimuth tracking loop presents to the line of sight azimuth angle in describing the mean value of the loop tracking response. In system design, B_L is a specified parameter, and must be large enough to allow the loop to track the expected time variations (i.e., motion) of the line of sight azimuth vector to the transmitter. Using B_L from (11.4.12) in (11.4.10) and (11.4.11), substituting into (11.4.3), and dividing by the coefficient of the second term yields the differential equation

$$0 = \rho\psi p(\psi) + \frac{dp(\psi)}{d\psi} + \sum_{\substack{j \geq 3 \\ \text{odd}}}^{\infty} C_j \frac{d^{j-1}}{d\psi^{j-1}} [\psi p(\psi)] + \sum_{\substack{j \geq 4 \\ \text{even}}}^{\infty} C_j \frac{d^{j-1}}{d\psi^{j-1}} [p(\psi)]$$

$$(11.4.13)$$

where

$$\rho = \left(\frac{n_s}{2B_L}\right)\left(\frac{n_s d/\lambda}{n_s + 4n_b}\right) \tag{11.4.14}$$

$$C_j = \left(\frac{2^{j-1}}{j!}\right)\left(\frac{2B_L}{n_s}\right)^{j-2}\left(\frac{n_s d/\lambda}{n_s + 4n_b}\right), \qquad j \text{ odd}$$

$$= \left(\frac{2^{j-1}}{j!}\right)\left(\frac{2B_L}{n_s}\right)^{j-2}, \qquad\qquad j \text{ even} \tag{11.4.15}$$

We see that the coefficients of the differential equation depend explicitly on the parameter n_s/B_L, which is the average signal count collected from the received optical signal beam over a time period $1/B_L$, that is, the response time of the loop. Note that the coefficients beyond $j = 3$ vary inversely with this parameter, and we would expect a decreasing importance in their effect as the signal count rate is increased.† In particular, for $n_s/B_L \gg 1$, we can approximate the system equation as

$$0 = \rho \psi p(\psi) + \frac{dp}{d\psi}, \qquad \frac{n_s}{B_L} \gg 1 \tag{11.4.16}$$

by neglecting higher order terms. The solution to (11.4.16) is easily found to be

$$p(\psi) = \frac{1}{(2\pi/\rho)^{1/2}} \exp\left[-\frac{\psi^2}{2(1/\rho)}\right] \tag{11.4.17}$$

under the initial condition that $p(\psi)$ integrate to unity. The above is the familiar Gaussian density with zero mean and variance $1/\rho$. Thus the azimuth tracking loop tends to track the azimuth angle with a zero mean (on the average it is pointed in azimuth at the transmitter) and a random spread, or mean squared jitter, about that mean given by the reciprocal of (11.4.14). Hence, the tracking jitter depends inversely on the detected count over the reciprocal of the loop bandwidth. The above is a first-order approximation to the true error density, and is essentially valid for small tracking angle errors ($\psi \ll 1$). Higher order approximations would require inclusion of more terms in the solution of (11.4.13). An equivalent result can be derived for the elevation angle as well, as the transmitter is being tracked by the receiver. The overall effect is to place a two-dimensional (circular) Gaussian error density over the azimuth and elevation angles to describe the tracking error. This model is useful for first-order tracking analysis (Problem 11.5).

† Here we make use of the results of the studies in Chapter 10 on pulse tracking to justify this statement.

REFERENCES

[1] McIntyre, C., Peters, W., Chi, C., and Wischnia, H., "Optical Components and Technology in Laser Space Communication Systems," *Proc. IEEE*, **58-10**, 1491 (October 1970).

[2] Lozins, N., "Pointing in Space," *Space Aeronautics*, 76–83 (August 1966).

[3] Thompson, W. T., "Passive Control of Satellite Vehicles," in *Guidance and Control of Aerospace Vehicles* (C. T. Leondes, ed.), McGraw-Hill Book Co., New York, 1964.

[4] Whitford, R. K., "Design of Altitude Control Systems for Earth Satellites 2313-0001-RV-000," Space Technology Laboratories Report, June 1961.

[5] Lindsey, W., *Synchronization Systems in Communication and Control*, Prentice-Hall, Englewood Cliffs, New Jersey, 1972, Chap. 7.

PROBLEMS

1. (a) An orbiting satellite circles the Earth every 2 hrs at an altitude of approximately 1000 miles. The satellite-Earth link has a pointing error totaling 50 μrad, and the uncertainty in the altitude is ± 10 miles. Determine a suitable look-ahead angle and beamwidth in order to maintain communications.

(b) Derive an equation for up and down look-ahead angles when both stations are in motion. Assume the stations move parallel to each other and each has separate pointing and velocity errors.

2. An optical receiver operates with a received count rate of 2×10^6 and a noise count per mode of 0.2. We assume an uncertainty field of view of 10^7 spatial modes and a desired resolution of 0.1 μrad. The acquisition probability is given by 0.98. Use Figure 11.6 and determine the maximum slewing rate for an azimuth search.

3. A point source optical transmitter operating at 10^{14} Hz is located somewhere within a 1° by 1° field of view. The receiver has a 3 cm by 3 cm square area.

(a) Find the number of diffraction limited spatial modes that must be searched.

(b) If a resolution of 50 arcseconds by 50 arcseconds is desired, find the value of Q in (11.2.6).

4. Derive (11.3.7) by determining the areas under an offset circle of radius r_0 that fall within symmetrically located quadrants.

5. Using the small error model for azimuth and elevation tracking, determine the probability that the receiver area will be aimed to within a distance of m miles (in any direction) of the transmitter located a distance L away. Assume the background count rate is $\beta\%$ of the signal rate, and $n_s/B_L = N$.

6. (a) A quadrant error sensor uses four detectors with unequal gains G_i, $i = 1, 2, 3, 4$. Determine the bias error produced when receiving a constant intensity field of I_0 over a focal plane area r_0.

(b) If $G_1 = G_2 \neq G_3 = G_4$, is there a best arrangement of the detectors for minimizing the bias error? Explain.

7. (a) A lobing detector has slit detectors located on the quadrant axis, as shown in Figure 11.1a. The lobing circle in the detector plane is offset by the amount y_0 along the elevation axis. Derive an expression for the distance between detector output pulses, assuming the lobing circle intersects all detectors.

(b) Repeat for a shift of x_0 along the azimuth axis.

(c) Show that the time shifts in (a) and (b) are linearly related to these offsets, x_0 and y_0, when the offset is small.

8. A receiver lens system focuses the received plane wave to a point (approximately) in the focal plane. A rotating reticle is placed just in front of the focal plane, as shown in Figure 11.17. The received field is a monochromatic field

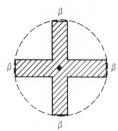

Figure 11.17. With problem 11.8.

of intensity I_0. The background has a constant radiance N_0. The filter following the detector has bandwidth equal to the frequency of reticle rotation. Determine the filtered SNR_p assuming an infinite bandwidth photodetector.

9. In an acquisition operation, heterodyning may be used during reception. If a strong oscillator signal is used for heterodyning, show that the single scan PAC_1 in (11.2.7) becomes

$$PAC_1 = \int_{-\infty}^{\infty} \int_{-\infty}^{\infty} dx\, dy\, \frac{\exp[-\frac{1}{2}(x-\lambda)^2 + y^2]}{2\pi}$$

$$\times \left[\int_{-\infty}^{x} \int_{-\infty}^{y} du\, dv\, \frac{\exp[\frac{1}{2}(u^2 + v^2)]}{2\pi} \right]^{Q-1}$$

where $\lambda = \dfrac{n_s}{n_L + n_b}$

n_s = signal count rate

n_b = background count rate

n_L = local oscillator count rate

[*Hint:* Use the fact that the counting statistics become Gaussian distributed when the count rate is high.]

Appendix A   Review of Probability Theory and Random Fields

A.1 RANDOM VARIABLES

Let x be a real random variable, having probability density $p_x(x)$. Then the probability that x is in an interval, or collection of intervals, I, is given by

$$\text{Prob}(x \in I) = \int_I p_x(x)\, dx \qquad (A.1.1)$$

The average (expected, mean) value of any function of x, $f(x)$ is defined by

$$E_x[f(x)] = \int_{-\infty}^{\infty} f(x) p_x(x)\, dx \qquad (A.1.2)$$

where E is called the expectation operator of x, and performs the integration shown. The subscript on E indicates the random variable whose density is used in the averaging (subscripts are omitted if no confusion exists). The operator defines the nth moment of x as

$$m_n = E_x[x^n] \qquad (A.1.3)$$

The first several moments are most important, denoted as

$$m_1 = E_x[x] = \text{mean of x}$$

$$m_2 = E_x[x^2] = \text{mean squared value of x} \qquad (A.1.4)$$

Also important is the variance, defined as

$$\sigma^2 = E_x[(x - m_1)^2] = m_2 - m_1{}^2 \qquad (A.1.5)$$

The characteristic function of x is

$$\psi_x(\omega) = E_x[e^{j\omega x}]$$

$$= \int_{-\infty}^{\infty} p_x(x) e^{j\omega x}\, dx \qquad (A.1.6)$$

Note that $\psi_x(\omega)$ is the Fourier transform of the density $p_x(x)$. Characteristic functions are important for generating the moments of x, since

$$m_n = \left(\frac{1}{j}\right)^n \left[\frac{\partial^n}{\partial\omega^n} \psi_x(\omega)\right]_{\omega=0} \tag{A.1.7}$$

Hence, moments of random variables can be generated by integrating with the probability density as in (A.1.3) or by differentiating with the characteristic function as in (A.1.7). Note that the Taylor series expansion of $\psi_x(\omega)$ yields the moments directly as coefficients:

$$\psi_x(\omega) = 1 + m_1 \frac{(j\omega)}{1!} + m_2 \frac{(j\omega)^2}{2!} + \cdots + m_n \frac{(j\omega)^n}{n!} + \cdots \tag{A.1.8}$$

Often it is easier to expand the log of the characteristic function:

$$\log \psi_x(\omega) = \chi_1 \frac{(j\omega)}{1!} + \chi_2 \frac{(j\omega)^2}{2!} + \cdots + \chi_n \frac{(j\omega)^n}{n!} + \cdots \tag{A.1.9}$$

The coefficients $\{\chi_i\}$ are called the semiinvariants of x and are related to the moments by

$$\chi_1 = m_1$$
$$\chi_2 = m_2 - m_1{}^2 \tag{A.1.10}$$
$$\chi_3 = m_3 - 3m_1 m_2 + 2m_1{}^2$$

The most important random variable is the Gaussian variable, whose density is given by

$$p_x(x) = \frac{1}{\sqrt{2\pi}\sigma} \exp\left[-\frac{(x - m_1)^2}{2\sigma^2}\right] \tag{A.1.11}$$

where m_1 is its mean and σ^2 is its variance. Its characteristic function is

$$\psi_x(\omega) = \exp\left(-j\omega m_1 - \frac{\sigma^2\omega^2}{2}\right) \tag{A.1.12}$$

A discrete random variable takes on values only at discrete points. Its probability density is given by

$$p_x(x) = \sum_{i=1}^{\infty} P(x_i)\, \delta(x - x_i) \tag{A.1.13}$$

where $\{x_i\}$ are the points of the density and $\delta(x - x_i)$ is the Dirac delta function, defined by

$$\int_I g(x)\, \delta(x - x_i)\, dx = g(x_i) \tag{A.1.14}$$

that the joint characteristic function is the two-dimensional Fourier
orm of $p(x, y)$. An important moment in (A.2.4) is $m_{11} = E(xy)$. The
les x and y are said to be *uncorrelated* if $m_{11} = E(x)E(y)$. The variables
d to be *independent* if the joint density factors as

$$p_{xy}(x, y) = p_x(x)p_y(y) \tag{A.2.6}$$

onditional density of x given $y = y$ is denoted

$$p_{x|y}(x|y) = \frac{p_{xy}(x, y)}{p_y(y)} \tag{A.2.7}$$

in be interpreted as the probability density of x when y has the value y.
onditional expectation operator is then

$$E_{x|y}[f(x)] = \int_{-\infty}^{\infty} f(x)p_{x|y}(x|y) \, dx \tag{A.2.8}$$

a function of y. The joint expectation in (A.2.3) is seen to be equivalent
sequence of expectations,

$$E_{x,y}[f(x, y)] = E_y[E_{x|y}[f(x, y)]] \tag{A.2.9}$$

in therefore be obtained by first conditional averaging, then averaging
the conditioning variable. Conditional moments and characteristic
ons can be obtained as in (A.1.3) and (A.1.6) using conditional proba-
s.

en dealing with sums of independent random variables, $z = x + y$,
te that the characteristic function of z becomes

$$\psi_z(\omega) = E_z[e^{j\omega z}] = E_z[e^{j\omega(x + y)}] = \psi_x(\omega)\psi_y(\omega) \tag{A.2.10}$$

therefore equal to the product of the characteristic function of each
Being a Fourier transform, the probability density of z is then the convo-
of the individual densities of x and y:

$$p_z(z) = \int_{-\infty}^{\infty} p_x(z - u)p_y(u) \, du \tag{A.2.11}$$

nces of Random Variables. Let $\mathbf{x} = (x_1, x_2, \ldots, x_N)$ be a sequence
ed set) of N real random variables. Many of the properties of $\mathbf{x}$ are
imply the N-dimensional extension of the properties of two random
les. In particular, the N-dimensional joint density $p_x(x_1, x_2, \ldots, x_N)$
used to derive joint probabilities of being in specific intervals. The

Here $g(x)$ is any well-defined continuous function
containing the point x_i. The coefficient $P(x_i)$ in (A
that $x = x_i$. That is,

$$P(x_i) = \text{Prob}(x = x_i)$$

Averaging with a discrete variable reduces to

$$E_x[f(x)] = \int_{-\infty}^{\infty} f(x)p_x(x)\,dx$$

$$= \sum_{i=1}^{\infty} f(x_i)P(x_i)$$

and appears as a weighted summation instead of a

Any function of a random variable is itself a
$z = f(x)$ is a random variable, and its probability d
that of x by the transformation

$$p_z(z) = \left[\frac{p_x(x)}{|df(x)/dx|}\right]_{x = f^{-1}(z}$$

where $f^{-1}(z)$ is the solution of $z = f(x)$ for x.

A.2 JOINT RANDOM VARIABLES

Let x, y be a pair of real random variables, having j
$p_{xy}(x, y)$. Then

$$\text{Prob}[x \in I_x \quad \text{and} \quad y \in I_y] = \int_{I_y} \int_{I_x} p_{xy}$$

The individual densities $p_x(x)$ can be obtained by inte

$$p_x(x) = \int_{-\infty}^{\infty} p_{xy}(x, y)\,dy; \qquad p_y(y) = \int_{-}^{}$$

The joint expectation operator becomes

$$E_{x,y}[f(x, y)] = \iint_{-\infty}^{\infty} f(x, y)p_{xy}(x, y)$$

for any function f of x and y. The (ij) joint mome

$$m_{ij} = E_{x,y}[x^i y^j]$$

and the joint characteristic function is

$$\psi_{x,y}(\omega_1, \omega_2) = E_{x,y}[\exp(j\omega_1 x + j$$

joint average over the sequences of any function $f(\mathbf{x})$ is given by

$$
E_{\mathbf{x}}[f(\mathbf{x})] = \int_X f(\mathbf{x}) p_{\mathbf{x}}(\mathbf{x})\, dx
$$

$$
= \int_{-\infty}^{\infty} \cdots \int_{-\infty}^{\infty} f(x_1, x_2, \ldots, x_N) p_{\mathbf{x}}(x_1, x_2, \ldots, x_N)\, dx_2 \cdots dx_N
$$

$$(A.2.12)$$

which we see is an N-fold integral over N-dimensional Euclidean space X.

A sequence $\mathbf{x}$ is said to be an *uncorrelated* sequence if every pair of components are uncorrelated. That is, (x_i, x_j) are uncorrelated for every $i \neq j$. The sequence is *independent* if components are pairwise independent.

The N-dimensional joint characteristic function is the N-dimensional extension of (A.2.5), and therefore becomes

$$
\psi_{\mathbf{x}}(\omega_1, \omega_2, \ldots \omega_N) = E_{\mathbf{x}}[\exp(j\omega_1 x_1 + j\omega_2 x_2 + \cdots + j\omega_N x_N)] \quad (A.2.13)
$$

A Gaussian sequence has the joint density given by

$$
p_{\mathbf{x}}(x_1, x_2, \ldots, x_N) = \frac{1}{(2\pi)^{N/2}\sqrt{|M|}} \exp\left[-\frac{1}{2} \chi^{\mathrm{Tr}} M^{-1} \chi \right] \quad (A.2.14)
$$

where

$$
\chi = \begin{bmatrix} x_1 - m_{x_1} \\ x_2 - m_{x_2} \\ \vdots \\ x_N - m_{x_N} \end{bmatrix}_{N \times 1}, \qquad
M = \begin{bmatrix} \mu_{11} & \mu_{12} & \cdots & \\ \mu_{21} & \mu_{22} & & \\ \vdots & & \ddots & \\ & & & \mu_{NN} \end{bmatrix} \quad (A.2.15)
$$

and m_{x_i} is the mean of x_i, $\mu_{ij} = E_{\mathbf{x}}[(x_i - m_{x_i})(x_j - m_{x_j})]$, and $|M|$ is the determinant of M. The matrix M is called the covariance matrix of the sequence. We see that if the sequence is uncorrelated, $\mu_{ij} = 0$, $i \neq j$, then the quadratic form in the exponent of (A.2.14) expands such that $p_{\mathbf{x}}(x_1 \cdots x_N)$ is a product of terms in each x_i. Hence, uncorrelated Gaussian sequences are also independent sequences.

We often must sum the components of a random sequence. Define $z = x_1 + x_2 + \cdots + x_N$ as the sum variable. If the sequence is independent, the characteristic function of z is then

$$
\psi_z(\omega) = \prod_{i=1}^{N} \psi_{x_i}(\omega) \quad (A.2.16)
$$

which is the product of the individual characteristic functions. The corresponding probability density becomes

$$
p_z(z) = p_{x_1}(x_1) \otimes p_{x_2}(x_2) \otimes \cdots \otimes p_{x_N}(x_N) \quad (A.2.17)
$$

where $\otimes$ denotes convolution. The sum density is therefore obtained by an $(N - 1)$-fold convolution of the individual densities. For a Gaussian sequence, z will always be a Gaussian random variable, whose mean is the sum of the means of each component. If the Gaussian sequence is independent, then z will have in addition a variance equal to the sum of the component variances. If the sequence is not Gaussian, then it has been shown that under relatively weak conditions, the density of z will converge to a Gaussian density as $N \to \infty$ (central limit theorem) [1].

Random Fields. Let v be a vector of scalar coordinates, representing a point in a Euclidean space V. Then $f(\mathbf{v})$ is a random *field* over V if, for every sequence of points $\mathbf{v}_1, \mathbf{v}_2, \ldots$, from V, $f(\mathbf{v}_1), f(\mathbf{v}_2), \ldots$ is a sequence of random variables. Loosely speaking, $f(\mathbf{v})$ is a random field if it represents a random variable at every $\mathbf{v}$ in V. When V is a one-dimensional scalar (such as time) $f(\mathbf{v})$ is called a random process.

Random fields are described by their N-dimensional density $p_{\mathbf{f}}(f_1, f_2, \ldots, f_N, f_i = f(\mathbf{v}_i)$ corresponding to a sequence of N points $(\mathbf{v}_1, \mathbf{v}_2, \ldots, \mathbf{v}_N)$ taken from V, for all integers N. First-order statistics of f correspond to the single random variable $f(\mathbf{v})$ at any $\mathbf{v}$. Such variables are described by the first-order density, first-order moments, and so on, which may be a function of $\mathbf{v}$. Second-order statistics involve joint densities between two points $(\mathbf{v}_1, \mathbf{v}_2)$ from V. The coherence function of f is defined as

$$R_f(\mathbf{v}_1, \mathbf{v}_2) = E[f(\mathbf{v}_1)f^*(\mathbf{v}_2)] \qquad (A.2.18)$$

where the asterisk denotes complex conjugate. The coherence function is a function of the points $\mathbf{v}_1$ and $\mathbf{v}_2$, and indicates the amount of correlation, or the degree of randomness, between two points of the field. A random field is often given special names depending on properties of this coherence function (see Section 1.4). When dealing with complex fields it is convenient to write

$$f(\mathbf{v}) = A(\mathbf{v}) \, e^{j\phi(\mathbf{v})} \qquad (A.2.19)$$

where $A(\mathbf{v})$ and $\phi(\mathbf{v})$ are the amplitude and phase at point $\mathbf{v}$. The coherence function is then

$$\begin{aligned} R_f(\mathbf{v}_1, \mathbf{v}_2) &= E[f(\mathbf{v}_1)f^*(\mathbf{v}_2)] \\ &= E[\exp((\ln A_1 + \ln A_2) + j(\phi_1 - \phi_2))] \end{aligned} \qquad (A.2.20)$$

where $A_i = A(\mathbf{v}_i)$ and $\phi_i = \phi(\mathbf{v}_i)$. It is convenient to use

$$\begin{aligned} l_{12} &= \ln A_1 + \ln A_2 \\ \phi_{12} &= \phi_1 - \phi_2 \end{aligned} \qquad (A.2.21)$$

so that (A.2.20) simplifies to

$$R_f(\mathbf{v}_1, \mathbf{v}_2) = E[\exp(l_{12} + j\phi_{12})] \qquad (A.2.22)$$

When the field amplitude and phase are independent,

$$R_f(\mathbf{v}_1, \mathbf{v}_2) = E[e^{l_{12}}]E[e^{j\phi_{12}}]$$
$$= \psi_{1_{12}}(-j)\psi_{\phi_{12}}[1] \qquad (A.2.23)$$

where $\psi(\omega)$ is the characteristic function of the variables involved. Hence, the field coherence factors into the product of characteristic functions for independent amplitude and phase. The result is particularly convenient when the log amplitude $\ln[A(\mathbf{v})]$ and phase $\phi(\mathbf{v})$ are considered *homogeneous*, Gaussian field variables. For then ϕ_{12}, l_{12} are Gaussian with moments

$$\begin{aligned}
\text{mean } l_{12} &= 2\bar{l}\\
\text{variance } l_{12} &= 2[C_a(0) + C_a(\rho)]\\
\text{mean } \phi_{12} &= 0\\
\text{variance } \phi_{12} &= 2[R_\phi(0) - R_\phi(\rho)]
\end{aligned} \qquad (A.2.24)$$

where $\bar{l} = E[\ln A(\mathbf{v})]$, $\rho = |\mathbf{v}_1 - \mathbf{v}_2|$, $C_a(\rho) = E[(\ln[A(\mathbf{v}_1)] - \bar{l})(\ln[A(\mathbf{v}_2)] - \bar{l})]$ and $R_\phi(\rho) = E[\phi(\mathbf{v}_1)\phi(\mathbf{v}_2)]$. Using (A.2.24) in (A.1.12) the field correlation then simplifies to

$$R_f(\rho) = \exp[2\bar{l} + C_a(0) + C_a(\rho) - R_\phi(0) + R_\phi(\rho)]$$
$$= [\exp(2\bar{l} + C_a(0) - R_\phi(0))][\exp(C_a(\rho) + R_\phi(\rho))] \quad (A.2.25)$$

Thus the variation of the field correlation with ρ depends only on the co-variance of the log amplitude and the correlation of the phase over ρ. Hence, measurement of field coherence reduces to measurements of log amplitude and phase correlations.

When the mean squared value of the complex field is normalized to unity at all points, then $E[f(\mathbf{v}_1)f^*(\mathbf{v}_1)] = R_f(0) = 1$. Therefore, from (A.2.25), $\exp[2(\bar{l} + C_a(0))] = 1$, which requires

$$\bar{l} = -C_a(0) \qquad (A.2.26)$$

In this case, $R_f(\rho)$ in (A.2.25) can be written as

$$R_f(\rho) = \exp[-\tfrac{1}{2}D(\rho)] \qquad (A.2.27)$$

where

$$D(\rho) = 2[C_a(0) - C_a(\rho)] + 2[R_\phi(0) - R_\phi(\rho)] \qquad (A.2.28)$$

The function $D(\rho)$ is called the *field structure function*, with the first bracket yielding the *log amplitude structure* function and the second the *phase structure* function. In intensity detection systems we are interested in the correlation of the amplitude $|f(\mathbf{v})|$, and only the log amplitude structure function is of interest. Note that the field correlation in (A.2.27) varies as $\exp[C_a(\rho) + R_\phi(\rho)]$, and therefore is directly related to the correlation distance of the structure function. In general, $R_\phi(\rho)$ falls off more slowly than $C_a(\rho)$ and the former usually defines the correlation region. Studies and measurements of structure functions have been pursued to some depth in the literature. [See References 3–8 in Chapter 6.)

REFERENCES

[1] Papoulis, A., *Probability, Random Variables, and Stochastic Processes*, McGraw-Hill Book Co., New York, 1965, Chap. 8.

Appendix B Expansions of Fields into Series

Throughout this text we often expand optical fields into a sum of orthonormal functions. We then treat this sum as being identically equal to the field itself, without formally establishing its validity. In this appendix we review the properties of orthonormal field expansions.

Let $\mathbf{v}$ be a point of an N-dimensional real Euclidean space V, and let $x_1, x_2, \ldots, x_N$ be the components of $\mathbf{v}$. A function $f(\mathbf{v})$ with domain V is called a field over V. We denote integrals of products of fields over V as

$$(f, g) = \int_V f(\mathbf{v})g^*(\mathbf{v})\, d\mathbf{v} \tag{B.1}$$

and integrals of squared fields as

$$\|f\|^2 \triangleq (f, f) = \int_V |f(\mathbf{v})|^2\, d\mathbf{v} \tag{B.2}$$

Note that this integral is equivalent to the energy over V. Let $\{\phi_i(\mathbf{v})\}$ be a complete orthonormal set over V, which means

$$\begin{aligned}(\phi_i, \phi_j) &= 1, \qquad i = j \\ &= 0, \qquad i \neq j\end{aligned} \tag{B.3}$$

and no field exists that is not in $\{\phi_i\}$ satisfying $(f, \phi_i) = 0$. A deterministic field of bounded energy can be approximated by a series of terms from $\{\phi_i\}$. Let $f(\mathbf{v})$ be such a field with $\|f\|^2 < \infty$, and denote

$$f_N(\mathbf{v}) = \sum_{n=1}^{N} c_n \phi_n(\mathbf{v}) \tag{B.4}$$

It can be shown that

(a) $\|f - f_N\|^2$ is minimized for any N if

$$c_n = (f, \phi_n) \tag{B.5}$$

(b) $$\lim_{n \to \infty} \|f - f_N\|^2 = 0 \tag{B.6}$$

Part (a) is shown by expanding (B.2):

$$\|f - f_N\|^2 = \|f\|^2 + \|f_N\|^2 - (f, f_N) - (f_N, f)$$
$$= \|f\|^2 + \sum_n |c_n|^2 - \sum_n c_n^*(f, \phi_n) - \sum_n c_n(\phi_n, f)$$
$$= \|f\|^2 - \sum_n |(f, \phi_n)|^2 + \sum_n |c_n - (f, \phi_n)|^2 \qquad (B.7)$$

Clearly (B.7) is minimized for any N if $c_n = (f, \phi_n)$, which are called the Fourier coefficients of the field. When used in (B.7) the result is

$$\|f - f_N\|^2 = \|f\|^2 - \sum_{n=1}^{N} |c_n|^2 \geq 0 \qquad (B.8)$$

Equation B.8 is called Bessel's inequality. Since $\{\phi_i\}$ is complete, and $\|f\|^2$ is bounded, increasing N can only decrease the right side of (B.8). The only violation would occur if $c_n = 0$ for all n beyond some N, or if $f(v)$ had an additive term orthogonal to the set $\{\phi_i(v)\}$. But this requires $(f, \phi_i) = 0$, $i \geq N$, violating the completeness assumption. Hence, part (b) in (B.6) follows. Thus the integrated squared difference between $f(v)$ and $f_N(v)$ must converge to zero as more terms are used in (B.4). We say that the infinite sum

$$f(v) = \sum_{n=1}^{\infty} c_n \phi_n(v) \qquad (B.9)$$

represents the field $f(v)$ in an integrated squared sense. This means that we can get arbitrarily close, in terms of integrated difference, by using enough terms. We point out that this does not necessarily mean that the sum on the right of (B.9) will exactly equal $f(v)$ on the left for every v. Theoretically, points in V can exist for which the two are not equal but the integrated difference is zero. (These points are said to constitute points of measure zero when integrating.) Thus, as long as we are primarily interested in eventually integrating (filtering) the field, use of the orthonormal expansion is mathematically identical to use of the field itself.

With (B.9), we see that (Parceval's theorem)

$$\|f\|^2 = \int_V |f(v)|^2 \, dv = \sum_{n=1}^{\infty} |c_n|^2 \qquad (B.10)$$

Since $\|f\|^2$ is the energy of the field we see that the sum of the squares of the components of the expansion is equal to the field energy. It is natural to interpret $|c_n|^2$ as the field energy in the nth component.

Random Fields. Let $f(\mathbf{v})$ be zero mean random field over V, having coherence function

$$R_f(\mathbf{v}_1, \mathbf{v}_2) = E[f(\mathbf{v}_1)f^*(\mathbf{v}_2)] \tag{B.11}$$

Let $\{\phi_i(\mathbf{v})\}$ again be a complete orthonormal set over V, and consider the expansion

$$f_N(\mathbf{v}) = \sum_{n=1}^{N} c_n \phi_i(\mathbf{v}) \tag{B.12}$$

where again

$$c_n = (f, \phi_i) \tag{B.13}$$

This is the random field equivalent to (B.4). However, here the $\{c_n\}$, being integrals of zero mean random fields, must be a sequence of zero mean, complex random variables. We can now show that if we can expand

$$R_f(\mathbf{v}_1, \mathbf{v}_2) = \sum_{i=1}^{\infty} \gamma_i \phi_i(\mathbf{v}_1)\phi_i^*(\mathbf{v}_2) \tag{B.14}$$

then it follows that

$$\lim_{N \to \infty} E[(f - f_N)^2] = 0 \tag{B.15}$$

This can be shown by expanding, substituting, and averaging, yielding

$$E[|f - f_N|^2] = E[|f(\mathbf{v})|^2] + E[|f_N|^2] - E(f, f_N) - E(f_N, f)$$

$$= E[|f(\mathbf{v})|^2] + \sum_{i=1}^{N}\sum_{j=1}^{N} \phi_i(\mathbf{v})\phi_j^*(\mathbf{v}) \int_V\int_V R_f(\mathbf{v}_1, \mathbf{v}_2)$$

$$\times \phi_i^*(\mathbf{v}_1)\phi_j(\mathbf{v}_2)\, d\mathbf{v}_1\, d\mathbf{v}_2$$

$$- \sum_{i=1}^{N} \phi_i^*(\mathbf{v}) \int_V R_f(\mathbf{v}, \mathbf{v}_1)\phi_i(\mathbf{v}_1)\, d\mathbf{v}_1$$

$$- \sum_{i=1}^{N} \phi_i(\mathbf{v}) \int_V R_f(\mathbf{v}, \mathbf{v}_1)\phi_i^*(\mathbf{v}_1)\, d\mathbf{v}_1 \tag{B.16}$$

Substituting from (B.14) then yields

$$E[|f - f_N|^2] = R_f(\mathbf{v}, \mathbf{v}) - \sum_{i=1}^{N} \gamma_i \phi_i(\mathbf{v})\phi_i^*(\mathbf{v}) \tag{B.17}$$

from which (B.15) follows. Thus the random field expansion in (B.12) converges in the mean squared sense, rather than an integrated squared sense, when using an orthonormal expansion with Fourier variables. This convergence is sometimes called convergence in probability, in that the random field $f_N(\mathbf{v})$ converges to $f(\mathbf{v})$ with a probability of unity. The only condition required is that the field coherence be expandable as in (B.14). It is possible to show by Mercer's theorem [1] that any continuous, bounded coherence function can be so expanded [2]. Many classes of discontinuous coherences can also be expanded. In communication systems, coherence functions encountered are generally continuous, and mean squared series expansions are valid.

It is convenient in expanding a field if the sequence of complex, random variables $\{c_n\}$ is pairwise uncorrelated. This requires

$$E(c_n c_m^*) = 0, \qquad n \neq m \tag{B.18}$$

Using (B.13) in (B.18), we equivalently require

$$\int_V \left[\int_V R_f(\mathbf{v}_1, \mathbf{v}_2) \phi_n(\mathbf{v}_2) \, d\mathbf{v}_2 \right] \phi_m^*(\mathbf{v}_1) \, d\mathbf{v}_1 = 0, \qquad n \neq m \tag{B.19}$$

Because of the completeness of the orthonormal set, this can occur for every $n \neq m$ if and only if

$$\int_V R_f(\mathbf{v}_1, \mathbf{v}_2) \phi_n(\mathbf{v}_2) \, d\mathbf{v}_2 = \gamma_n \phi_n(\mathbf{v}_1) \tag{B.20}$$

for some positive constant γ_n. Thus the random coefficients will be uncorrelated only if the orthonormal set has each member satisfying (B.20). The latter is an integral equation of the Fredholm type, and the $\{\phi_n(\mathbf{v})\}$ satisfying it are called its *eigenfunctions* and the associated $\{\gamma_n\}$ are its *eigenvalues*. The resulting expansion with these orthonormal functions is called a *Karhunen–Loeve* (KL) expansion. Thus an orthonormal expansion will be a KL expansion (i.e., have uncorrelated field coefficients) if the orthonormal functions are the eigenfunctions of the integral equation in (B.20). Note that for a KL expansion

$$E[|c_n|^2] = \gamma_n \tag{B.21}$$

obtained by using (B.20) in (B.19) with $m = n$. Thus the eigenvalues of (B.20) are the mean squared values of the KL coefficients. Last, we note that the mean field energy becomes

$$E \int_V |f(\mathbf{v})|^2 \, d\mathbf{v} = E \sum_{n=1}^{\infty} |c_n|^2$$

$$= \sum_{n=1}^{\infty} \gamma_n \tag{B.22}$$

Hence, the sum of the eigenvalues is the average field energy, and each γ_n can be considered the average energy contributed by the nth field component.

In this way, the importance of a particular field component can be related to the size of its eigenvalue. This notion is applied in several places in the text. Equation B.22 is a form of Parseval's theorem.

REFERENCES

[1] Courant, R. and Hilbert, D., *Methods of Mathematical Physics*, Vol. I, John Wiley and Sons (Interscience), New York, 1953.

[2] Selin, I., *Detection Theory*, Princeton University Press, Princeton, New Jersey, 1965, Chap. 7.

Appendix C Laguerre Polynomials

The generalized Laguerre polynomial $L_n^\alpha(x)$ is defined as [1]

$$L_n^\alpha(x) = \sum_{k=0}^{n} (-1)^k \binom{n+\alpha}{n-k} \frac{x^k}{k!} \tag{C.1}$$

where n is a nonnegative integer, called the order of the polynomial, $\alpha > -1$ is called the index, and

$$\binom{n+\alpha}{n-k} = \frac{\Gamma(n+\alpha+1)}{\Gamma(n-k+1)\Gamma(\alpha+k+1)}$$

$$\Gamma(y+1) = \int_0^\infty x^y e^{-x} dx \tag{C.2}$$

When the index α is itself an integer,

$$\binom{n+\alpha}{n-k} = \frac{(n+\alpha)!}{(n-k)!(\alpha+k)!} \tag{C.3}$$

The polynomials $L_n^\alpha(x)$ can also be defined by the Rodriguez formula:

$$L_n^\alpha(x) = e^x \frac{x^{-\alpha}}{n!} \frac{d^n}{dx^n} (e^{-x} x^{n+\alpha}) \tag{C.4}$$

Laguerre polynomials are the solutions $f(x, n, \alpha)$ to the recursion equation

$$(n+1)f(n+1) + (x-\alpha-2n-1)f(n) + (n+\alpha)f(n-1) = 0, \qquad n \geq 1 \tag{C.5}$$

the differential equation

$$x\frac{d^2 f(x)}{dx^2} + (\alpha+1-x)\frac{df(x)}{dx} + (n+\alpha)f(x) = 0, \qquad n \geq 0 \tag{C.6}$$

and the integral equation

$$e^{-x/2} x^{\alpha/2} f(x) = \frac{(-1)^n}{2} \int_0^\infty J_\alpha(\sqrt{xy}) e^{-y/2} f(y) \, dy \tag{C.7}$$

The simplest class of Laguerre polynomials consists of those of zero index, written as $L_n^0(x) = L_n(x)$. For these we have the simpler expansion

$$L_n(x) = 1 - \binom{n}{1}\frac{x}{1!} + \binom{n}{2}\frac{x^2}{2!} + \cdots + (-1)^n \frac{x^n}{n!} \qquad (C.8)$$

Tabulations of the coefficients and plots of $L_n(x)$ are available [2].

When the index α is not an integer, the gamma function must be evaluated in (C.2) for each coefficient. When $\alpha = \pm\frac{1}{2}$, the series has the integral expressions:

$$L_n^{1/2}(x) = \frac{e^x}{n!\sqrt{\pi x}} \int_0^\infty e^{-t}t^n \sin(2\sqrt{xt})\, dt \qquad (C.9)$$

$$L_n^{-1/2}(x) = \frac{e^x}{n!\sqrt{\pi}} \int_0^\infty e^{-t}t^{n-1/2} \cos(2\sqrt{xt})\, dt \qquad (C.10)$$

Laguerre polynomials are orthogonal with weight $x^\alpha e^{-x}$ over the interval $(0, \infty)$. That is,

$$\int_0^\infty x^\alpha e^{-x} L_n^\alpha(x) L_m^\alpha(x)\, dx = 0, \qquad n \ne m$$

$$= \frac{\Gamma(n + \alpha + 1)}{n!}, \qquad n = m \qquad (C.11)$$

This means the function set

$$l_n^\alpha(x) = \frac{n!}{\Gamma(n + \alpha + 1)}\, x^{\alpha/2}\, e^{-x/2} L_n^\alpha(x), \qquad n \ge 0 \qquad (C.12)$$

is orthonormal over $[0, \infty)$. The above are called the *Laguerre functions*, or the *Laguerre wave functions*. The wave functions are complete, and therefore can be used to expand any integrably squared function, $s(x)$, over $(0, \infty)$:

$$s(x) = \sum_{n=0}^\infty c_n l_n(x) \qquad (C.13)$$

where

$$c_n = \int_0^\infty s(x) l_n(x)\, dx \qquad (C.14)$$

and the convergence is in the sense discussed in Appendix B. Laguerre polynomials differentiate into Laguerre polynomials, according to the rule

$$\frac{dL_n^\alpha(x)}{dx} = -L_{n-1}^{\alpha+1}(x) \qquad (C.15)$$

and recursively reproduce as

$$L_n^{\alpha-1}(x) = L_n^{\alpha}(x) - L_{n-1}^{\alpha}(x) \tag{C.16}$$

Since a Laguerre polynomial is always positive when its argument is negative, (C.15) and (C.16) also show that the polynomial $L_n^{\alpha}(-x)$, $x > 0$, is a monotonically increasing function of x, α, and n.

C.1   ASYMPTOTIC BEHAVIOR

Large n,

$$L_n^{\alpha}(x) \approx \frac{\Gamma(n + \alpha + 1)}{n!} e^{x/2}(Nx)^{-\alpha/2} J_\alpha(2\sqrt{Nx}), \qquad n \to \infty, \quad N = n + \frac{\alpha+1}{2} \tag{C.1.1}$$

$$\approx \frac{1}{\sqrt{\pi}} e^{x/2} n^{\alpha/2 - 1/4} x^{-\alpha/2 - 1/4} \cos\left(2\sqrt{nx} - \frac{\alpha\pi}{2} - \frac{\pi}{4}\right), \qquad n \to \infty \tag{C.1.2}$$

Large x,

$$L_n^{\alpha}(x) = (-1)^n \frac{x^n}{n!}, \qquad x \to \infty \tag{C.1.3}$$

Small x,

$$L_n^{\alpha}(x) \approx \binom{n+\alpha}{n} - \binom{n+\alpha}{n-1}x \tag{C.1.4}$$

C.2   BOUNDS

$$|L_n^{\alpha}(x)| \leq \frac{\Gamma(\alpha + n + 1)}{n!\,\Gamma(\alpha + 1)} e^{x/2}, \qquad \alpha \geq 0, \quad x \geq 0 \tag{C.2.1}$$

$$|L_n(x)| \leq e^{x/2} \tag{C.2.2}$$

$$|L_n^{\alpha}(x)| \leq \left[2 - \frac{\Gamma(\alpha + n + 1)}{n!\,\Gamma(\alpha + 1)}\right] e^{x/2}, \qquad -1 < \alpha < 0, \quad x \geq 0 \tag{C.2.3}$$

$$|L_n^{\alpha}(x)| \leq L_n^{\alpha}(-x) \tag{C.2.4}$$

C.3 RELATIONSHIP TO OTHER FUNCTIONS

Hypergeometric series

$$L_n^\alpha(x) = \binom{n + \alpha}{n}_1 F_1(-n; \alpha + 1; x) \tag{C.3.1}$$

Bessel Functions

$$L_n^\alpha(x) = \frac{e^x x^{-\alpha/2}}{n!} \int_0^\infty e^{-t} t^{n+\alpha/2} J_\alpha(2\sqrt{tx})\, dt \tag{C.3.2}$$

$$(ax)^{-\alpha/2} J_\alpha(2\sqrt{ax}) = e^{-a} \sum_{n=0}^\infty \frac{a^n}{\Gamma(n + \alpha + 1)} L_n^\alpha(x)$$

$$x > 0, \qquad a > 0, \qquad \alpha > -1 \tag{C.3.3}$$

Hermite Polynomials

$$H_{2n}(x) = (-1)^n 2^{2n} n! L_n^{-1/2}(x^2) \tag{C.3.4}$$

$$H_{2n+1}(x) = (-1)^n 2^{2n+1} n! x L_n^{1/2}(x^2) \tag{C.3.5}$$

Exponential Function

$$e^{-ax} = (a + 1)^{-\alpha-1} \sum_{n=0}^\infty \left(\frac{a}{a+1}\right)^n L_n^\alpha(x), \qquad 0 \le x < \infty \tag{C.3.6}$$

Gamma Function

$$e^x x^{-\alpha} \Gamma(\alpha, x) = \sum_{n=0}^\infty \frac{L_n^\alpha(x)}{n+1}, \qquad 0 < x < \infty, \quad \alpha > -1 \tag{C.3.7}$$

Jacobi Polynomials

$$L_n^\alpha(x) = \lim_{\beta \to \infty} P_n^{(\alpha, \beta)}\left(1 - \frac{2x}{\beta}\right) \tag{C.3.8}$$

Exponential Integral

$$-e^x Ei(-x) = \sum_{n=0}^\infty \frac{L_n(x)}{n+1}, \qquad 0 < x < \infty \tag{C.3.9}$$

Erf Functions

$$\frac{\sqrt{\pi}}{\text{Erf}[\sqrt{x}]} = e^{-x} \sum_{n=1}^\infty \frac{\Gamma(\tfrac{1}{2} + n) L_{n-1}(x)}{n!} \tag{C.3.10}$$

Power Series

$$x^v = \Gamma(v + \alpha + 1)\Gamma(v + 1) \sum_{n=0}^{\infty} \frac{(-1)^n L_n^\alpha(x)}{\Gamma(n + \alpha + 1)\Gamma(v - n + 1)}, \quad \begin{matrix} 0 < x < \infty \\ \alpha > -1 \end{matrix}$$

(C.3.11)

Contour Integrals

$$L_n^\alpha(x) = \frac{1}{2\pi j} \oint \left(1 + \frac{x}{z}\right)^n e^{-z} \left(1 + \frac{z}{x}\right)^\alpha \frac{1}{z} \, dz,$$

(C.3.12)

$$= \frac{x^{-x}e^x}{2\pi j} \oint \left(\frac{z}{z-x}\right)^n \frac{z^\alpha}{z-x} e^{-z} \, dz$$

(C.3.13)

Delta Function

$$\delta(x - m) = e^{-m/2} \sum_{n=0}^{\infty} L_n(m)\left[e^{-x/2}L_n(x)\right]$$

$$= \sum_{n=0}^{\infty} l_n(m)l_n(x)$$

(C.3.14)

C.4 SUM FORMULAS

$$\sum_{m=0}^{\infty} \frac{m!}{\Gamma(m + \alpha + 1)} L_m^\alpha(x)L_m^\alpha(y)$$

$$= \frac{(n + 1)!}{\Gamma(n + \alpha + 1)(x - y)}\left[L_n^\alpha(x)L_{n+1}^\alpha(y) - L_{n+1}^\alpha(x)L_n^\alpha(y)\right] \quad \text{(C.4.1)}$$

$$\sum_{m=0}^{n} \frac{\Gamma(\alpha - \beta + m)}{\Gamma(\alpha - \beta)m!} L_{n-m}^\beta(x) = L_n^\alpha(x)$$

(C.4.2)

$$\sum_{m=0}^{n} L_m^\alpha(x) = L_n^{\alpha+1}(x)$$

(C.4.3)

$$\sum_{m=0}^{n} L_m^\alpha(x)L_{n-m}^\beta(x) = L_n^{\alpha+\beta+1}(x + y)$$

(C.4.4)

$$\sum_{k=0}^{\infty} \frac{(2n - 2k)!(2k)!\, L_{2k}^{2\alpha}(2x)}{\Gamma(\alpha + 1 - k)(n - k)!} = \frac{n!}{\Gamma(n + \alpha + 1)}\left[L_n^\alpha(x)\right]^2$$

(C.4.5)

$$\sum_{k=0}^{\infty} \frac{L_{n-k}^{\alpha+2k}(x + y)(xy)^k}{\Gamma(k + \alpha + 1)k!} = \frac{n!}{\Gamma(n + \alpha + 1)} L_n^\alpha(x)L_n^\alpha(y)$$

(C.4.6)

$$\sum_{k=0}^{\infty} \frac{(-1)^k}{k!} y^k L_n^{\alpha+k}(x) = e^{-y} L_n^{\alpha}(x+y) \tag{C.4.7}$$

$$\sum_{i_1+i_2+\cdots+i_k=n} L_{i_1}^{\alpha_1}(x_1) L_{i_2}^{\alpha_2}(x_2) \cdots L_{i_k}^{\alpha_k}(x_k)$$

$$= L_n^{\alpha_1+\alpha_2+\cdots+\alpha_k+k-1}(x_1+x_2+\cdots+x_k) \tag{C.4.8}$$

$$\sum_{k=0}^{n} (-1)^{n-k} \binom{n}{k} \left(1+\frac{\beta}{\alpha}\right)^k L_k(\beta x) = \frac{\beta^n}{\alpha^n} L_n[(\alpha+\beta)x] \tag{C.4.9}$$

$$\sum_{m=0}^{n} \binom{n}{m} k^m (1-k)^{n-m} L_m(x) = L_n(kx) \tag{C.4.10}$$

$$\sum_{m=0}^{n} \binom{n}{m} (k-1)^{n-m} L_m(kx) = k^n L_n(x) \tag{C.4.11}$$

C.5   FUNCTIONAL RELATIONS

$$\frac{\partial}{\partial x} [L_n^{\alpha}(x) - L_{n+1}^{\alpha}(x)] = L_n^{\alpha}(x) \tag{C.5.1}$$

$$\frac{\partial}{\partial x} L_n^{\alpha}(x) = -L_{n-1}^{\alpha+1}(x) \tag{C.5.2}$$

$$x \frac{\partial}{\partial x} L_n^{\alpha}(x) = n L_n^{\alpha}(x) - (n+\alpha) L_{n-1}^{\alpha}(x) \tag{C.5.3}$$

$$= (n+1) L_{n+1}^{\alpha}(x) - (n+\alpha+1-x) L_n^{\alpha}(x) \tag{C.5.4}$$

$$x L_n^{\alpha+1} = (n+\alpha+1) L_n^{\alpha}(x) - (n+1) L_{n+1}^{\alpha}(x) \tag{C.5.5}$$

$$= (n+\alpha) L_{n-1}^{\alpha}(x) - (n-x) L_n^{\alpha}(x) \tag{C.5.6}$$

$$L_n^{\alpha-1}(x) = L_n^{\alpha}(x) - L_{n-1}^{\alpha}(x) \tag{C.5.7}$$

$$(n+1) L_{n+1}^{\alpha}(x) - (2n+\alpha+1-x) L_n^{\alpha}(x) + (n+\alpha) L_{n-1}^{\alpha}(x) = 0 \tag{C.5.8}$$

$$e^x \int_x^{\infty} e^{-t} L_n^{\alpha}(t)\, dt = L_n^{\alpha}(x) - L_{n-1}^{\alpha}(x) \tag{C.5.9}$$

$$\int_0^x L_m(t) L_n(x-t)\, dt = \int_0^x L_{m+n}(t)\, dt = L_{m+n}(x) - L_{m+n+1}(x) \tag{C.5.10}$$

$$\Gamma(\alpha+\beta+n+1) \int_0^x (x-t)^{\beta-1} t^{\alpha} L_n^{\alpha}(t)\, dt$$

$$= \Gamma(\alpha+n+1)\Gamma(\beta) x^{\alpha+\beta} L_n^{(\alpha+\beta)}(x), \qquad \text{Re } \alpha > -1, \qquad \text{Re } \beta > 0 \tag{C.5.11}$$

C.6 GENERATING FUNCTIONS

$$(1 - z)^{-(\alpha + 1)} \exp\left[\frac{xz}{z - 1}\right] = \sum_{n=0}^{\infty} L_n^{\alpha}(x)z^n, \qquad |z| < 1 \qquad \text{(C.6.1)}$$

$$e^{-\alpha z}(1 + z)^{\alpha} = \sum_{n=0}^{\infty} L_n^{\alpha - n}(x)z^n, \qquad |z| < 1 \qquad \text{(C.6.2)}$$

$$J_{\alpha}(2\sqrt{xz})e^z(xz)^{-\alpha/2} = \sum_{n=0}^{\infty} \frac{L_n^{\alpha}(x)}{\Gamma(n + \alpha + 1)} z^n, \qquad \alpha > -1 \qquad \text{(C.6.3)}$$

$$\frac{(xyz)^{-\alpha/2}}{1 - z} \exp\left[\frac{-z(x + y)}{1 - z}\right] I_{\alpha}\left(\frac{2(xyz)^{1/2}}{1 - z}\right)$$

$$= \sum_{n=0}^{\infty} \frac{n!\, L_n^{\alpha}(x)L_n^{\alpha}(y)z^n}{\Gamma(n + \alpha + 1)}, \qquad |z| < 1 \qquad \text{(C.6.4)}$$

$$e^x x^{-\alpha}\Gamma(\alpha, x) = \sum_{n=0}^{\infty} \frac{L_n^{\alpha}(x)}{n + 1}, \qquad \alpha > -1, \quad x > 0 \qquad \text{(C.6.5)}$$

REFERENCES

[1] Laguerre, E., "Sur l'integrale $\int_x^{\infty} e^{-x}\, dx/x$," *Bull. Soc. Math. Fr.*, 1879.

[2] Abramowitz, M. and Stegun, I., *Handbook of Mathematical Functions*, National Bureau of Standards, Washington, D.C., 1965, Chap. 22.

[3] Gradsteyn, I. and Ryshik, I., *Tables of Series, Products, and Integrals*, Academic Press, New York, 1965, p. 1037.

[4] Head, J. W. and Wilson, W., *Laguerre Functions: Tables and Properties*, Monograph No. 183R, British Broadcasting Corp., June 1956.

[5] Lebedev, N., *Special Functions and Their Applications*, Prentice-Hall, Englewood Cliffs, New Jersey, 1965, Chap. 4.

Appendix D   Random Channels

In optical communications it is often necessary to deal with the properties of a random channel between the transmitter and receiver. This occurs for example when attempting to communicate over a turbulent medium or over an underwater scattering channel. The principle effect is that the channel produces random fields in the system model, and subsequent communication analysis must take into account the associated field statistics. To assess performance we are generally interested in the spatial coherent properties inherent in the optical field as it arrives at the receiver. This information is contained in the field spatial coherence function describing the receiver field. Also, the field intensity (irradiance) function over the optical detector area allows us to determine the amount of collected field power. In this appendix we review some basic field relations that aid in determining these functions when dealing with random channels.

D.1   COHERENCE RELATIONS

Consider a monochromatic random field defined at points $\mathbf{r}$ in a plane $\mathscr{C}$ (Figure D.1). The field over $\mathscr{C}$ is therefore coherence separable, and we concentrate only on its spatial coherence function. We write this as

$$R_{\mathscr{C}}(\mathbf{r}_1, \mathbf{r}_2) = E[f_{\mathscr{C}}(\mathbf{r}_1)f_{\mathscr{C}}^*(\mathbf{r}_2)] \tag{D.1}$$

where $f_{\mathscr{C}}(\mathbf{r})$ is the random complex field envelope produced at a point $\mathbf{r}$, and $\mathbf{r}_1$ and $\mathbf{r}_2$ are points in $\mathscr{C}$. From Huygen's propagation rule, the field envelope produced at a point $\mathbf{q}$ on a plane $\mathscr{D}$, due to the propagation of the field from $\mathscr{C}$ over a random medium, is given by

$$f_{\mathscr{D}}(\mathbf{q}) = \int_{\mathscr{C}} f_{\mathscr{C}}(\mathbf{r})g(\mathbf{q}, \mathbf{r}) \frac{e^{j(2\pi/\lambda)p}}{\lambda p} \, d\mathbf{r} \tag{D.2}$$

where λ is the field wavelength, p is the distance from $\mathbf{r}$ in $\mathscr{C}$ to $\mathbf{q}$ in $\mathscr{D}$, and $g(\mathbf{q}, \mathbf{r})$ is a random, complex multiplicative factor to account for the alteration in field amplitude and phase over the path from $\mathbf{r}$ to $\mathbf{q}$ due to the medium.

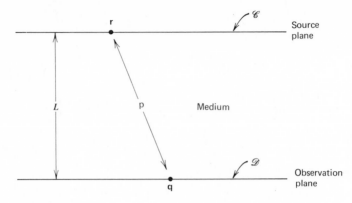

Figure D.1

[In free space $g(\mathbf{q}, \mathbf{r}) = 1$.] We may consider the plane $\mathscr{C}$ as a source plane in which a field is radiated toward $\mathscr{D}$, and plane $\mathscr{D}$ as a response, or observable, plane upon which the source field impinges. The spatial coherence function of the field over the observable plane $\mathscr{D}$ is then

$$R_{\mathscr{D}}(\mathbf{q}_1, \mathbf{q}_2) = E[f_{\mathscr{D}}(\mathbf{q}_1)f_{\mathscr{D}}^*(\mathbf{q}_2)]$$

$$= \int_{\mathscr{C}} \int_{\mathscr{C}} R_{\mathscr{C}}(\mathbf{r}_1, \mathbf{r}_2)M(\mathbf{q}_1, \mathbf{r}_1; \mathbf{q}_2, \mathbf{r}_2)\frac{e^{j(2\pi/\lambda)(p_1 - p_2)}}{\lambda^2 p_1 p_2}\, d\mathbf{r}_1\, d\mathbf{r}_2 \quad (D.3)$$

where p_i is the distance from $\mathbf{r}_i$ to $\mathbf{q}_i$ and

$$M(\mathbf{q}_1, \mathbf{r}_1; \mathbf{q}_2, \mathbf{r}_2) = E[g(\mathbf{q}_1, \mathbf{r}_1)g^*(\mathbf{q}_2, \mathbf{r}_2)] \quad (D.4)$$

The function M in (D.4) is due entirely to the medium and is, in a sense, the coherence function of the medium response. Equation D.3 describes how the spatial coherence function in one plane is related to that over another plane. In essence, (D.3) describes how a coherence function "propagates" through a random medium. The irradiance of the field at a point $\mathbf{q}$ in $\mathscr{D}$ then follows as

$$I_{\mathscr{D}}(\mathbf{q}) = R_{\mathscr{D}}(\mathbf{q}, \mathbf{q}) \quad (D.5)$$

and can be determined directly from (D.3). Thus the field irradiance propagates in a manner similar to that of the coherence function. We point out that (D.3) is also valid for nonrandom fields defined over the source plane $\mathscr{C}$, with (D.1) replaced by the product of the field at each pair of points. When the separation between the planes, L, is large enough, the Fresnel approximations:

$$p \approx L + \frac{|\mathbf{q} - \mathbf{r}|^2}{2L} \quad (D.6)$$

can be applied, and (D.3) simplifies to

$$R_{\mathscr{D}}(\mathbf{q}_1, \mathbf{q}_2) = \left(\frac{1}{\lambda L}\right)^2 \int_{\mathscr{C}} \int_{\mathscr{C}} R_{\mathscr{C}}(\mathbf{r}_1, \mathbf{r}_2) M(\mathbf{q}_1, \mathbf{r}_1; \mathbf{q}_2, \mathbf{r}_2)$$

$$\times e^{j(2\pi/\lambda L)(|\mathbf{q}_1 - \mathbf{r}_1|^2 - |\mathbf{q}_2 - \mathbf{r}_2|^2)} \, d\mathbf{r}_1 \, d\mathbf{r}_2 \qquad (D.7)$$

In optical communication models, the Fresnel approximation is generally valid, and the simplification in (D.7) is usually used for link analysis.

We obtain a further simplification when the random field over the source plane $\mathscr{C}$ is incoherent. In this case $R_{\mathscr{C}}(\mathbf{r}_1, \mathbf{r}_2) = I_{\mathscr{C}}(\mathbf{r}_1)\,\delta(\mathbf{r}_1 - \mathbf{r}_2)$, where $I_{\mathscr{C}}(\mathbf{r}_1)$ is the source field irradiance function over $\mathscr{C}$. Inserting into (D.7), and integrating out the delta function, produces

$$R_{\mathscr{D}}(\mathbf{q}_1, \mathbf{q}_2) = \frac{e^{j\psi}}{(\lambda L)^2} \int_{\mathscr{C}} I_{\mathscr{C}}(\mathbf{r}) M(\mathbf{q}_1, \mathbf{q}_2, \mathbf{r}) e^{j(2\pi/\lambda L)(\mathbf{r} \cdot \mathbf{q}_1 - \mathbf{q}_2)} \, d\mathbf{r} \qquad (D.8)$$

where now

$$\psi = \frac{\pi}{\lambda L}\left(|\mathbf{q}_2|^2 - |\mathbf{q}_1|^2\right) \qquad (D.9)$$

and

$$M(\mathbf{q}_1, \mathbf{q}_2, \mathbf{r}) = M(\mathbf{q}_1, \mathbf{r}_1; \mathbf{q}_2, \mathbf{r}_2)|_{\mathbf{r}_1 = \mathbf{r}_2 = \mathbf{r}} \qquad (D.10)$$

Equation D.8 indicates how the irradiance function of an incoherent source field generates a coherence function over the observing plane, after propagating through the medium. The parameter ψ is a phase factor associated with the complex coherence. Note that (D.8) has the appearance of a two dimensional Fourier transform, and $R_{\mathscr{D}}(\mathbf{q}_1, \mathbf{q}_2)$ is therefore proportional to the transform of the product of the source irradiance and the function in (D.10) describing the medium. In free space, $M = 1$, and $R_{\mathscr{D}}(\mathbf{q}_1, \mathbf{q}_2)$ is proportional to the transform of the source plane irradiance. Thus a "wide" source irradiance function (spread over a large area) produces a "narrow" coherence function at $\mathscr{D}$, while a confined source irradiance would produce a highly correlated random field (wide coherence function). From (D.8) we see that the effect of the medium can only decrease the field coherence at $\mathscr{D}$, which is equivalent to effectively spreading the source irradiance. Thus a random medium appears to increase further the spatial extent of an incoherent source.

To better understand the medium function in (D.10), consider the special case of (D.7) when the source plane $\mathscr{C}$ contains a single point source at $\mathbf{r}_s$ of unit intensity. This point source can be represented in the source plane as a field with coherence function

$$R_{\mathscr{C}}(\mathbf{r}_1, \mathbf{r}_2) = \delta(\mathbf{r}_1 - \mathbf{r}_s)\,\delta(\mathbf{r}_1 - \mathbf{r}_2) \qquad (D.11)$$

Using (D.11) in (D.7) yields

$$R_{\mathscr{D}}(\mathbf{q}_1, \mathbf{q}_2) = \frac{e^{j[\psi + (2\pi/\lambda L)(\mathbf{r}_s \cdot \mathbf{q}_1 - \mathbf{q}_2)]}}{(\lambda L)^2} M(\mathbf{q}_1, \mathbf{q}_2, \mathbf{r}_s) \qquad (D.12)$$

Hence the function in (D.10) can be defined as the scaled coherence function produced over $\mathscr{D}$ by a point source at $\mathbf{r}$ in $\mathscr{C}$. That is, M describes the coherence function in the observable plane due to a point source in the medium. The integrand of (D.8) therefore represents a multiplication of the source irradiance at each point of $\mathscr{C}$ by the coherence function of the medium from that point. The function $M(\mathbf{q}_1, \mathbf{q}_2, \mathbf{r})$ is often referred to as the *point source coherence function* of the medium. Its value lies in the fact that, once determined, the coherence function response to any source irradiance distribution can be theoretically determined by evaluating (D.8). Since a point source would produce at $\mathscr{D}$ a constant coherence function (plane wave) in free space, the spatial extent of $M(\mathbf{q}_1, \mathbf{q}_2, \mathbf{r}_s)$, as a function of $\mathbf{q}_1$ and $\mathbf{q}_2$, is an indication of the effect of the medium. The larger this extent, the less the medium has affected the coherency of the field. As $M(\mathbf{q}_1, \mathbf{q}_2, \mathbf{r}_s)$ becomes narrower, the coherency region of the observed field is reduced, and the field appears to have been generated from a large incoherent source. Thus, the medium "spreads" the point source into an extended incoherent source.

Various types of medium are defined by the properties of its point source coherence function. If $M(\mathbf{q}_1, \mathbf{q}_2, \mathbf{r})$ is a function of only the difference $(\mathbf{q}_1 - \mathbf{q}_2)$, for each $\mathbf{r}$, the medium is said to be *spatially homogeneous*. Note that the coherence function in (D.8) is always itself homogeneous if the medium is homogeneous. If, in addition, $M(\mathbf{q}_1, \mathbf{q}_2, \mathbf{r})$ depends only on $|\mathbf{q}_1 - \mathbf{q}_2|$, the medium is said to be *isotropic*.

Now let us examine optical detection of a field after transmission through a medium. Consider an optical receiver whose aperture lies in the observation plane $\mathscr{D}$. The impinging field at $\mathscr{D}$, due to a field from a source plane $\mathscr{C}$, has the coherence function $R_{\mathscr{D}}(\mathbf{q}_1, \mathbf{q}_2)$, as described previously. We wish to determine the intensity distribution produced in the detector (focal) plane of the receiver (Figure D.2) due to the impinging random field. The coherence function over the aperture $\mathscr{A}$ of the receiver is then

$$R_{\mathscr{A}}(\mathbf{q}_1, \mathbf{q}_2) = R_{\mathscr{D}}(\mathbf{q}_1, \mathbf{q}_2)P(\mathbf{q}_1)P(\mathbf{q}_2) \qquad (D.13)$$

where $P(\mathbf{q})$ is the aperture pupil function. [For a simple aperture $P(\mathbf{q}) = 1$, $\mathbf{q} \in \mathscr{A}$, and zero elsewhere]. If the receiving optics are constructed so as to allow for Fraunhoffer diffraction, the intensity in the detector plane a distance f_c behind the aperture can be obtained directly from (D.7). We consider the receiver aperture plane as the source plane, the receiver detector plane as the

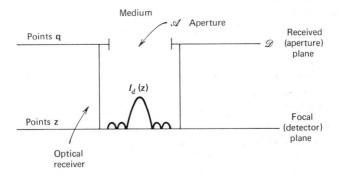

Figure D.2

observation plane, $L = f_c$, and the function M is set equal to one (free space propagation within the receiver). Hence

$$I_d(\mathbf{z}) = \frac{1}{(\lambda f_c)^2} \int_{\mathscr{D}} \int_{\mathscr{D}} R_{\mathscr{D}}(\mathbf{q}_1, \mathbf{q}_2)P(\mathbf{q}_1)P(\mathbf{q}_2)e^{j(2\pi/\lambda f_c)(\mathbf{z} \cdot \mathbf{q}_1 - \mathbf{q}_2)}d\mathbf{q}_1\, d\mathbf{q}_2 \quad (D.14)$$

where $\mathbf{z}$ is a point in the detector plane, and $I_d(\mathbf{z})$ is the intensity at $\mathbf{z}$. Equation D.14 therefore describes the intensity distribution over the detector plane. Collected receiver power is obtained by integrating $I_d(\mathbf{z})$ over the actual photodetector area. When the received field is homogeneous [i.e., $R_{\mathscr{D}}(\mathbf{q}_1, \mathbf{q}_2) = R_{\mathscr{D}}(\mathbf{q}_1 - \mathbf{q}_2)$] then (D.14) can be rewritten, using the change of variable $\mathbf{u} = \mathbf{q}_1 - \mathbf{q}_2$, as

$$I_d(\mathbf{z}) = \frac{1}{(\lambda f_c)^2} \int_{\mathscr{D}} R_{\mathscr{D}}(\mathbf{u})H(\mathbf{u})e^{j(2\pi/\lambda f_c)(\mathbf{z} \cdot \mathbf{u})}\, d\mathbf{u} \quad (D.15)$$

where

$$H(\mathbf{u}) = \int_{\mathscr{D}} P(\mathbf{u} + \mathbf{q}_2)P(\mathbf{q}_2)\, d\mathbf{q}_2 \quad (D.16)$$

Thus the detector irradiance distribution is the scaled Fourier transform of the product of the impinging coherence function and the function $H(u)$ in (D.16). The latter is called the *operating transfer function* (OTF) of the receiving system and depends only upon the pupil function of the receiving lens. For a simple aperture, the receiver OTF corresponds to the area of overlap of two offset pupil functions located in the aperture plane [1]. Conversely, $I_d(\mathbf{z})$ can be obtained from a convolution of the transform of the received coherence function and the transform of the receiver OTF. If the received coherence function is widely extended (corresponding to a highly correlated received field from a point source) the irradiance distribution is given approximately by the transform of the receiver OTF, which is generally concentrated over a narrow region in the detector plane. Hence the receiver is effectively imaging the point source. On the other hand, if the coherence function is extremely narrow (e.g., because of a highly random medium) the irradiance distribution is itself spread, or blurred, over the detector plane, and the source image is lost.

In the typical communication applications, a point source transmitter is located in a homogeneous medium and is attempting to communicate with a receiver. The detector irradiance is therefore obtained by using (D.12) in (D.15). In this case the irradiance distribution is related to the Fourier transform of the product of the medium point source coherence function and the receiver OTF. For a widely spread incoherent source (such as a noisy sky background) in the random medium, (D.8) must be evaluated and used in (D.15). In general, large incoherent sources tend to produce uniformly spread irradiance at the detector, while the spreading of a point source will depend upon the degree of randomness of the medium.

D.2 SCATTER CHANNELS

In optical communications the most important type of random medium encountered is the scattering channel. Because of the small optical wavelengths, particulates in the transmission medium, such as water drops, aerosols, molecules, and clouds, act as individual point scatters to the optical field. The overall effect is to produce a random scattering of the optical beam, and in general a reduction of the field power and a distortion of the coherent structure of the wave.

Scatter channels have been studied from several viewpoints, including variations in refractive index [2], point scattering [3, 4, 5], and radiation transport theory [6]. The basic analysis procedure, called *Mie scattering theory*, is to determine the scattering effect of a single point scatter of a given size by finding electromagnetic boundary value solutions. The solution is then averaged over a scatterer size and location distribution. This technique has been used to determine the unit power point source coherence function for an isotropic, homogeneous scattering medium, and analytical solutions have been developed that conform well with measured data, and give accurate insight into the behavior of the scattered field. We omit the derivation, and give the point source coherence function solution as

$$M(\rho, \mathbf{r}_s) = e^{-\alpha_a L} \exp[-\alpha_s L(1 - \beta(\rho))] \tag{D.17}$$

where L is the distance from the point source $\mathbf{r}_s$ to a normal observing plane, and ρ is the distance between two points in the plane. The parameters α_a and α_s are called the *absorption* and *scattering* coefficients, respectively, and give the power loss per unit distance due to these effects. The function $\beta(\rho)$ indicates the variation in coherence as a function of ρ, and is given by

$$\beta(\rho) = \int_0^\pi \left[\int_0^1 J_0\left(\frac{u2\pi\rho \tan \theta}{\lambda}\right) du \right] \sigma(\theta) \sin \theta \, d\theta \qquad (D.18)$$

where J_0 is the Bessel function and $\sigma(\theta)$ is the normalized, circularly symmetric, scattering function describing the angular distribution of scattered power from a scattering volume, and is normalized such that

$$\int_0^\pi \sigma(\theta) \sin \theta \, d\theta = 1 \qquad (D.19)$$

Absorption loss is due to field absorption by the scatters, while scattering loss is due to the scattering of the field in other directions. These parameters are also a property of the scattering medium, and depend upon the cross section and location distribution of the scatterers. Since $M(\rho, \mathbf{r}_s)$ is a coherence function in ρ, $M(0, \mathbf{r}_s)$ gives the irradiance at any observation point. It is easy to show that

$$M(0, \mathbf{r}_s) = e^{-\alpha_a L} \qquad (D.20)$$

That is, the intensity at a point is reduced by only the absorption loss when collected from all directions. On the other hand,

$$M(\infty, \mathbf{r}_s) = e^{-(\alpha_a + \alpha_s)L} \qquad (D.21)$$

implying points widely separated have their correlation reduced to a constant value. Since a constant coherence function corresponds to a plane wave, (D.21) indicates that the plane wave received from the point source has its power level reduced by both the absorption and scattering losses. This means that a detector with a narrow field of view looking at only the point source would see the received point source field with power reduced by this value.

Limiting forms to the coherence function in (D.17) can be obtained by expanding the Bessel function in (D.18) for large and small values of ρ. This leads to

$$M(\rho, \mathbf{r}_s) = e^{-\alpha_a L} e^{-(\rho/r_0)^2}, \qquad \rho \ll \frac{r_0}{(\alpha_s L)^{1/2}}$$

$$= e^{-(\alpha_a + \alpha_s)L}, \qquad \rho \gg \frac{r_0}{(\alpha_s L)^{1/2}} \qquad (D.22)$$

where

$$r_0 = \frac{\lambda}{(\alpha_s L)^{1/2}\theta_{rms}} \qquad (D.23)$$

$$\theta_{rms}^2 = \int_0^\pi \theta^2 \sigma(\theta) \sin \theta \, d\theta \qquad (D.24)$$

Here θ_{rms} is the root mean squared scatter angle in radians of a scattering volume. The value of θ_{rms} depends on frequency and scatterer size. Its value typically ranges from a few degrees for water scattering to as much as 60 deg. for cloud scattering. Note that M is essentially Gaussian in shape for small ρ with a spread of approximately r_0. The latter parameter is called the *coherency distance* of the medium, and is seen to be inversely related to propagation distance into the medium, the rms scattering angle, and frequency. The value of r_0 is almost always less than the wavelength of the field and is generally on the order of microns.

We observe three basic regions of interest concerning this point source coherence function. For points (values of ρ) much less than the coherence length apart the the field is coherent. A receiving aperture whose dimensions are of this order of magnitude would therefore observe a coherent field, and only the absorption loss occurs. For points whose separation is approximately r_0, a loss in field coherence begins to appear, in addition to the absorption loss. Finally, for points separated by much more than the coherence length, the field appears spatially incoherent, except for the residual plane wave field that has been extinguished by both the absorption and scattering losses. Receiving apertures whose dimensions are much larger than the coherence length would therefore observe the combination of an incoherent (scattered) field and an extinguished coherent plane wave field.

REFERENCES

[1] J. W. Goodman, *Introduction to Fourier Optics*, McGraw-Hill Book Co., New York, 1968. Section 6-3.

[2] V. I. Tatarski, *Wave Propagation in a Turbulent Medium*, McGraw-Hill Book Co., New York 1961, Chapter 7.

[3] H. M. Heggestad, "Optical Communication Through a Multiple-Scattering Media," Technical Report 472, Massachusetts Institute of Technology, Cambridge, Mass., November 1968.

[4] R. F. Lutomirski and H. T. Yura, "Propagation of an Optical Beam in an Inhomogeneous Medium," *Appl. Opt.*, **10**, 1954 (July 1971).

[5] W. H. Wells, "Loss of Resolution in Water as a Result of Multiple Scattering," *J. Opt. Soc. Am.*, **59**, No. 6, 686 (June 1969).

[6] D. Arnush, "Underwater Light Propagation with Small Angle Scattering," *J. Opt. Soc. Am.*, **62**, No. 9, 1109 (September 1972).

Index